The Praeger Handbook of Environmental Health

The Praeger Handbook of Environmental Health

VOLUME 1
FOUNDATIONS OF THE FIELD

Robert H. Friis, Editor

AN IMPRINT OF ABC-CLIO, LLC
Santa Barbara, California • Denver, Colorado • Oxford, England

Copyright 2012 by ABC-CLIO, LLC

All rights reserved. No part of this publication may be reproduced, stored in a retrieval system, or transmitted, in any form or by any means, electronic, mechanical, photocopying, recording, or otherwise, except for the inclusion of brief quotations in a review, without prior permission in writing from the publisher.

Library of Congress Cataloging-in-Publication Data

The Praeger handbook of environmental health / Robert H. Friis, editor.
    p. cm.
  Includes bibliographical references and index.
  ISBN 978-0-313-38600-8 (hardback) — ISBN 978-0-313-38601-5 (ebook)
 1. Environmental health—Handbooks, manuals, etc. I. Friis, Robert H.
 RA565.P73   2012
 613'.1—dc23

2012003306

ISBN: 978-0-313-38600-8
EISBN: 978-0-313-38601-5

16 15 14 13 12   1 2 3 4 5

This book is also available on the World Wide Web as an eBook.
Visit www.abc-clio.com for details.

Praeger
An Imprint of ABC-CLIO, LLC

ABC-CLIO, LLC
130 Cremona Drive, P.O. Box 1911
Santa Barbara, California 93116-1911

This book is printed on acid-free paper ∞
Manufactured in the United States of America

Every reasonable effort has been made to trace the owners of copyright materials in this book, but in some instances this has proven impossible. The editor and publisher will be glad to receive information leading to more complete acknowledgments in subsequent printings of the book and in the meantime extend their apologies for any omissions.

Contents

Introduction: Current Status of Environmental Health, ix
  Robert H. Friis

CHAPTER 1
Ecology and Environmental or Ecosystem Health, 1
  Joanna Burger

CHAPTER 2
Human Exposure Assessment, 27
  Derek G. Shendell

CHAPTER 3
The Environment and Endangered Species/Species Extinctions, 45
  John E. Fa

CHAPTER 4
Environmental Noise and Health, 69
  Irene van Kamp, Wolfgang Babisch, and A. L. Brown

CHAPTER 5
Environmental Epigenetics, 95
  Mihalis I. Panayiotidis, Aglaia Pappa, and Dominique Ziech

CHAPTER 6
Climate Change and Health, 111
  Jenny Griffiths

CHAPTER 7
Oceanic Pollution, 133
  Stephen B. Weisberg and Karen E. Setty

CHAPTER 8
Fundamentals of Environmental Epidemiology, 155
Marc A. Strassburg

CHAPTER 9
Using Epidemiology as a Tool to Study
Environmental Health, 181
Keith B. G. Dear and Anthony J. McMichael

CHAPTER 10
Solutions to the Growing Solid Waste Problem, 197
Robert H. Friis

CHAPTER 11
Population Trends and the Environment, 215
Colin D. Butler

CHAPTER 12
International Environmental Law, 233
Margaret Alkon

CHAPTER 13
Biomarkers Used in Environmental Health, with a Focus
on Endocrine Disrupters, 259
*Tanja Krüger, Mandana Ghisari, Manhai Long,
and Eva Cecilie Bonefeld-Jørgensen*

CHAPTER 14
Environmental Health Risk Assessment, 281
Ryan G. Sinclair, Kristen Gunther, and Rhonda Spencer-Hwang

CHAPTER 15
The Environmental Health Policy-Making Process, 305
Barry L. Johnson

CHAPTER 16
Significant Environmental Health Statutes
and Key Regulations, 323
Barry L. Johnson

Chapter 17
The Role of State and Local Public Health Departments in Environmental Enforcement, 347
Mark B. Horton, Alison F. Dabney, Cindy A. Forbes, Patrick Kennelly, Richard A. Kreutzer, Gregg W. Langlois, Carl Lischeske, Robert Schlag, and Gary H. Yamamoto

Chapter 18
Risk Communication and Environmental Health: Principles, Strategies, Tools, and Techniques, 367
Vincent T. Covello

Chapter 19
The Role of Health Professionals in Protecting Environmental Health, 391
Robert M. Gould

Chapter 20
Renewable Energy, 409
Brian C. Black and Richard E. Flarend

Chapter 21
Eco-Friendly Transportation and the Built Environment, 427
David A. Sleet, Rebecca Naumann, Grant Baldwin, T. Bella Dinh-Zarr, and Reid Ewing

Chapter 22
Lighting and Astronomy, 441
Christian B. Luginbuhl, Constance E. Walker, and Richard J. Wainscoat

Chapter 23
Green Living: Reducing the Individual's Carbon Footprint, 455
Clinton J. Andrews and Robert H. Friis

Chapter 24
Educating the Environmental Health Workforce, 475
Maureen Lichtveld and Christine Rosheim

About the Editor and Contributors, 503

Index, 509

Introduction: Current Status of Environmental Health

Robert H. Friis

Introduction

Environmental health is a topic of ongoing fascination for contemporary society, as illustrated by the continuing flow of media reports on environmental disasters such as earthquakes and tsunamis, floods, oceanic oil spills, drinking water contamination by toxic chemicals, species extinctions, and outbreaks of infectious and communicable diseases linked to poor sanitation. One issue that has captured the attention of environmentalists is global warming, a controversial subject debated with great passion by government officials, policy makers, scientists, economists, and the business community. The focus of the debate has been the severity of warming, its causes, methods for prevention, and the cost of reducing carbon emissions. The specter of global warming was heralded in summer 2010 by devastating floods in Pakistan and record-breaking heat waves such as those experienced in Russia, which normally has a much cooler summer temperature. Extreme heat waves, a kind of temperature anomaly, are believed to be the consequence of global warming, as are rising ocean levels caused by melting of the polar ice caps and glaciers; this phenomenon threatens human habitation in some low-lying coastal zones.

Governmental bodies and health-related agencies have responded to environmental threats caused by global warming and other human impacts by proposing a number of policy initiatives. For example, in the United States President Barack Obama has called for a reduction in carbon dioxide emissions (linked to global warming and climate change) by using clean energy sources to produce 80 percent of domestic electricity by the year 2035.[1] Policy makers in the European Union have identified four strategic priorities related to the environment: climate change, protection of nature and biodiversity, conservation of natural resources, and the linkage of the environment with health and quality of life.[2] Given the transnational scope of many environmental issues and the growing global interconnectedness of modern society, actions that address threats to the environment require cooperation across political boundaries, be they international, national, or local.[3]

In recent years the media have portrayed several compelling, environmentally related events: oil spills, global warming, and natural disasters. During 2010 the catastrophic spill of nearly five million barrels of crude oil into the Gulf of Mexico from BP's Deepwater Horizon oil-drilling platform threatened marine life, the fishing industry, and tourism in the southeastern United States. Television monitors located one mile beneath the gulf's surface tracked the progress of attempts to staunch the flow of gargantuan volumes of oil. For several months local residents and fishermen stood by helplessly as crude oil drifted into ecologically sensitive coastal nesting areas for birds and endangered turtles. The event called into question the wisdom of drilling for crude oil at sea, where unintentional releases of oil can have devastating environmental consequences.

Adverse impacts on the environment are not caused exclusively by anthropogenic sources; natural forces such as volcanoes and earthquakes also affect the environment. For example, the massive eruption of the Eyjafjallajokull volcano in Iceland in April 2010 interrupted air traffic over northern Europe. Severe earthquakes in 2010 (affecting Chile, China, and Haiti) displaced thousands of people, caused widespread loss of life, and produced substantial economic losses. A tsunami in Sri Lanka (2004) killed an estimated 200,000 people; in Japan (March 2011) an earthquake followed by a tsunami caused significant mortality and extreme environmental devastation, which was coupled with explosions at a nuclear reactor and the subsequent dispersal of radioactivity into the surrounding region.

The Impact of the Environment on Health

Environmental factors play a significant role in human health, illness, and mortality. Regrettably, the adverse impacts of environmental exposures and conditions are likely to continue unabated or even increase in the future, despite many improvements. All human beings are affected in some way by exposure to environmental hazards associated with lifestyle: at work, at home, during recreation, or as passengers or drivers on the roadways.

From the early 1900s until the present century, life expectancy in the United States has almost doubled. This improvement in life expectancy has come about, in part, because of positive environmental changes that occurred during the past century. These positive changes include better sanitation, cleaner water, improved air quality, attention to the safe use of chemicals, and prevention of unsafe behaviors.[4] For example, in recent years lead has been taken out of gasoline, potential carcinogens have been removed from pharmaceuticals, some toxic chemicals have been eliminated from products used in the home and workplace, and pregnant women and other consumers have been advised not to consume fish that might contain high levels of mercury.

According to the World Health Organization, preventable environmental exposures were responsible for about one-quarter of global disease during the middle of the first decade of the twenty-first century.[5] Some estimates place the portion of deaths worldwide caused by environmental factors at 40 percent.[6] Exposures to potentially hazardous agents such as microbes, toxic chemicals and metals, pesticides, and ionizing radiation account for many of the forms of environmentally associated morbidity (acute and chronic conditions, allergic responses, and disability) and mortality that occur in today's world. These environmentally related determinants are believed to be important for the development of chronic diseases such as cancer, although most chronic diseases are thought to be the result of complex interactions among environmental and genetic factors.[7]

In the United States environmental factors are implicated in many of the leading causes of death, including the five leading causes: diseases of the heart (heart disease), malignant neoplasms (cancer), cerebrovascular diseases (stroke), chronic lower respiratory diseases, and accidents (unintentional injuries).[8] Similar trends in the causes of death associated with environmental risk factors are evident in many other developed countries.

Air pollution, both indoor and outdoor, has been researched extensively as a risk factor for heart disease and related cardiovascular conditions, cancer, stroke, and respiratory diseases—four of the leading causes of mortality in the United States. Examples of sources of indoor pollution are cigarette smoke, chemicals contained in household cleaning materials, dust mites and cockroaches, and emissions from building materials and furniture. The components of outdoor air pollution include particles from vehicle exhaust and toxic gases such as carbon monoxide. Examples of the hazards associated with air pollution are increased risk of heart attacks and respiratory infections from exposure to secondhand cigarette smoke and particle pollution.[9, 10]

Unintentional injuries, the fifth leading cause of death, are linked to environmental concomitants. Often the public, educators, and health officials do not consider unintentional injuries to be a dimension of environmental health. Contrary to this point of view, injury researchers argue that features of the environment, for example, the design and maintenance of highways, structures, and occupational settings, are implicated in unintentional injuries, as are people's risky behaviors.

Some of the potentially toxic chemicals and elements identified as possible human carcinogens include organic pesticides, solvents, cadmium, and arsenic as well as other chemicals that may be present in the air, foods, and water supply. Researchers attribute approximately three-quarters of the cases of cancer and cancer deaths to environmental factors.[11] Rachel Carson's groundbreaking work *Silent Spring*, published in 1962, helped to focus the public's attention on the harmful effects of pesticides.[12] A major concern of environmental professionals is the negative consequences of exposure to toxic chemicals, including pesticides such as DDT and persistent organic chemicals (POPs), for example, dioxins and PCBs. DDT, dioxins, and PCBs are now banned in the United States and many other countries.

Vulnerable populations such as women and children are at heightened risk for the adverse impacts of environmental conditions. The World Health Organization (WHO) points out that the toll on children from environmentally associated conditions is particularly severe.[13] The WHO estimates that more than ten million children succumb yearly to conditions to which the environment makes a contribution and that three million of these children are under five years of age. The environmental factors that are implicated in children's deaths include air pollution and unsafe water, poor sanitary conditions, and toxic chemicals and metals. Disease vectors (e.g., insect vectors) infect children with serious vector-borne diseases such as malaria, which takes the lives of almost two million children each year. Worldwide, many children suffer from diarrheal diseases, unintentional injuries, and respiratory conditions. Unclean water and food and an unsanitary environment are responsible for outbreaks of diarrheal conditions. Respiratory diseases such as asthma and pneumonia are linked to the indoor use of biomass cooking fuels. Some research findings suggest that women's exposures to environmental toxins during pregnancy may be related to adverse pregnancy and birth outcomes (e.g., miscarriage, growth restriction in utero, and low birth weight) and may negatively affect children (e.g., by causing cognitive and motor delays) of exposed women.[14]

Trends in Environmental Health

There are several definitions for the term *environmental health*; moreover, the field of environmental health has a broad scope.[15] The World Health Organization has developed a widely adopted definition:

> Environmental health addresses all the physical, chemical, and biological factors external to a person, and all the related factors impacting behaviours. It encompasses the assessment and control of those environmental factors that can potentially affect health. It is targeted towards preventing disease and creating health-supportive environments.[16]

Organization of the Handbook

Table I.1 provides an alphabetical listing of current topics in the field of environmental health. Although that list is comprehensive, it is not exhaustive. The goal is to address the majority of these topics in this handbook, which is organized into four volumes that span the basic foundations of environmental health, environmental disease agents, water and air quality and solid waste disposal, and miscellaneous topics. Each volume covers several of the content areas and associated subtopics shown in the table. In a few instances, some topics are presented in more than one volume.

Volume 1, *Foundations of the Field*, provides an introduction to the environmental health field. Topics presented are ecology, environmental policy and regulations, exposure and risk assessment, communication about the environment, global climate change, the human population and environmental quality, and development of the environmental health workforce. Volume 2, *Agents of Disease*, covers food safety, heavy metals and other toxic elements, infectious and vector-borne diseases, radiation, and toxic chemicals and substances. Volume 3, *Water, Air, and Solid Waste*, presents information on these three topics. The focus of volume 4, *Current Issues and Emerging Debates*, is on occupational environments, unintentional injuries, and vulnerable populations.

TABLE I.1
Current Major Themes, Topics, and Subtopics in the Environmental Health Field

Broad Theme or Topic	Examples of Subtopics
Air quality	Asthma
	Cardiovascular disease
	Exposure to secondhand cigarette smoke
	Health effects of goods movement
	Impact of outdoor air pollution
	Indoor air pollution and use of biomass fuels
	Mortality
	Specific air pollutants: carbon monoxide, diesel exhaust, particles, and ozone
Communication/workforce development	Educating the environmental health workforce
	Methods of risk communication
	Roles of health professionals
Ecology	Carrying capacity of the earth
	Land use planning
	Preservation of endangered animals/species extinctions
	Smart growth
Environmental factors and chronic disease	Breast cancer
	Parkinson's disease
	Reproductive effects
Environmental policy and regulations	Environmental justice
	Environmental policy
	Ethics and the environment
	Governmental roles in environmental quality
	International environmental laws
Exposure and risk assessment	Environmental epigenetics
	Epidemiology
	Molecular methods and biomarkers
	Toxicology
Food safety	Contamination of food
	Microbial agents present in food
	Prevention of foodborne illness
Global climate change/global warming	Extreme weather events
	Rainforest destruction
	Use of fossil fuels
Green living	Alternative and clean energy sources
	Energy and lifestyle
	Sustainability
	Transportation alternatives

(continued)

TABLE I.1
(Continued)

Broad Theme or Topic	Examples of Subtopics
Heavy metals and other toxic elements found in the environment	Arsenic Cadmium Lead
Human population and environmental quality	Air quality in developing countries Effects of the built environment Overpopulation and crowding The social environment Urban environmental health
Infectious, communicable, and vector-borne diseases related to environmental factors	Antibiotic-resistant bacteria Arthropod-borne diseases Emerging infectious diseases Zoonoses
Occupational environments as a subset of the general environment	Biomonitoring Occupational diseases Safety and health
Physical environmental factors	Ionizing and nonionizing radiation Light pollution Noise
Toxic chemicals and substances	Dioxin Endocrine disruptors Pesticides Phthalates Polychlorinated biphenyls
Unintentional injuries and their relationship to the environment	Falls Motor vehicle and traffic-related injuries and deaths Sports-associated injuries
Vulnerable populations and environmental impacts	Birth outcomes among women of childbearing age Disproportionate effects upon children: lead poisoning, childhood asthma
Waste disposal	Agricultural waste Hazardous materials management Liquid and solid waste disposal Toxic waste sites
Water quality	Adequate supplies of safe drinking water Oceanic pollution Safety of beaches and recreational water Water disinfection

The authors who contributed to this handbook are outstanding authorities in their fields. They have been selected from the ranks of academia, government, industry, and environmental organizations. The *Praeger Handbook of Environmental Health* is a valuable resource for students, educated laypersons, and environmental professionals who are seeking a unique and comprehensive collection of information on the key topics in environmental health.

Conclusion

Environmental health is a topic of central importance in the twenty-first century. Despite many positive advances in the control of some toxic chemicals and other environmental hazards, the current century has been witness to continuing degradation of the environment and concomitant challenges to the health of the planet and its inhabitants. Particularly affected by environmental hazards are less-developed countries undergoing industrialization, regions experiencing high population growth, and economically disadvantaged sections of first world countries.

Although much progress has been made in controlling and remediating adverse environmental conditions, the environment continues to be under threat from human activities. This handbook provides an overview of environmentally associated hazards, their assessment, and methods for their remediation.

Notes

1. F. J. Frommer, "Obama Seeks New Path to Environmental Goals" [editorial], *Associated Press Online*, January 27, 2011, http://www.aarp.org/politics-society/environment/news-01-2011/obama_seeks_new_path_to_environmental_goals.html (accessed March 14, 2011).
2. European Environment Agency (EEA), *The European Environment—State and Outlook 2010: Synthesis* (Copenhagen, Denmark: EEA; 2010).
3. L. Huong, *Environmental Governance* (Nairobi, Kenya: United Nations Environment Programme, 2009).
4. National Institutes of Health, *The New Environmental Health*, NIH publication no. 02-5081 (Research Triangle Park, NC: NIEHS, n.d.), http://www.niehs.nih.gov/health/assets/docs_p_z/the_new_environmental_health.pdf (accessed October 5, 2011).

5. World Health Organization, *Almost a Quarter of All Disease Caused by Environmental Exposure* (Geneva: World Health Organization, 2006), http:www.who.int/mediacentre/news/releases/2006/pr32/en/index.html (accessed March 8, 2011).
6. National Institute of Environmental Health Sciences, "Forum: Killer Environment," *Environmental Health Perspectives* 107 (1999): A62–A63.
7. P. G. Butterfield, "Upstream Reflections on Environmental Health: An Abbreviated History and Framework for Action," *Advances in Nursing Science* 25 (1) (2002): 32–49.
8. A. M. Minino, J. Xu, and K. D. Kochanek, "Deaths: Preliminary Data for 2008," *National Vital Statistics Reports* 59 (2) (2010): 1–9.
9. American Lung Association, *Particle Pollution* (Washington, DC: State of the Air, 2010), http://www.stateoftheair.org/2010/health-risks/health-risks-particle.html (accessed March 14, 2011).
10. American Lung Association, *Outdoor Smoking Bans* (Washington, DC: Fighting for Air, 2010), http://www.lungusa.org/associations/states/california/press-room/alacinthenews/los-angeles-times-outdoor.html (accessed March 14, 2011).
11. R. V. Snowden, *ACS Report Addresses Environmental Pollutants and Cancer Risk* (American Cancer Society, 2009), http://www.cancer.org/cancer/news/news/acs-report-addresses-environmental-pollutants-and-cancer-risk (accessed March 14, 2011).
12. R. Carson, *Silent Spring* (New York: Houghton Mifflin, 1962).
13. World Health Organization, *The Environment and Health for Children and Their Mothers*, Fact Sheet No. 284 (Geneva: World Health Organization, 2005), http://www.who.int/mediacentre/factsheets/fs284/en/index.html (accessed March 9, 2011).
14. E. Harrison, J. Partelow, and H. Grason, *Environmental Toxicants and Maternal and Child Health: An Emerging Public Health Challenge* (Baltimore, MD: Women's and Children's Health Policy Center, Johns Hopkins Bloomberg School Public Health, 2009).
15. R. H. Friis, *Essentials of Environmental Health*, 2nd ed. (Sudbury, MA: Jones and Bartlett Learning, 2012).
16. World Health Organization, *Environmental Health*, http://www.who.int/topics/environmental_health/en/ (accessed February 22, 2010).

1

Ecology and Environmental or Ecosystem Health

Joanna Burger

Introduction

The public, scientists, managers, and public policy makers are interested in maintaining healthy ecosystems, both for ecosystem protection and for the benefits they provide society, including goods and services, medicinal products, and religious and cultural benefits. Ecosystems, whether natural, degraded, or developed, have always faced biological, physical, chemical, and radiological stressors, but many of these have increased in magnitude and frequency since the Industrial Revolution. Industrialization, including the development of chemical weapons and weapons of mass destruction during and after the Second World War, has led to massive habitat destruction and extensive contamination of the land and water around factories. Although species assemblages and communities have adapted or adjusted to these stressors, the cumulative effect has ranged from minor to devastating. In some cases, however, the chemical and physical stressors caused by industrial or military development have resulted in protection of ecosystems that were minimally affected, such as around many of the Departments of Defense and Energy sites in the United States, where large tracts of undisturbed buffer lands were maintained.[1]

This chapter examines the features that are important to maintaining healthy ecosystems, ecosystem disruptions (natural and anthropogenic), recovery and resiliency, and the effect of ecosystem disruptions on human health. Though entire books can be written about ecology[2] and each of these topics, the intent of this chapter is to provide a broad overview of the ecological concepts that govern environmental health, where environmental health refers to the maintenance of ecosystems with appropriate and dynamic structure and functioning. Environmental health then encompasses all organisms living within ecosystems, from microbes to people. Humans are just one part of ecosystems, albeit one we are particularly interested in.

Some authors use *environmental health* to refer to aspects of human health that relate to the environment or ecosystems (e.g., exposure to lead, mercury, radionuclides, or occupational exposures), as opposed to internal diseases (cancer, diabetes, stroke, heart disease) that might or might not have an environmental component. In this chapter *environmental health* refers to ecosystems, and in some ways can be referred to as ecosystem or ecological health. *Ecology* in its broadest sense refers to the interactions and relationships between the physical and biotic environments. An *ecosystem* is a defined spatial system and can range from a small mud puddle or rotting log to the earth.

Defining Ecosystem Health

Defining ecosystem health is like defining human health: We all know what being healthy is, but it is difficult to write a short definition that covers all the essential aspects. I define *ecosystem, ecological,* or *environmental* health (in this case, all have the same meaning) as the condition in which an ecosystem has appropriate structure and functioning that can provide the goods and services people require,[3] where *appropriate* relates to local geographical, geophysical, and meteorological conditions. A healthy system is the one that was there before human intervention, although clearly nearly all ecosystems on Earth have been impacted in some way by humans. A healthy ecosystem is dynamic, resilient, and self-correcting.[4] That is, the system can recover from natural stressors, such as fires, hurricanes, drought or floods, and disease.

Defining other terms used in ecology is less difficult (see Table 1.1). Ecology covers different levels of biological organization (molecular, cellular, individual, populations), different spatial scales (up to landscape and global), different interactions (behavior, competition, predation, disease, population dynamics), and different processes (food webs, productivity

TABLE 1.1
Common Terms Used by Ecologists

Biomass: The amount of biological material in a defined space (e.g., the biomass of a forest).

Biome: An easily recognized habitat type, such as tropical rain forest, desert, temperate forest.

Carnivore: An animal that eats meat.

Cascading effects: A range of secondary and tertiary effects on organisms or processes. For example, adverse effects on oceanic invertebrates can affect small fish, fish that compete with the small fish, fish that eat the small fish, and larger fish that eat the fish that eat the small fish.

Community: The organisms living within a given, defined geographic region, usually involving some degree of interactions.

Competition: When two or more individuals use the same resource that is in short supply.

Decomposers: Organisms (primarily bacteria and fungi) that break down dead material by the process of absorption (releasing their enzymes into dead bodies to degrade and digest them).

Detritivore: Organism that decomposes organic matter (e.g., microbes).

Eco-cultural attributes: The interdependencies that people place on the cultural or religious resources that have a necessary natural resource component.

Eco-cultural dependency webs: Assessments of the interlocking system of people and biota that is sustainable over time, particularly for the purposes of risk assessment and characterization.

Ecological evaluations: Evaluating aspects of the environment and ecosystems. Ecological evaluations form the basis for determining status and trends of biological, physical, or chemical/radiological conditions; conducting environmental impact assessments; selecting remediation and restoration options; performing remedial actions should remediation fail; managing ecosystems and associated wildlife; and assessing the efficacy of long-term stewardship.

Ecological risk: The risk (probability of harm) to ecosystems (or specific biota or groups of biota) from biological, physical, or chemical/radiological stressors.

Ecology: The study of ecosystems, including all species, interactions, and processes.

Ecosystem: The interacting system of a biological community and the abiotic (nonbiological) environment. The spatial extent of an ecosystem is defined (e.g., a particular forest).

Energy transfer: The movement of energy through the food chain, from primary producers (plants) to herbivores and carnivores.

(continued)

TABLE 1.1
(Continued)

Food web: The complex energy sequence of links (or trophic levels) that starts with herbivores and ends with top-trophic-level carnivores. Webs can be simple and involve few organisms, or complex, involving hundreds of organisms.

Fragmentation: The breaking up of habitat to create smaller patches (plots), and the possible isolation of patches (e.g., small woodlot surrounded by houses).

Habitat: A recognizable ecosystem, such as a forest, an old field, or a grassland.

Herbivore: An animal that eats plants.

Landscape scale: Consideration of larger regional-scale analysis of relationship among habitat and ecosystem types (e.g., amount and relationship among developed and undeveloped lands, forests, lakes, and corridors).

Life stages (or life cycle): The different forms of an organism. For example, a frog has an egg stage (normally aquatic), a tadpole stage (normally aquatic), and an adult frog stage (usually terrestrial, or terrestrial and aquatic); a butterfly has an egg stage, a caterpillar (usually feeds on plant material), a chrysalis, and an adult butterfly (usually feeds on nectar from flowers). Each phase may differ markedly in time period, habitat use, and trophic level.

Niche: The role that an organism plays in its environment.

Omnivore: An animal that eats both plants and animals.

Population: An ecological term referring to the individuals of the same species living within a defined area and potentially exposed to the same hazards.

Predation: The killing and consuming of one organism by another. Examples of predation include a wolf killing and eating a deer, a snake killing and eating a mouse, a lion killing and eating a gazelle, and a human killing (either directly or by surrogate) and eating a cow.

Productivity: The biomass produced at each trophic level in a given period of time (e.g., per year).

Species diversity: The number of species in an ecosystem or defined region. It can refer to a list of species or can be computed using a formula that takes into account both numbers of species and their abundance.

Sustainability: The handling and conservation of natural resources and the orientation of technological and institutional change, so as to ensure the continuous satisfaction of human needs for present and future generations. It involves meeting the needs of the present without compromising the ability of future generations to meet their own needs.

Top-level predators: Predators that are at the top of the food chain and are thus expected to bioaccumulate the highest levels of contaminants.

Trophic level: The level within the food chain at which an organism functions. Major trophic levels are producers (plants), herbivores (plant eaters), carnivores (meat eaters) and omnivores (eating at different levels, usually plants and animals).

and energy flow, nutrient cycling), among others. Although initially ecologists focused on individual and population interactions, recent attention has focused on community dynamics, landscape and global dynamics (e.g., regional development along coasts, water resources, global warming), and geophysical aspects (e.g., mercury cycling). The field has moved from being descriptive to developing predictive models.[5]

Ecosystem Structure and Function

The basis of environmental health is an understanding of the structure and functions of healthy ecosystems. Ecosystems are composed of the physical environment and a wide range of species, including plants, animals, and microorganisms. The relationships among these parts, and how these parts function together, result in ecosystems, which range from degraded to healthy. A distinction can be made between environmental health and ecosystem integrity. Although ecologists can argue about the meanings of each term, *environmental health* generally refers to the services it provides (clean air, water, and species diversity), whereas *ecosystem integrity* refers to the possession of structure and function approximating an undisturbed ecosystem. Both can be endpoints of remediation and of restoration following disturbances.

Ecosystem Structure

Ecosystem structure refers to the component parts of the ecosystem and the relationship among them. In its simplest sense, it is the physical environment (rocks, soil, gradients, and nutrients) and the microbes, plants, and animals that make up the system. A simple measure of structure is species diversity: the number of different species that make up the ecosystem. Since some ecologists have argued that species diversity is the regulator of ecosystem stability,[6] the more species, the more stable the system, and the less likely it is to be affected by physical or biological stressors, the loss or gain of species, or invasions by exotic species. Although species diversity clearly impacts ecosystem dynamics, it is just one factor among many,[7] such as predators,[8] food webs,[9] and species interactions such as competition.[10]

Any consideration of ecosystem structure needs to include understanding of different levels of biological organization, including cells, individuals, species, populations, communities, ecosystems, and landscape. Individuals of a species are linked into populations, which in turn are linked with spatially mediated interactions to other populations and communities, which lead to understanding how species and environments interact on a landscape scale.[11] Such linkages lead directly to considerations of how ecosystems function.

Understanding the structure of ecosystems is not only a matter of determining species diversity or listing the different trophic levels that occur within the system, but of understanding the physical structure of the community. For example, a tropical rain forest has ground-level vegetation, herbs that are taller, short and tall shrubs, vines that may reach to the tree tops, and trees that range from seedlings to 300-foot-tall trees. There are different layers to a tropical forest. Temperate forests also have different layers, but there are not as many, nor are the trees as tall. Similarly, the number of tree species in an acre of tropical rain forest is ten or more times greater than in a temperate forest. These differences mean that there can be a greater number of different animal species (and niches) in a tropical forest than in a temperate one.

In contrast, the boreal forest has few layers because the trees are so dense that there is little sun penetration to the forest floor, and few plants can live in low light. Tundra environments have no trees or shrubs, and the vegetation that grows in the wind-swept land is only a few inches tall. The low plant diversity and lack of different structural layers results in a lower number of animal species (and niches). Overall species diversity and structural complexity decrease from the equator to the Arctic.

Ecosystem Functions

The functions of ecosystems include nutrient cycling, energy cycling, productivity, food webs, predator–prey relationships, and overall dynamic complexity, among others, and should be sustainable.[12] One of the most important concepts in ecology and in the maintenance of ecosystem health is energy production and transfer. Food web dynamics is a central theme of ecology[13] and of understanding the role ecosystems play in human health.[14] Managing any ecosystem, including evaluation, protection, restoration, remediation, or holding in stewardship, requires understanding of food web dynamics. It is absolutely essential for protecting both ecological and human health.

Productivity and energy cycling are the primary functions of ecosystems, and all other functions derive from these. All energy comes from the sun; the conversion of solar radiation into chemical energy by photosynthesis is the starting point of energy flow within every system. Autotrophs (organisms that can convert solar radiation into biomass) use carbon dioxide and water to produce carbohydrates using chlorophyll (see below). The movement of energy through ecosystems is unidirectional and noncyclic (the energy produced is ultimately used up and does not cycle).

Nutrient cycling refers to how nutrients (carbon, water, nitrogen, phosphorous, sulfur, and magnesium, among others) move through ecosystems. Nutrient cycling and energy transfer go hand in hand. As plants are grazed, chemical energy in the form of carbohydrates, fats, and proteins is transferred to herbivores, and nutrients move through the system. In the final step in the food chain, dead bodies are reduced by decomposers to the nutrients that made up these bodies, making them available for recycling within the system.

Primary Productivity in Different Ecosystems

Determining productivity in different ecosystems is difficult, because it involves assessing the productivity of all the components of the system, from plants to top-level predators. Farmers have been determining productivity for centuries in terms of harvest yields: so many bushels of wheat, corn, or rice; bushels of apples; baskets of tomatoes; gallons of milk; or pounds of beef or pork. Ecologists also have been "harvesting" above-ground vegetation to determine productivity of a given ecosystem. Although this provides a useful comparative method among crop types or habitats, it does not provide information on primary productivity (including both above- and below-ground productivity). It is quite difficult to determine total primary productivity; imagine digging up several trees to determine root biomass in addition to trunks, branches, and leaves.

Some general productivity estimates have been made in different ecosystems.[15] There are three terms that are essential for our understanding of productivity: 1) biomass (the material that is present in an ecosystem,

TABLE 1.2

Variations in Primary Productivity, Biomass, and Transient Time for Some Typical Biomes

Biome	Primary Productivity (grams/m²/year)	Biomass (grams/m²)	Transient Time (years)
Tropical forest	1,800	45,000	22.5
Temperate forest	1,250	30,000	25
Boreal forest (pine trees)	800	20,000	?
Temperate grassland	500	1,500	3.0
Desert scrub	70	700	10
Swamp and marsh	2,500		6
Algae beds and reefs	2,000		1
Estuaries	1,800		1
Lakes	500		0.04
Open ocean	125		0.02

Source: After E. T. Odum, *Fundamentals of Ecology* (Philadelphia, PA: Saunders, 1957); E. Komandy, *Concepts of Ecology* (Upper Saddle River, NJ: Prentice Hall, 1996); and others.

2) primary productivity (the amount produced by plants), and 3) transient time (the time for nutrients to move through the system).

For plants, *biomass* includes what one can see in the system (above-ground stems, leaves, fruits, flowers) and below-ground roots. While people see only the above-ground parts, it is important to remember that for some plants there is more biomass below ground than above ground.

Primary productivity differs in different ecosystems, but some generalizations are useful (see Table 1.2). In general, productivity is higher in ecosystems with grass (much more above-ground material than below-ground roots) than in those with trees (more above-ground material). This is because the above-ground material in grass is produced anew every year or two, while trees remain and produce only their leaves or needles annually. This means that *transient time* is lower for trees than for grass; the nutrients tied up in trees remain there for many dozens or hundreds of years, while the above-ground material in grass dies and decays each year (or every few years for some grasses). Productivity is highest in swamps and marshes because of the high nutrient content and presence of algae, herbs, and grasses that grow rapidly. Standing biomass is greatest in tropical forests because the trees are older and taller, do not lose their leaves as often, and have the longest growing season (all year).

The Food Web and Energy Transfer

Food Webs and Trophic Levels

Food webs are diagrams of the connections and linkages among organisms in the transfer of energy within ecosystems. They start with plants, go through herbivores, and end with top-trophic-level carnivores. A food web starts with a species that eats no other species and ends with a species that is eaten by no other species. Webs can be simple and involve few organisms or complex, involving hundreds of organisms. Initially ecologists used the term *food chain*, because they linked only a few organisms. The complexity of species interactions was soon acknowledged, and more complex food webs were constructed with dozens and then hundreds of linkages.[16]

Food webs reflect the reality of feeding relationships: who is eating what. It is critical to understand the complexities of food webs to protect or manage ecosystems. And it is important to know the effect of removal of a given species from the food web within an ecosystem.[17] Some species are critical to food webs, and without them the food web may collapse. For example, removing small prey fish (such as anchovies) from a food web may cause the collapse of a local fishing industry. Similarly, overfishing horseshoe crabs (*Limulus polyphemus*) in Delaware Bay may have resulted in the decline of commercial and recreational fisheries, as well as the decline of shorebirds.[18]

Trophic level refers to the main categories of producers (plants) and consumers (animals) and includes autotrophs (organisms that produce carbohydrates from incoming solar radiation); herbivores, which eat plants; carnivores, which eat other animals; and omnivores, which eat plants and animals. Carnivores are often divided into top-level predators (e.g., a mountain lion) and others (e.g., a skunk, raccoon, or domestic cat). Because energy is lost at every trophic level (or link in the food web), fewer organisms can be maintained at each higher trophic level. Thus, in any given ecosystem there are fewer top-level predators than other carnivores, and far fewer carnivores than herbivores. The number of trophic levels is limited because of the decreasing availability of energy, resulting from the inefficiencies in energy transfer from one level to the next.

10 The Praeger Handbook of Environmental Health

Energy Flow

Energy flow through food webs is the most important aspect of ecosystem functioning. Energy flow through ecosystems can be summarized as follows: Producers convert solar energy into biomass, herbivores eat plants, carnivores eat herbivores, top-level carnivores eat other carnivores, omnivores eat anything, decomposers reduce dead bodies to their nutrient components, and these nutrients are then reused by plants through root uptake (see Figure 1.1, top). Food chains are constructed by naming the species that fit into each category and showing the linkages among them.

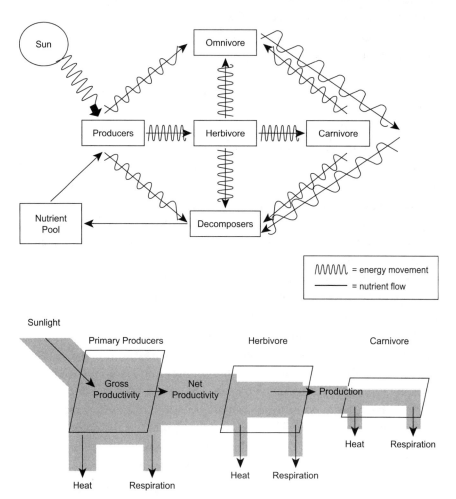

Figure 1.1 Top. Diagram of energy and nutrient cycling in ecosystems. Bottom: Diagram of energy flow from the sun to carnivores, and through the detritus food chain.

As energy flows through the system, it is lost through heat and respiration (to maintain the organism) and is converted into biomass (see Figure 1.1, bottom). The efficiency of organisms in each trophic level is fairly similar; that is, how much biomass can be produced by eating a certain amount of food is relatively constant. The efficiency of plants (producers) ranges from 5 to 20 percent, of herbivores from 8 to 16 percent, and of carnivores from 5 to 22 percent. This suggests that the same amount of energy is used to maintain the body for each trophic level.

The basic energy relationships result in important facts of ecosystems: 1) energy is lost with each transfer from one organism to another, 2) the number of links in the food chain is limited, 3) the number of trophic levels is limited, and 4) more organisms can be maintained by eating plants than by eating animals. By eating meat, people are increasing the number of links in the food chain and decreasing the available food. For example, 17,850 pounds of grain can maintain forty-two children who weigh 50 pounds for one year. The same 17,850 pounds of grain, if cycled through beef (resulting in 2,250 pounds of beef), can sustain only two children weighing 50 pounds. Humans make a trade-off when they decide to eat meat rather than grain, and there is a cost to the ecosystem.

Population Dynamics and Species Interactions

Energy and nutrient flow does not happen in the air, but in organisms, and it is the interactions among individuals, species, and populations that help define the ecosystem. Maintaining ecological health requires understanding both the factors that contribute to population stability and the interactions of species within communities.

Population Dynamics

Darwin and early biologists recognized that, if unchecked by nature, populations would continue to increase until there was no room left. Fish females each lay hundreds of eggs, yet the oceans are not crowded with fish; insects each lay hundreds of eggs, yet the fields are not black with hordes of insects; sea turtle females each lay a hundred eggs, but sea turtles are all currently endangered; snakes can lay up to tens or dozens of eggs, yet the earth is not overrun with them; and human females can produce 150 viable eggs or more, yet women never have this many offspring.

Populations normally increase and at some point level off, and they do not continue to increase. The leveling off point is where the population is limited, usually by resources (food, space, mates, nesting or breeding sites) or interactions with other species (competition, predation). The leveling off point is often called the "carrying capacity" of the habitat or environment; it is the population level that can be supported by a given ecosystem.

The ability of a population to recover or increase is dependent upon two factors internal to the species itself: the reproductive potential of the individual and the number of years an individual can breed. Reproductive potential is a function of age of first reproduction, the number of young that can be produced, and the number of years individuals reproduce. Thus an insect that breeds when it is only a few weeks old, lays dozens or hundreds of eggs at one time, but lives only a few years may have a lower reproductive potential than a fish that breeds when it is a few months old, lays dozens or hundreds of eggs at one time, but lives for dozens of years. In contrast, a seabird that doesn't breed until it is ten years old (e.g., an albatross), lays only one egg at a time, but can live to be forty or fifty years old has a greater potential to produce more offspring that a songbird that lives only three or four years, breeds at one year of age, and lays only three or four eggs at a time. For animals, productivity is the number of young produced that reach breeding age, not the number of eggs or young initially produced. In nature populations are limited by both intrinsic factors (age to first breed, number of eggs produced, years to reproduce) and extrinsic factors that include habitat suitability, weather constraints (storms, temperature), predators, and competitors.

Competition and Predation

Two of the most important species interactions controlling population dynamics and food chain relationships are competition and predation. *Competition* occurs when two or more organisms use the same resource that is in short supply. *Predation* is one organism killing another for food. Both predation and competition affect not only the organisms directly involved, but also other organisms in the food web.[19]

When a predator eats a prey organism, there are secondary effects, such as removal of the prey organism from the food web, which 1) prevents other predators from eating it; 2) prevents that prey organism from eating other animals or plants; 3) prevents that organism from competing with others in its same trophic level; and 4) provides more space within the habitat for other organisms to flourish. These secondary effects are called "cascading effects": the effect of one predation event having numerous effects on the food web.[20]

Competition has similar effects. Competition between organisms often results in one individual or species obtaining a greater share of the resources, decreasing the population of one species at the expense of another. Competition is most severe among similar organisms because they use the same resource that is in short supply. That is, two species of songbirds that both eat seeds are in competition, whereas a songbird that eats seeds and one that eats worms are not.

Top-Down or Bottom-Up Processes

The movement of nutrients, detritus, prey, and consumers among habitats is a feature of food webs and population dynamics. Bottom-up effects occur when a change in the behavior or population levels of organisms at the base of the food chain (limited by resources, habitat, or food) affects energy transfer and population levels higher up. Top-down effects occur when changes in organisms at the top of the food web (grazers or predators) affect population dynamics and behavior of organisms below them.[21] Though it was once thought that primary producers were influencing ecosystem dynamics, it is now clear that predators and competitors can influence productivity and population dynamics of the organisms around them.[22] That is, removing a particular predator from a system allows predators who are competitors to increase, as well as influencing the population levels of their prey (which, in turn, affects the plants they eat).

Ecosystem Vulnerability to Stressors

All levels of biological organization (organisms, populations, communities, ecosystems, landscapes) are vulnerable to biological, physical, and chemical/radiological stressors. Biological stressors such as diseases can destroy individuals or populations, and invasive species can replace native ones and change community dynamics. Both processes can lead to species extinctions and have community-level effects. Physical disruptions are caused by storms, floods, earthquakes, volcanoes, and other natural disasters that can modify the structure of ecosystems. Chemical and radiological stressors can be natural, such as naturally occurring mercury in seawater or arsenic in some soil that gets into the water.

Species or species assemblages, however, evolved with stressors and have mechanisms for adapting, although not necessarily to rapid or massive human-induced stressors. The mechanisms that allow species to cope with natural stressors (e.g., intrinsic genetic variability) also allow species to recover eventually from anthropogenic stressors. There are three types of adaptation to disturbances: 1) of species, through natural selection (a long process), 2) because of the plastic behavior of individuals, and 3) involving adjustments of assemblage structure and function. An example of the latter is a switch in prey type because preferred prey was eliminated or decreased due to a disturbance (such as a storm or chemical). The latter two responses may preserve ecosystem health in the face of disturbance. For example, the devastation of a natural fire is similar to that of a fire following a controlled burn or a bombing; ecosystems can usually recover.

Ecosystems (and the species and populations within them) experience some degradation in response to a wide range of stressors, both natural and anthropogenic. The degree of degradation depends upon the vulnerability of the species composing that ecosystem; deleterious effects in some organisms have a greater effect on overall ecosystem integrity than others. For example, the devastation of hardwood trees in the northeastern United States by gypsy moths can have a greater effect on forest health than an insect that affects a ground cover species, because the hardwoods are the dominant species.

With time, ecosystems and individual species begin to recover; given enough time, the same or a similar ecosystem can develop. The time needed to recover depends on the strength of the stressor, whether there are multiple stressors, and the individual species effects, among other factors. Physical disruptions, however, have the potential to cause additional degradation, particularly if soil is removed, habitat is destroyed (e.g., capping), and roads are built into the habitat. Roads allow nuisance, exotic, and introduced species to move into intact habitats. For example, roads through an otherwise pristine environment allow foxes, raccoons, and other predators to move into interior forest regions, causing increased predation rates on eggs in turtle and snake nests, small rodents or lizards, and neotropical migrants that breed only in interior forests.

Ecosystem Health and Human Health

Ecosystem disruption and degradation, the opposite of ecosystem health and well-being, can affect human health in a number of ways: 1) loss of food, fibers, medicines, hunting and fishing opportunities, and gathering opportunities; 2) loss of recreational, subsistence, or cultural practices; 3) loss of the protective value of intact ecosystems, buffering of catastrophic events, and maintenance of stable ecosystems; 4) acute or chronic illness and disease;, and 5) stress, cultural, or other societal issues, particularly connected to habitat loss or contamination. All of these issues directly and indirectly affect human health and can have chronic and acute, as well as lethal, effects. A number of examples of ecosystem disruptions and their effects on ecosystem and human health are given in Table 1.3.

TABLE 1.3
Effects of Ecosystem Degradation on Ecological and Human Health*

Degradation Type	Effect on Human Health
Lower Productivity	
Direct Effects	Less food, fiber, medicines, cultural products (less total biomass produced)
	Lower-quality food, fiber, medicines, cultural products
	Potential for small- or large-scale famine in some communities or regions
	Loss of medicines or other products from specific biota
Indirect Effects	Fewer hunting and fishing opportunities for recreational, cultural, or religious purposes
	Fewer suitable places for recreational, cultural, or religious activities that require pristine environments
	Loss of the aesthetic or existence values of seeing healthy and pristine environments
	Potential shortening of links in food webs

(continued)

TABLE 1.3
(Continued)

Degradation Type	Effect on Human Health
Loss of Species Diversity	
Direct Effects	Loss of particular food items, such as species of fish, game, or herbs
	Potential loss of critical food sources, such as fish (tuna), shrimp, fish oil, or other foods from nature
	Economic losses due to collapse of fish stocks or other species
	Loss of currently used medicinal and other cultural products
	Disruption of food webs, creating cascading effects
	Loss of recognizable species assemblages (such as "shorebirds," "whales," or "sea turtles")
	Potential loss of endangered, threatened, or species of special concern
Indirect Effects	Loss of aesthetic or existence values of seeing diverse assemblages of species
	Fear of possible collapse of productive food webs in particular habitats or biomes
	Loss of plant or animals species that might (in the future) provide as yet undiscovered foods, nutrients, or medicines
Fragmentation	
Direct Effects	Loss of habitat for natural ecosystems that provide wild foods
	Loss of hunting, fishing, and other recreational, cultural, and religious environments
	Loss of buffer lands for preservations or national parks; preservation of coastal habitats from storms and hurricanes
	Loss of endangered, threatened, and species of special concern
	Loss of recreational, religious, or cultural lands
	Loss of buffer lands to protect natural habitats from human development
	Loss of archeological or cultural sites
Indirect Effects	Possible effects on species that require specific home range sizes that are heretofore unknown
	Stress of knowing habitats are being lost; cultural and religious habitats are destroyed
	Loss of commercially viable species, possible starvation or food stress of human populations dependent upon native species (such as fish)
	Unknown cascading effects

(continued)

TABLE 1.3
(Continued)

Degradation Type	Effect on Human Health
Invasive Species	
Direct Effects	Loss of species, changes in species composition, changes in food webs
	Loss of natural foods for humans; loss of recreational and subsistence hunting, fishing, and gathering because of competition of invasive species with native species
Indirect Effects	Human stress of not knowing the long-term effects of invasive species on food, fiber, medicines, and other ecosystem services
Chemical/Radiological Contamination	
Direct Effects	Loss of clean drinking water, clean air, clean soil, and unpolluted lakes, rivers, and oceans
	Loss of food, fiber, medicines, and other products because they are contaminated
	Lethal and sublethal effects on humans and other populations of plants and animals living adjacent to the sites
	Increased mutations or other genetic changes in humans and other biota
	Loss of fish, wildlife, herbs, roots, berries, medicinal plants, and fiber due to contamination
	Loss of habitat for plants and animals (longer-term effects from radiological contamination and heavy metals)
	Loss of habitat for people for recreational, cultural, religious, or archeological purposes
Indirect Effects	Emotional, religious, and cultural stresses of knowing that habitats can no longer be visited or used
Global Destabilization (Warming)	
Direct Effects	Changes in ecosystem structure and function
	Decreases in food production or spatial distribution of food, creating hardships and distribution problems for people and communities
	Increases in diseases in humans and other organisms due to increased temperatures and longer summer seasons
	Losses due to increases in insect populations (in foods, some other products)
Indirect Effects	Changes in the quantity and interspersion of habitat types
	Cascading effects on food webs

*These are meant only as examples. The degradation types can occur through loss of habitat (e.g., development), physical disruptions (e.g., road building, soil removal), biological disruptions (e.g., diseases, invasive species), and chemical/radiological stressors.

Goods and Services

Humans are absolutely dependent on healthy ecosystems for food, fiber, medicines, herbs, clean water and clean air, hunting and fishing opportunities, and a host of other ecosystem goods and services.[23] However, there are a range of other goods and services that are less obvious, such as protecting the coastline from hurricanes and tidal floods (mangroves and saltmarshes) and providing windbreaks on the prairies (trees; see below). In addition there are aesthetic, religious, and existence values that ecosystems provide, and these fall into the category of "eco-cultural attributes" (the interdependencies that people have with cultural or religious resources that have a necessary natural resource component.[24] Eco-cultural attributes tie together the Western view of the goods and services ecosystems can provide with the aesthetic, religious, and existence values so critical for people who are close to the land, including American Indian nations.

Disease, contamination, and other stressors that affect natural ecosystems can quickly transfer to anthropogenic ecosystems, such as farms, dairies, and fisheries. A disease outbreak in wild ungulates can transfer to cattle, and vice versa. Diseases that affect weeds in hedgerows can move to farm fields. Insect outbreaks in forests and abandoned fields can move to farm fields, and conversely, loss of pollinators (e.g., the recent loss of bees in many places) can affect commercial crops, fields, and forests. When deer, woodchucks, and other animals run out of wild foods, they turn to agricultural fields and suburban gardens or simply move into these environments.

Loss of Recreational, Subsistence, and Other Cultural Attributes

Ecosystem disruptions result in the loss of recreational, subsistence, medicinal, and cultural opportunities. Loss of habitat has the effect of rendering these activities completely impossible on lost land, but degradation due to natural causes (hurricanes, floods, fires, earthquakes), as well as anthropogenic causes (human-caused fires, contamination, destruction because of development), also decreases the value of the land for these purposes. If a natural area is cut in half, the ecosystem is usually degraded by a greater degree, because the minimum space for healthy populations of some plants and animals may no longer be available. For example, cutting a forest in half may result in loss of neotropical migrants because the amount of interior forest is no longer sufficient; foxes, large snakes, or other large animals may no longer have sufficient territory; and invasive plants and animals will have a greater ability to penetrate the forest interior (decreasing nesting success for ground-nesting birds and increasing rates of disease, predation, and competition). Similarly, people will no longer have the recreational, religious, medicinal, or other opportunities a large forest presented.

The interactions and relationships between ecosystem goods and services and human health have been woven together by Harris and Harper into eco-cultural dependency webs that model and describe the interlocking system of people and biota that is sustainable over time.[25] Eco-cultural attributes add the aesthetic and religious components to the totality of human dependence and coexistence within ecosystems that leads to sustainability.

Protective Value

Intact ecosystems protect other ecosystems, as well as people, from physical and catastrophic stressors, such as volcanoes and earthquakes, fires, hurricanes, and tornadoes. Salt marshes and mangroves that border coasts protect the interior land from high tides, storms, and hurricanes. Without salt marshes and mangroves, mainland areas would quickly be eroded by natural forces. Large streams and rivers provide natural firebreaks. Small woodlots and forests provide windbreaks for storms moving across the prairies.

Intact ecosystems have robust species diversity, and that diversity protects the whole ecosystem from disease and other stresses. For example, if one plant or animal species dies out in that ecosystem, other species can take over that function. If the system has only one or two species that fill a general role in that system, then loss of one species could result in the collapse of that system. Loss of one species is a greater problem in systems that have naturally low species diversity, such as tundra systems, because there are fewer other species to take up the role of a species that disappears. Without the arctic fox, tundra ecosystems might be overrun with lemmings. Similarly, in the northeastern United States, the loss of natural predators of white-tailed deer, such as mountain lions and wolves (and even sufficient human hunters), has resulted in an overabundance of deer. Backyard gardens and other plants provide sufficient food for deer populations to increase even more.

Acute and Chronic Illnesses and Death

The aspect of ecosystem health that most concerns people, in addition to habitat and species loss, is the intersection of contamination and human health. This is the most obvious effect of ecosystem degradation on people, largely because direct exposure of people to contaminated water, air, and food can lead to diseases and mortality (of biota, as well as humans). Classic examples of acute, chronic, and lethal effects of chemicals include 1) IQ deficits and learning problems for children exposed to lead from lead paint and lead in gasoline;[26] 2) neurobehavioral effects from mercury and PCBs, particularly derived from high levels in fish;[28] 3) exposure to asbestos, even in drinking water;[27] 4) dioxins and the Agent Orange used during the Vietnam War; 5) endocrine disruptors, whose effects are often contested for humans, but which have clear and devastating effects in animals (and animal models);[29] and 6) respiratory diseases caused by particulate matter in the air (e.g., smog), among others. There are also some acute exposures that can lead to immediate illness or even death, such as respiratory disease from marine toxins like red tides, which are caused by runoff leading to plankton blooms that can make people and biota sick or cause mortality.[30]

Chemical and radiological exposures are not the only kind of ecosystem anthropogenic disruption that can cause illness and death. Disruptions to ecosystems can result in changes in the environment that cause increases in disease in both biota and humans. For example, during mild winters some insect pests do not die out, resulting in higher populations the following year, higher rates of plant (or animal) diseases (and thus lower crop yields), and higher incidences of human disease (such as Lyme disease or West Nile disease).

Stress, Cultural, or Other Societal Issues

Illness and death caused by chemicals (or microbes) is clearly recognizable as environmentally caused and due to ecosystem degradation and contamination. However, people experience stress when they lose habitats or species, or when they are unable to conduct their recreational, religious, medicinal, or cultural activities in habitats, and such stress can lead directly to illness. This effect is most clearly recognized among Native American tribes that have had their tribal lands contaminated or destroyed by development or activities of developers, decreasing their ability to depend on the diversity of plants and animals (including fish) traditionally hunted and gathered for food, medicines, and religious ceremonies.[31]

Conclusion

Finally, and perhaps most important, ecosystems are not healthy if they are not sustainable. The definition of *sustainability* approved by the United Nations Food and Agriculture (FAO) Council in 1988 is the handling and conservation of natural resources and the orientation of technological and institutional change so as to ensure the continuous satisfaction of human needs for present and future generations.[32] A more accepted definition of sustainability is meeting the needs of the present without compromising the ability of future generations to meet their own needs.[33] Ideally, the use or extractions from an ecosystem should not compromise the biological structure and functioning of the system, or it is not sustainable on a long-term basis.[34]

Because ecosystems are defined spatially, it is usually the case that nutrients and energy move into or out of the system. That is, in natural systems, animals and seeds move between systems, moving nutrients and energy. In agricultural systems, fertilizers, seeds, and water are added, and energy and nutrients are removed during harvesting. Ideally, food webs are not subsidized from elsewhere.[35] However, ecosystems (at any level of definition) are interconnected, and they function together such that in a region or larger area they are self-sustaining. In an ultimate sense, the earth functions as an ecosystem, since only energy (via sunlight) enters, and otherwise nutrients and energy are cycling within the system.

Ecosystems are sustainable only if they exhibit appropriate structure and function, can continue these functions into the future; are resilient and able to recover from natural events and catastrophes (such as fires or floods); and continue to provide the goods, services, and eco-cultural attributes people need. This definition implies a system in which nutrients and energy continue to cycle, with the appropriate species diversity, species abundances, and trophic-level relationships. Disruption of any of these aspects will eventually lead to decreases in ecosystem health and in human health, because we depend on these systems to provide all our resource and energy needs, as well as environments for recreational, medicinal, cultural, aesthetic, and religious activities.

Notes

1. J. Burger, T. M. Leschine, M. Greenburg, et al., Shifting priorities at the Department of Energy's bomb factories—protecting human and ecological health, *Environmental Management* 31(2) (2003): 157–167.
2. E. T. Odum, *Fundamentals of ecology* (Philadelphia, PA: Saunders, 1957); M. L. Cain, W. D. Bowman, and S. D. Hacker, *Ecology* (Sunderland, MA: Sinauer Associates, 2008); T. M. Smith and R. L. Smith, *Elements of ecology* (San Francisco: Benjamin Cummings, 2008); M. C. Molles, *Ecology: Concepts and applications* (New York: McGraw-Hill, 2009).
3. A. B. Leitao and J. Ahern, Applying landscape ecological concepts and metrics in sustainable landscape planning, *Landscape and Urban Planning* 59 (2002): 65–93; J. Burger, Environmental management: Integrating ecological evaluation, remediation, restoration, natural resource damage assessment and long-term stewardship on contaminated lands, *Science of the Total Environment* 400 (2008): 6–19; J. Burger, M. Gochfeld, K. Pletnikoff, et al., Ecocultural attributes: Evaluating ecological degradation: Ecological goods and services vs. subsistence and tribal values, *Risk Analysis* 28 (2008): 1261–1271.
4. J. Cairns Jr. (ed.), *Rehabilitating damaged ecosystems* (Boca Raton, FL; CRC Press, 1995).
5. P. A. Keddy, Assembly and response rules: Two goals for predictive community ecology, *Journal of Vegetation Science* 3 (1992): 157–164.
6. G. E. Hutchinson, Homage to Santa Rosalia, or why are there so many kinds of animals? *American Naturalist* 93 (1959): 145–159; R. H. MacArthur, Patterns of species diversity, *Biological Reviews* 40 (1965): 510–533.
7. R. M. May, *Stability and complexity in model ecosystems* (Princeton, NJ: Princeton University Press, 1973).
8. A. A. Berryman, The origins and evolution of predator-prey theory, *Ecology* 73 (1992): 1330–1335; A. R. Ives, B. J. Cardinale, and W. E. Snyder, A synthesis of subdisciplines: Predatory-prey

interactions, and biodiversity and ecosystem functioning, *Ecology Letters* 8 (2005): 102–116.
9. F. Briand, Environmental control of food web structure, *Ecology* 64 (1983): 253–263.
10. D. Tilman, Competition and biodiversity in spatially structured habitats, *Ecology* 75 (1994): 2–16.
11. J. B. Dunning, B. J. Danielson, and H. R. Pulliam, Ecological processes that affect populations in complex landscapes, *Oikos* 65 (1992): 169–175; R. T. T. Forman and M. Godron, *Landscape ecology* (New York: Springer-Verlag, 1986).
12. Odum, *Fundamentals of ecology*; R. T. Payne, Food web complexity and species diversity, *American Naturalist* 100 (1966): 65–75; C. T. Hunsaker, R. L. Graham, G. W. Suter II, et al., Assessing ecological risk on a regional scale, *Journal of Environmental Management* 14 (1990): 325–332.
13. S. D. Fretwell, Food chain dynamics: The central theory of ecology, *Oikos* 50 (1987): 291–301.
14. J. Burger, The effect on ecological systems of remediation to protect human health, *American Journal of Public Health* 97 (2007): 1572–1578.
15. Odum, *Fundamentals of ecology*; E. Kormondy, *Concepts of ecology* (Upper Saddle River, NJ: Prentice Hall, 1996).
16. Briand, Environmental control of food web structure; S. L. Pimm, J. H. Lawton, and J. E. Cohen, Food web patterns and their consequences, *Nature* 350 (1991): 669–674; S. J. Hall and D. Raffaelli, Food-web patterns: Lessons from a species-rich web, *Journal of Animal Ecology* 60 (1991): 823–841.
17. R. J. William and N. E. Martinez, Simple rules yield food webs, *Nature* 404 (2000): 180–183.
18. L. J. Niles, H. P. Sitters, A. D. Dey, et al., Status of the red knot (Calidris canutus rufa) in the western hemisphere," *Studies in Avian Biology* 36 (2008): 1–185.
19. J. Roughgarden, Competition and theory in community ecology, *American Naturalist* 122 (1983): 583–601.
20. T. L. DeVault, O. E. Rhodes, and J. A. Skivik, Scavenging by vertebrates: Behavioral, ecological, and evolutionary perspectives on an

important energy transfer pathway in terrestrial ecosystems, *Oikos* 102 (2003): 225–234; M. C. Emmerson and D. Raffaelli, Predator-prey body size, interaction strength and stability of a real food web, *Journal of Animal Ecology* 73 (2004): 399–409; G. C. Trussell, P. J. Ewanchuk, and C. M. Matassa, The fear of being eaten reduces energy transfer in a simple food chain, *Ecology* 87 (2006): 2979–2984.

21. S. A. Levin, T. M. Powell, and J. H. Steele, *Patch dynamics* (Berlin: Springer-Verlag, 1993).
22. D. A. Spiller and T. W. Schoener, Effects of top and intermediate predators in a terrestrial food web, *Ecology* 75 (1994): 182–196; G. A. Polis and D. R. Strong, Food web complexity and community dynamics, *American Naturalist* 147 (1996): 814–846.
23. R. S. deGroot, M. A. Wilson, and R. M. J. Boumans, A typology for the classification, description and valuation of ecosystem functions, goods, and services, *Ecological Economics* 41 (2002): 393–408; J. Burger, M. A. Carletta, K. Lowrie, et al., Assessing ecological resources for remediation and future land uses on contaminated lands, *Environmental Management* 34 (2004): 1–10.
24. Burger et al., Ecocultural attributes.
25. S. G. Harris and B. L. Harper, Using eco-cultural dependency webs in risk assessment and characterization of risks to tribal health and cultures, *Environmental Science and Pollution Research* 2 (2000): 91–100.
26. D. C. Bellinger and A. M. Bellinger, Childhood lead poisoning: The torturous path from science to policy, *Journal of Clinical Investigation* 116 (2006): 853–857.
27. P. M. Cook, G. E. Glass, and J. H. Tucker, Asbestiform amphibole minerals: Detection and measurement of high concentrations in municipal water supplies, *Science* 185 (1974): 853–855.
28. Institute of Medicine (IOM), *Seafood choices: Balancing benefits and risks* (Washington, DC: National Academy Press, 2006).
29. National Research Council (NRC), *Hormonally-active agents in the environments* (Washington, DC: National Academy Press, 1999).
30. Massachusetts, *Red tide (paralytic shellfish poisoning)* (Boston: Health and Human Services, 2010).
31. B. L. Harper, A. D. Harding, T. Waterhous, and S. G. Harris, *Traditional tribal subsistence exposure scenario and risk assessment*

guidance manual (U.S. Environmental Protection Agency, 2008), EPA-STAR-J1-R831-46, http://www.hhs.oregonstate.edu/ph/sites/default/files/xposure_Scenario_and_Risk_Guidance_Manual.
32. F. Cena, The farm and rural community as economic systems, in *Rural planning from an environmental systems perspective*, edited by F. B. Golley and J. Gellot (New York: Springer, 1999), 229–286.
33. United Nations (UN), Report of the World Commission on Environment and Development, General Assembly resolution 42/187 (December 11, 1987), http://www.un.org/documents/ga/res/42/ares42-187.htm (accessed February 2010).
34. H. Cabezas, C. W. Pawlowski, A. L. Mayer, and N. T. Hoagland, Simulated experiments with complex sustainable systems: Ecology and technology, *Resources, Conservation and Recycling* 44 (2005): 279–291.
35. G. A. Polis, W. B. Anderson, and R. D. Holt, Toward an integration of landscape and food web ecology: The dynamics of spatially subsidized food webs, *Annual Review of Ecology Systems* 28 (1997): 289–316.

Acknowledgments

I thank M. Gochfeld, M. Greenberg, C. W. Powers, J. Clarke, B. Goldstein, C. Chess, and D. Wartenberg for stimulating discussions about ecological and human health. Partial funding was provided by Consortium for Risk Evaluation with Stakeholder Participation (CRESP; Department of Energy; DE-FG 26-00NT 40938, DE-FC01-06EW07053), NIEHS (ESO 5022), Nuclear Regulatory Commission (38-07-S02M02), EPA, Trust for Public Lands, Wildlife Trust, and New Jersey Department of Environmental Protection. The interpretations reported herein are the sole responsibility of the author and do not represent the views of any funding agency.

2

Human Exposure Assessment

Derek G. Shendell

Introduction

Worldwide, people across age groups, gender, and race/ethnicity live in urban, suburban, and rural communities of varying levels of development, geographic size, and population density. People live and clean, cook, eat, learn, play, pray, provide care, research, sleep, study, travel, work, and worship every minute of every day during their lives. From an environmental health perspective, it is important to have a fundamental understanding of the potential exposure agents (hazards) of concern and the various factors that may make a person or community more, or less, susceptible and vulnerable. An identified hazard, or mixture of agents, poses risks of adverse acute and/or chronic human health effects, as well as ecological health and environmental quality (negative aesthetic) effects. This chapter introduces human exposure assessment and cross-references other chapters in other volumes of this handbook.

Human and ecological exposure assessment is one of the four core components of the traditional quantitative risk assessment (QRA) process. QRA is more recently discussed as having six components, the four core and the additional components of risk management and risk communication. Exposure assessment is the third step of QRA, and the first step is hazard identification; that is, what are the exposure agents of concern? The second step of QRA is dose response, and the fourth step is risk characterization, which incorporates steps one through three. In addition, the science of exposure assessment has received renewed attention in the private and public (government and nonprofit advocacy) sectors at the federal, state, and local levels for research, emergency planning, and response to natural and human disasters; consulting services; interventions; and old, abandoned hazardous waste site cleanups. (For additional information, see volume 1, chapters 14–19 and 24, volume 3, chapters 22–25, and volume 4, chapters 21–22 and 24.)

This chapter provides an overview of human exposure assessment principles for policy, practice, research, and services. Concepts and definitions outlined include sources of pollution-emitting agents/hazards; the hazards themselves, which may be biological, chemical, ergonomic, physical, psychosocial, or radiological; and subpopulations known or thought to be more susceptible and vulnerable to environmental and occupational exposures, as well as health disparities.

Core Concepts and Definitions

Exposure is actual contact between a person (or animal, plant, fish, etc.) and an agent (pollutant, hazard), or the vector of a known agent. The agent would have been released or emitted from one or more sources and reach one or more environmental media through transport pathways. The exposure or actual contact occurs through one or more defined routes, at or over a defined period of time in some geographic area, such as a defined location or microenvironment (space) in the community (environmental exposure), or at a workplace (occupational exposure). See volume 3, chapter 22, on geographic information systems in urban waste generation. Needham and others[1] provided an easily understandable flowchart of the exposure assessment process in their article about the development of the U.S. National Children's Study.

Another way to conceptualize exposure is to consider the relative contributions from three aspects of exposure, that is, the contributions of ambient (outdoor) exposure and nonambient exposure (indoors, whether inside a home, school, workplace, stores, mode of transport, etc.) to total personal exposure. See Figure 2.1 and the figure and accompanying text in Wilson and Brauer.[2]

There have traditionally been four levels of measurements conducted in exposure science. First, measurements from models based on various input data parameters may offer crude-to-detailed estimates; sensitivity

Human Exposure Assessment 29

DOSE MAKES THE POISON ... BUT WHAT MAKES THE DOSE?

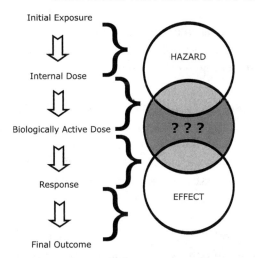

Questions to Consider:

* What are the agents of concern?

* To what concentration are individuals exposed, and in which media?

* How do different subjects process the agent upon exposure, thereby yielding the internalized dose?

* How do different physiological systems deal with agents after entry into the bloodstream?

* Do the subjects implement or receive any preventive measures or treatments?

* What physical and/or mental effects do subjects experience?

Example: People drinking water from the Broad Street Pump in London during the mid-1800s all got exposed to *Vibrio cholerae* bacteria, but not all developed cholera. Of those who did develop the disease, some became extremely ill and died while others experienced mild symptoms and survived. Possible reasons for this disparity include individual pathology; variable concentration of *V. cholerae* in any given sample of water; and handling of water (e.g., boiling) prior to ingestion.

Figure 2.1 A Venn diagram attempting to relate human exposure to dose—as the frequency, duration, and magnitude of exposure—and to potential adverse acute and chronic health outcomes within toxicological and epidemiologic research. The gray area in the middle of this Venn diagram represents the fact that most of the underlying biological and physiological mechanisms, including genetics, and psychosocial factors, are not yet known from research studies for each agent of concern whether considered individually or in mixtures. (Created by Alexandra C.H. Nowakowski and Derek G. Shendell in February 2008 at UMDNJ-SPH)

analyses concerning their accuracy (validity) and precision (repeatability) are also conducted. These measurements involve experts in engineering, mathematics, statistics, and physical sciences. Second, measurements within laboratory experiments such as acute and chronic toxicity tests on animals may be conducted, but these are indirect measurements or estimates of human exposure due to the required extrapolations from the animal to humans plus considerations of gender and age or developmental stage. Third, direct measurements under controlled environmental chamber conditions attempting to simulate indoor microenvironments may be conducted. Finally, direct field measurements at community and occupational locations, at area or personal levels, may be conducted to more closely characterize actual exposure. If the monitoring is conducted at one or more metropolitan area (city or county) central monitoring locations, the measurement can be viewed as an estimated exposure over (for) a defined geographic area/region. (See also volume 1, chapters 8–9 for a discussion of environmental epidemiology and its use of measurements of exposure.)

If monitoring is conducted outside a private home or apartment building for air or at a point in the distribution system for water, the measure is an even better estimate of exposure, but still not as rigorous as personal monitoring techniques. Personal monitoring techniques have been developed and are continuously being refined, to assess 1) the "personal cloud" of air breathed indoors and outdoors; 2) food consumed via daily or weekly diaries and duplicate diet samples; 3) water and cold and hot beverages based on tap water consumed; 4) dust and soil particles consumed during hand-to-mouth or hand-to-object (or food-to-mouth) behaviors; and 5) biomarkers of exposure and/or effect. (See also volume 1, chapter 13, on biomarkers.)

These various exposure measurements may be quantitative or qualitative. The types and units of measurement of quantitative measures in samples of air, whether indoors or outdoors, in water, in soil and sediments, in surface dusts, in urine, and in blood have a *numerator* with a measurement of the weight (e.g., in grams, micrograms, or nanograms) or number (e.g., of coliform bacteria, pollen spores, particles of a specified size fraction, or lead) of a targeted agent or mixture of agents and a *denominator* defining a normalized weight or volume (e.g., liter or 100 milliliters of water sampled, cubic meter of air sampled, deciliter of blood volume sampled). Semiquantitative data may come from recorded diaries (e.g., human time-location-activity patterns, food/nutrition records, etc.). Qualitative data may come from using technician walk-through observational surveys, in-person questionnaires with consenting participants, and focus group techniques involving a designated person to transcribe discussions.

Exposure measurements are conducted by government agencies (federal, state, regional) and by the private sector to evaluate/monitor compliance with laws such as the Clean Air Act and its amendments, the Clean Water Act and its amendments, and the Safe Drinking Water Act and its amendments, as well as by clinicians and researchers in basic and applied sciences. In addition, exposure measurements and modeled estimates of exposure at various geographic levels (spatial resolution) have recently been used by federal and state agency collaborations initiating an environmental public health tracking network and in developing indicators of the effects of and (human and ecological) population adaptations to global climate change. See Table 1 in English et al.[3] (See also volume 1, chapters 1–3, 6–9, 15–18, and 23 as well as volume 3, chapter 11, for details on and examples of these uses of exposure assessment.)

Agents of Exposure

There are four main categories of exposure agents, which may also be classified as carcinogenic or noncarcinogenic. There are numerous specific examples, and most have natural and human sources, both indoors and outdoors, and/or sources related to human activity indoors. Note that the latter include agents with outdoor sources that are transported inside in water; in air via natural and/or mechanical ventilation; in soil and sediment on shoes; and in dust on clothing and other objects, including personal items such as bags and work equipment. The four categories are detailed below.

1. Biological (see volume 2, chapters 1–6, and volume 3, chapters 1–6, and 22–24)
   - Bacteria, including metabolic products and cell-wall components
   - Mold (fungi, mildew, spores)
   - Pollen from various trees, flowers, and plants
   - Protozoa
   - Viruses

   Note that some species of agents have human (engineered in the laboratory) sources. Note also that biological sources may create toxins and/or chemical toxicants due to metabolic processes; for example, microbial volatile organic compounds (MVOCs). The presence of these agents—that is, above the analytical detection limit—and levels they are found in—that is, measured concentrations per unit of environmental media sampled (e.g., in grams, milligrams, micrograms, milliliters, and liters)—depends on several factors:
   - Design and construction (e.g., materials selected) of interior spaces, which can affect the presence of and sources of food and of moisture required to live and grow

- Presence of natural and/or mechanical ventilation with particle filtration and possibly even removal of gases via activated carbon/charcoal filters
- "Sinks" such as carpets, which can capture and retain pollution
- Various sources of these agents both indoors and outdoors

2. Chemical (see volume 2, chapters 7–21 and 25, and volume 3, chapters 1–25)
   a. Organic (containing carbon and hydrogen atoms), with or without metals.
      i. Organic chemicals, including many volatile toxic air contaminants or hazardous air pollutants. They are typically known as volatile and semivolatile organic compounds (VOCs and SVOCs, respectively) in air (indoors and/or outdoors), on particles or in aqueous coatings of aerosols in the air, or on suspended particles or within humic substances in the water and sediments.
      ii. Persistent organic chemicals, referred to by named categories that represent many different congeners of configurations of atoms: for example, polychlorinated biphenyls, dioxins and DDT (for more details, see volume 2, chapters 15–19).
   b. Nonorganic (does not contain carbon), with nitrogen, sulfur, hydrogen, oxygen, and/or metals.
      i. Carbon dioxide (CO_2)
      ii. Carbon monoxide (CO) (see volume 2, chapter 12)
      iii. Metals, including heavy metals and transition elements on a chemistry periodic table. Metals can exist in elemental form as solids, liquids, or vapors (gas), and can be attached to or adsorbed to particles as well as in salt complexes with water. Metals may be found as organic compounds containing carbon and hydrogen, including metal-salt complexes in aqueous solution, or as inorganic compounds, including when mined in solid form. Metals include arsenic, boron, copper, cadmium, chromium, lead, magnesium, manganese, and mercury. Mercury is a metal posing issues in its organic and inorganic forms. (See volume 2, chapters 7–11.)
      iv. Particulate matter, or particles, exist in varying size ranges (PM_x) and usually not a perfect sphere in shape. The size refers

to their aerodynamic diameter in micrometers or microns (μm) under a microscope or, more commonly in the present, as counted/measured by laser-based technology. Size fractions most commonly referred to by scientists, engineers, and policy makers from local to global geographic scales over varying time frames are:

- Total suspended particles (TSP), generally ≤15μm
- Respirable particles (PM_{10})
- Coarse fraction of respirable particles ($PM_{2.5-10}$)
- Fine particles ($PM_{2.5}$), which are primary emissions or secondary in origin due to the formation of acid aerosol particles based on nitrates (HNO_3) and sulfates (H_2SO_4)
- Ultrafine particles ($PM_{0.1}$), including nanoparticles that may be designed as perfect spheres but still may vary in composition and thus potential toxicity (Research in this area is still nascent.)

  v. Nitrogen oxides (NO_x), including nitric oxide (NO), nitrogen dioxide (NO_2), and nitrous oxide (N_2O)

3. Physical (see volume 1, chapter 4, and volume 2, chapters 8 and 24).
   a. Asbestos (six types naturally occurring and mined in modern times), especially if in fibrous (friable) and not bound (nonfriable) form
   b. Light (natural or artificial, including fluorescent and incandescent), including ultraviolet (UV) rays such as UVA, UVB, and/or UVC
   c. Noise (frequency, in hertz [Hz]) and loudness, in A-weighted decibels (dBA)

4. Radiological (see volume 2, chapters 22–23). Note that the human sources engineered in the laboratory for research and development, energy resources, or consumer products used in various outdoor and indoor microenvironments can be located both indoors and outdoors.
   a. Electromagnetic fields and other forms of nonionizing radiation
   b. Ionizing radiation, e.g., X-rays
   c. Nuclear power waste fuel rods
   d. Radon gas, which is emitted from a natural source (i.e., underlying geology or rock formations) into outdoor air and water (think well water) and infiltrates indoor microenvironments such as homes and schools via three major pathways:

i. Through cracks in a building's envelope where walls meet
   ii. Through open windows (natural ventilation)
   iii. Through cracks in a building's foundation (soil-to-concrete, etc.) Also, radon is emitted from its source into indoor microenvironments. In homes radon is typically measured passively over twenty-four to forty-eight hours in picocuries per liter of air sampled—as a function of three main factors:
   iv. The particular location's underlying geology (rock formations)
   v. The slope of the land used for the construction site
   vi. The type, condition, and age of the building foundation

In occupational settings, ergonomics may be a subset of physical agents. In community (environmental) and occupational (workplace) settings, another category of agents, psychosocial stress, is of increasing interest to research, policy, programs, and practice.

Environmental Media

Known environmental media relevant to human exposure assessment are listed below. In addition to quantitative measurements (metrics, indicators) in environmental media and human samples (biomarkers of exposure and/or effect), it should be noted that interdisciplinary researchers in environmental public health, urban planning, and public policy have recently started conducting qualitative, semiquantitative, and quantitative measurements of the built or physical environment to inform, with demographic data, land use/zoning decisions as well as environmental impact assessments/statements (in the United States, required by law) and health impact assessments.[4] (See volume 1, chapters 14–16, volume 3, chapter 11, and volume 4, chapters 14–16, 18–20, and 24–25.)

1. Air quality (see volume 3, chapters 7–19):
   a. Indoors, in community locations (child care, community centers, homes, schools, etc.) and occupational settings, either industrial (think factory) or nonindustrial (think call centers and high-rise office buildings, and cleaning crews)
   b. Outdoors, in community locations and occupational settings (think ports, railroads and roadways, and construction and maintenance workers)

2. Dust:
   a. Indoors on surfaces of homes such as countertops, desks, floors, tables, and window sills, as well as in vehicles on the dashboard, seats, and steering wheel
   b. Outdoors in fields, at construction and demolition sites, and on asphalt or concrete surfaces including roads and door entry ways
3. Food, in particular agricultural products/crops, including fruits and vegetables
4. Sediments near surface water resources such as rivers and streams
5. Soils in rural, agricultural, and urban/suburban areas
6. Water resources (which then may impact groundwater aquifers) through rainout and washout as well as acidification by increased levels of carbon dioxide:
   a. Surface water resources including brooks, creeks, lakes, reservoirs, rivers, and streams
   b. Underground or groundwater resources, both unconfined aquifers below the water table and confined or artesian aquifers.
   (See volume 3, chapters 1–6, and volume 4, chapter 23.)

In addition, various bodily fluids and parts can be sampled (i.e., small samples taken by noninvasive, nonintrusive, nondisruptive means or, less commonly for exposure assessments, by an invasive procedure for a bone or skin graft). These samples are then analyzed for identified biomarkers of exposure and/or effect (typically preclinical) as one way to characterize and quantify personal exposure. Biomarkers of exposure and/or effect are potentially obtained from

1. bile produced by the liver,
2. blood (including from the mother's placenta, to estimate prenatal exposure of the fetus late during the third trimester),
3. breast milk,
4. feces via the gastrointestinal tract,
5. fingernails,
6. hair,
7. saliva,
8. sweat,

9. tears,
10. toenails, and
11. urine.

In the United States, the Centers for Disease Control and Prevention's (CDC) National Center for Health Statistics (NCHS), in collaboration with the National Center for Environmental Health (NCEH), leads a national effort through the National Health and Nutrition Examination Survey (NHANES) to update a developed National Report Card on Human Exposure to Environmental Chemicals ("report card").[5] Briefly, NHANES is a survey and set of measurements in mobile field laboratory clinics conducted about every two years on a probability (random) sample of the civilian, noninstitutionalized U.S. population. Participants' blood and urine samples are analyzed. To date, there have been four "report cards": 1999 NHANES data (first report out in 2001, with 27 chemicals assessed); 1999–2000 NHANES data (second report out in 2003, with 116 chemicals assessed); 1999–2002 NHANES data (third report out in July 2005, with 148 chemicals assessed); and 1999–2004 NHANES data (fourth report out in December 2009, with 212 chemicals assessed; plus "Updated Tables" released in July 2010, with 2005–2006 NHANES data for 51 chemicals within the fourth report and six more chemicals assessed for the first time).

Pathways, Including Fate and Transport (Agents from Sources)

How an agent or a mixture of agents travels from one location to another, or from the source(s) of emissions to a specific microenvironment and/or into specific environmental media with which humans may come in contact, is a process called an *exposure pathway*. In general, pathways are determined by a combination of factors:

1. The chemical and physical properties of the agent, or mixture of agents (if known)
2. The weather—temperature, humidity, wind speed and direction, amount of sunlight versus cloud cover
3. Topography of the land as well as the altitude relative to sea level

There are three categories or types of processes, listed below. They are especially relevant to air quality (both indoors and outdoors) as well as to water quality (of both surface water resources and/or groundwater aquifers).

1. Chemical processes:
   a. Absorption (into) and adsorption (attach to)
   b. Ionization
   c. Oxidation/reduction reactions ("redox" reactions in chemistry terminology)
   d. Photolysis, direct or indirect/sensitizing (light hits, then a reaction with target agent occurs)
2. Biological processes:
   a. Bioaccumulation—the tendency of an organic chemical agent to accumulate in an organism. Thus, it is lipophillic ("fat loving") or hydrophobic ("water hating"), but no other mechanistic information is available.
   b. Bioconcentration—accumulation of chemical agents (organic and/or inorganic) present in water by an aquatic organism by dietary and nondietary means
   c. Biodegradation, with or without sunlight
   d. Biomagnification—accumulation of chemical agents, particularly organic compounds, up the food chain, e.g., from plankton and algae to smaller fish and then into larger fish and then into humans
3. Physical processes:
   a. Sorption, or the attachment of one or a mixture of pollutants to sediment or soil particles that are then resuspended when dry and windy conditions persist
   b. Mixing
   c. Volatilization (from water to air or from sediment to air, or from soil to air)

Routes (Agents Contact Targets)

There are three primary routes of exposure relevant to humans during typical daily life.

1. Breathing or inhalation through the mouth and/or nose, generally as a function of a person's level of activity
2. Dermal or through the skin—whether intact or through open cuts—as well as eyes (ocular)
3. Drinking liquids or blended foods and eating solid foods, both known as ingestion

Three other highly relevant routes of exposure to humans are worth noting:

4. Across the placenta (in blood) from a mother to the developing fetus
5. Breast milk from a mother to an infant/toddler (unless exclusively fed using water- or milk-based formulae)
6. Intravenously for prescription or illicit drugs

There are other routes of exposure used in toxicology, that is, in toxicity studies on animals (e.g., mice, rats, hamsters, rabbits, guinea pigs, and dogs) in the laboratory, including controlled environmental exposure chambers. These routes of exposure—and the others listed above—are direct and intentional; for example, a known, applied dose is administered, typically with a needle but also with a mask if inhaled. These are:

7. Intraperitoneal or direct exposure into the stomach cavity
8. Intramuscular or direct exposure into a muscle
9. Subcutaneous or direct exposure from underneath a skin fold

Time and Space: Human Activity Patterns

Issues concerning air, food, soil, and water quality; exposure; and human health or environmental quality (aesthetics, ecological health, visibility) must be examined in both space and time. Time relates to the frequency and duration of exposure. Time can be characterized in two ways.

1. Acute, or one-time, exposures
2. Chronic exposures:
   a. At intervals or intermittently (think about your daily commute to work or school)
   b. Continuous (think about daily exposure to air or water pollution)
   c. Episodic (think about a volcanic eruption)

Along with quantity, frequency and duration define dose. There are two general categories of dose relevant to human exposure assessment research:

1. Internal dose (via pharmacokinetics)
2. Biologically active dose (via pharmacodynamics)

Space is typically thought of in terms of geography relative to altitude, using these terms:

1. Local (town, city, or small county)
2. Regional (multiple counties)
3. State
4. Region (multiple states)
5. National
6. International—more than one country, within or across a region or continents
7. Global—the entire planet may be affected (think impacts of global climate change)

The concepts of time and space help explain a portion of inter- (between) and intra- (within) individual variability in exposure and dose at any geographic level during a defined time period. In general, twenty-four-hour, forty-eight-hour, or weekly diaries at ten-, fifteen-, or thirty-minute resolution are used.

Other Factors Affecting Human Exposure

Other factors affecting human exposure may be at an individual level and relate to differences, or variability, between individuals (inter-individual variability) or within an individual over time and space (intra-individual variability). In addition, it is important to note the role of natural variation and its effect on ambient conditions (weather), psychosocial factors (e.g., behaviors and mood), and physiological parameters (e.g., blood pressure and heart rate). Furthermore, there is an increasing recognition of the potential role of gene–environment interactions in toxicological and epidemiologic research studies, particularly in laboratory and clinical settings, in understanding how exposure relates to dose and adverse effects.[6]

(For more information, refer to Bates et al., p. 1618. See also see volume 1, chapters 5 and 13.) These factors are also known in epidemiology as *covariates* or *potential confounding factors*. These factors are discussed below with respect to human exposure in the context of susceptibility—that is, defining subpopulations of interest—and vulnerability, that is, defining specific groups within an identified subpopulation relatively more at risk to exposure and adverse effects. (See volume 4, chapters 9–11.)

Susceptibility and Vulnerability Factors

Many publications list a number of the commonly known susceptibility and vulnerability factors. For example, there are two tables in an article in the peer-reviewed journal *Environmental Health Perspectives* by deFur et al. available for free on the Internet[7]; table 1 is "Examples of Specific Vulnerability Factors" and table 4 is "Factors Contributing to Vulnerability." As illustrative examples, vulnerability factors are discussed with respect to two commonly identified subpopulations, the youngest and the oldest. Factors listed below can be considered in terms of both susceptibility and vulnerability.

Using children as an example of a susceptible subpopulation in societies worldwide, the following are known vulnerability factors predisposing children to increased exposure and associated risk of adverse acute and chronic health effects:

1. Age (including prenatal, neonatal [0–28 days after birth], postneonatal [28–365 days after birth], infant, toddler, preschool, primary and secondary school age groups)
2. Crawling near ground, indoors or outdoors
3. Gender
4. General dependency on adults (caregivers/parents/guardians/other family)
5. Heightened skin sensitivity
6. Increased breathing rate (resting and active) per pound or kilogram body weight
7. Increased food consumption per pound or kilogram body weight
8. Increased water and other fluid consumption per pound or kilogram body weight

9. Immature immune system
10. Immature central nervous system, including the blood–brain barrier
11. Longer lifespan to still develop chronic illnesses
12. Pica behavior per curiosity (hand-to-object or food-to-mouth)
13. Race/ethnicity
14. Socioeconomic status (SES), which comprises a) education (in this case, of the mother, the father, and/or the individual child depending on age); b) employment (in this case, of the mother and/or father, characterized by industry and/or occupation); and c) income (in this case, the reported family or household annual income)
15. Specific microenvironments in which they spend relatively more time, for example, child care centers and schools as well as parks, recreation and sports facilities, and workplaces (see volume 4, chapters 5–7, 10, and 18–20).
16. Time-location-activity patterns, and thus frequency of contact with environmental media

Using older adults (elderly, senior citizens) as another example of a susceptible subpopulation in societies worldwide, following are known vulnerability factors predisposing them to increased exposure and associated risk of adverse acute and chronic effects:

1. Age
2. Gender
3. Genetics
4. Health insurance, including (in the United States) coverage and benefits from Medicaid and Medicare
5. Health status, including co-morbidity with diagnosed chronic diseases
6. Medications currently used by prescription and purchased over the counter
7. Potentially increased general dependency on adult caregivers/parents/guardians/other family for assistance with or to conduct activities of daily living such as exercise
8. Relatively weaker and/or already compromised immune system
9. Race/ethnicity

10. SES (i.e., the three attributes with respect to the individual in question and other family members living with the person at the time, or, if dependent, then with respect to the family and caregivers)
11. Specific microenvironments in which they spend relatively more time, such as their homes (living independently or with assistance), senior centers, or community centers
12. Time-location-activity patterns, and thus frequency of contact with environmental media

Factors Pertaining to Environmental Media in Indoor Microenvironments

These factors may affect measurements of exposure in various indoor microenvironments in the community and in occupational settings (see volume 4, chapters 1–4, 9, 13, 17, and 20):

1. Occupant density or the number of individuals per unit of surface area or room volume
2. Temperature of water from bathroom (shower, sink, and tub) and kitchen (sink) faucets
3. Temperature of and relative humidity in the air indoors
4. Use of air cleaning devices for bacteria, fungi or mold, particles, and pollen (from grass, flowers, trees), with or without the generation of ozone
5. Ventilation (natural; mechanical, including percent outdoor air versus percent recycled air in supply air and any filtration of particles and/or gases with activated carbon/charcoal filters)
6. Water faucet filtration of particles, including protozoa (a biological agent) and/or gases with activated carbon/charcoal filters

Factors Affecting Environmental Media Indoors and Outdoors

These factors may affect measurements of exposure in various outdoor—and thus, as explained above, indoor—microenvironments in community and occupational settings:

1. Albedo, or the reflectance off of lightly colored surfaces (think roofs painted white versus black or brown shingles on most homes)

2. Location or proximity of point, area (including nonpoint to surface water resources), and mobile sources (on-road and off-road including in surface waters)
3. Security issues surrounding surface water resources and drinking water and/or wastewater treatment plants, whether owned and operated by the public or private sectors
4. Shading due to strategic placement of or naturally occurring bushes, plants, and trees
5. Shading due to other building structures
6. Temperature of water in surface water resources or drinking water distribution systems
7. Temperature of and relative humidity in the air outdoors
8. "Urban heat islands" due to built environment factors, for example, asphalt-paved roads
9. Wind speed and direction, both prevailing (on average) and maximum gusts

Conclusion

Human and ecological exposure assessment via qualitative, semiquantitative, and quantitative measurements conducted with valid (accurate), precise (replicable) protocols and tools are essential to risk assessment and management, including physical and educational interventions to reduce or prevent further exposure in community and occupational settings. The adverse human and ecological health, and environmental quality, implications of exposure to one or a mixture of biological, chemical, physical (including ergonomic), radiological, and/or psychosocial agents (hazards) can be acute or chronic in nature. Details and real-world examples from the laboratory and the field are provided in other volumes and chapters of this handbook.

Notes

1. L. L. Needham, H. Özkaynak, R. M. Whyatt, et al., Exposure assessment in the National Children's Study: Introduction, *Environmental Health Perspectives* 113 (8) (2005): 1077.

2. W. E. Wilson and M. Brauer, Estimation of ambient and non-ambient components of particulate matter exposure from a personal monitoring panel study, *Journal of Exposure Science and Epidemiology* 16 (3) (2006): 265.
3. P. B. English, A. H. Sinclair, Z. Ross, et al., Environmental health indicators of climate change for the United States: Findings from the state environmental health indicator collaborative, *Environmental Health Perspectives* 117 (11) (2009): 1675.
4. B. Jakubowski and H. Frumkin, Environmental metrics for community health improvement, *Preventing Chronic Disease* 7(4) (2010), http://www.cdc.gov/pcd/issues/2010/jul/09_0242.htm (accessed July 25, 2011).
5. Centers for Disease Control and Prevention, *National report on human exposure to environmental chemicals*, 2009, http://www.cdc.gov/exposurereport/ (accessed July 26, 2011).
6. M. N. Bates, J. W. Hamilton, J. S. LaKind, et al., Workgroup report: Biomonitoring study design, interpretation, and communication—lessons learned and path forward, *Environmental Health Perspectives* 113 (11) (2005): 1615–1621.
7. P. L. deFur, G. W. Evans, E. A. C. Hubal, et al., Vulnerability as a function of individual and group resources in cumulative risk assessment, *Environmental Health Perspectives* 115 (5) (2007): 818 (table 1), 822 (table 4).

3

The Environment and Endangered Species/Species Extinctions

John E. Fa

Introduction

Biological diversity refers to the wealth of life forms found on Earth: millions of different plants, animals, and microorganisms, the genes they contain, and the intricate systems they form. However, life on Earth contains much greater variety than that measured by species alone. A single species may contain different races or breeds, and differences may also exist at the individual level. Species come together to form communities, which in turn combine in ecosystems. Many species survive in only one specific ecosystem, and thus discussions involving biological diversity must at a minimum define the concept by designating at least three distinct levels: genetic, species, and ecosystem. Human cultural diversity could also be considered part of biodiversity because human cultures represent "solutions" to the problems of survival in particular environments.

Genetic Diversity

Genes are the biochemical packages passed on by parents to determine the physical and biochemical characteristics of the offspring. *Genetic diversity* can refer to the variation found in genes within species. Although most of the genes are the same, subtle variations occur in some. The expression of these variations may be obvious, such as size and color; they may also be invisible, for example, susceptibility to disease. Genetic variability allows species to adapt to changes in their environment and can be manipulated to produce new breeds of crop plants and domestic animals.

Species Diversity

Species are usually recognizably different in appearance, allowing an observer to distinguish one from another, but sometimes the differences are extremely subtle. *Species diversity*, or species richness, usually measures the total number of species found within one geographical area. The problem in measuring species diversity is that it is often impossible to enumerate all species in a region.

Ecosystem Diversity

An *ecosystem* consists of communities of plants and animals and the nonliving elements of their environment (e.g., soil, water, minerals, and air). The functional relationships between the communities and their environment are frequently complex, through the mechanisms of major ecological processes such as the water cycle, soil formation, nutrient cycling, and energy flow. These processes provide the sustenance required by living communities and so lead to a critical interdependence. Two different phenomena are frequently referred to under the heading of ecosystem diversity: the variety of species within different ecosystems (the more diverse ecosystems contain more species) and the variety of ecosystems found within a certain biogeographical or political boundary.

How Many Species Are There?

It is difficult to quantify the world's genetic diversity. Estimates have varied from 2 to 100 million species, with a best estimate of somewhere near 10 million. Diversity at the species level is somewhat better known, but current estimates of the number of species only serve to underscore our degree of ignorance. The problem defining the limits of current knowledge of species diversity is compounded by the lack of a central database or list of the world's species. A recent compilation[1] has shown that since the beginnings of the science of taxonomy, scientists have identified and named about 1.4 million species of living organisms out of an estimated

10 million. Of the described forms, around 1.03 million are animals and 248,000 are higher plants. The best studied and most completely known groups are birds and mammals (roughly 9,000 and 4,000 species, respectively), although together they account for less than 1 percent of all known species. About 80 percent or more of all species of birds, mammals, reptiles, amphibians, and fishes have been described. Insects, however, are still little known even though they account for a high proportion of all known species. The number of insect species was estimated by C. B. Williams in 1969 to be around three million. By 1988 Stork estimated that insects comprised 57 percent of the total named species and one group of insects, the beetles, comprised 25 percent.[2] Using knockdown insecticides, Erwin[3] astonished the ecological world with his estimate of how many insects live in tropical forests. Working in Barro Colorado Island, Panama, Erwin sampled nineteen individual Luehea trees (family Tiliaceae) at different seasons. He collected 9,000 beetles belonging to more than 1,200 species. Erwin extrapolated from these data to estimate that 13.5 percent of these beetles (about 162 species) live only in Luehea trees. He then deduced that since roughly 50,000 species of tree live in tropical rain forests around the world, and "guácimo colorado" represents an average tropical tree, there must be about 1.8 million beetle species specializing on single-tree species. Because some beetles live on more than one tree species (Erwin thought about 2.7 million of them), a total of 10.8 million species of canopy beetles was possible. If all these estimates are correct, as many as 30 million species of insects are possible. Each step in Erwin's calculation of species diversity is so speculative that many scientists do not accept this estimate and keep to the earlier figure of 10 million total species.[4] Two other approaches have been employed in estimating biological diversity. The first, biological rules, involves determining how many species are involved in biological relationships. May estimated that in Britain and Europe there are about six times more fungi than plant species.[5, 6] Thus, if this ratio applies throughout the world, there may be as many as 1.6 million fungus species growing on the world's 270,000 plant species. A second method, associated specialists, assumes that if each species of plant and insect has at least one species of specialized bacteria, protist, nematode, and virus, estimates of the number of species should be multiplied by five, a grand total of 25 million using traditional estimates, or 150 million if Erwin's estimates are accepted.

New species are still being discovered—even new birds and mammals. On average, about three new species of birds are found each year. Although many assume nearly all mammalian species are known to scientists, since 1993, 408 new mammalian species have been described, around 10 percent of the previously known fauna.[7] Other vertebrate groups are far from being completely described: an estimated 40 percent of freshwater fishes in South America have not yet been classified. In addition, environments such as soil and the deep sea are revealing an unsuspected wealth of new species. Scientists believe that the deep sea floor may contain as many as a million undescribed species. Hydrothermal vent communities discovered less than two decades ago contain more than 20 new families and subfamilies, some 50 genera, and 100 new species.

Where Is Biological Diversity Found?

Biological diversity occurs in all habitats, because genetic diversity has allowed life to adapt to different environments. However, species are not spread evenly over the earth, and biological diversity is greater in some areas than in others. Some habitats, particularly tropical forests among terrestrial systems, possess a greater number or density of species than others. For example, a 13.7-square-kilometer area of the La Selva Forest Reserve in Costa Rica contains almost 1,500 plant species, more than the total found in the 243,500 square kilometers of Great Britain, while Ecuador harbors more than 1,300 bird species, or almost twice as many as the United States and Canada combined. Global biodiversity generally follows four clear patterns: 1) species diversity increases toward the tropics for most groups of organisms; 2) patterns of diversity in terrestrial species are paralleled by patterns in marine species; 3) species diversity is affected by local variation in topography, climate, and environment; and 4) historical factors are also important.

In general, the number of species (more cautiously called inventory of species) declines as one moves away from the equator, north or south. Examples of these latitudinal gradients, as the effect is known, abound. Since first described by Alfred Russell Wallace in 1878, around fourteen different hypotheses have been proposed to explain this phenomenon. It has been found in plants and animals in both aquatic as well as terrestrial environments. There are exceptions, such as marine algae in the North/

Central American Pacific, but these are indeed departures from the general pattern. Latitudinal gradients in species richness have also been found for fossil forms, in the Foraminifera for data stretching for 70 million years and in flowering plants in a data set of around 110 million years.

Concerns were raised about whether increasing species diversity along the tropics was an artifact of disproportionate interest to temperate area scientists describing the patterns in tropical areas. This worry was dispelled by comparing similar species in studied communities. For instance, one hectare of Atlantic rain forest in Bahia Province, the richest in the world, has 450 tree species.[8] In other more typical tropical forests, one hectare can have around 200–300 tree species. A test of latitudinal gradient in species richness comes from a study that meticulously compared the avifauna of a ninety-seven-hectare tropical forest in Peru with a similar-sized area in the temperate region of North America.[9] The study's authors believed that even though gamma diversities (the number of species in a region) have been shown to differ considerably between tropical and other areas, others have conjectured that the alpha diversity (the diversity of species in a particular geographical point) could be the same. The results showed that although in the tropical forest there were 160 bird species, the richest temperate forest had four to five times fewer species. Such wealth occurs although the number of individual birds per hectare is quite similar in the two biomes.

A corollary to the observed increases in species richness toward the equator is the latitudinal gradient in geographic range size, or Rapoport's rule.[10] Rapoport's rule has been demonstrated for a large variety of different taxa, although some exceptions have been found. The rule essentially describes how species' ranges at the polar end of a continent are larger than those at the equatorial end. An explanation for Rapoport's rule for mammals is that an individual animal at the polar end of a continent experiences a much wider range of climatic conditions than one at the equator. Thus unlike tropical species, the more temperate species cannot specialize on a narrow set of climatic conditions. In addition, other research has demonstrated that more northerly species were also more generalist, which correlates with the greater variability of habitats found in those environments. Theories to explain species richness in the tropics include the notion that tropical climates are more stable than temperate zones; that tropical communities are older and therefore have had more time to evolve; or that the warm temperatures and high humidity in the tropics provide favorable conditions for many species that are unable to survive in the temperate areas. Other theories suggest that the ever-present populations of pests, parasites, and diseases prevent any species or group from dominating communities and that because tropical regions are highly productive, they can provide a greater resource base that can support numerous species.

Loss of Biological Diversity

No species will exist forever. Species evolve, and some forms disappear over time. Extinction usually refers to the total disappearance of all individuals of a species. However, the meaning of the word *extinct* can vary according to the context in which it is used.[11] A species can be said to be extinct in the wild if individuals of a species remain alive only in captivity or similar situations. A species can also be locally extinct if it is no longer found in an area that it previously inhabited. *Ecologically extinct* species is a term that focuses on species that are found in such low numbers in a community that their impact is insignificant.

The fossil record shows that since life originated four billion years ago, the vast majority of species that have existed are now extinct. By the early nineteenth century, geologists had unearthed so many extinct species that as much as 82 percent of all species known to science were extinct. By the mid-twentieth century, it was estimated that probably more than 99 percent of known tetrapods from the mid-Mesozoic became extinct without leaving any descendants in our age.[12] Periods of mass extinctions have taken place due to sudden changes in sea level, climate (including the impact of a colliding comet), and volcanic activity. During these periods of sudden change, those species that were better adapted to new circumstances than others left descendants. Species that evolved fast enough to colonize new habitats in time and space survived and flourished.

Two broad processes influence the dynamics of populations and cause extinction: deterministic (or cause and effect) relationships, for example, glaciation or direct human interventions such as deforestation, and stochastic (chance or random) events, which may act independently or influence variation in deterministic processes. These deterministic and stochastic processes have been referred to as accidents and population interactions.[13] In other words, species may disappear for no predictable reason, or both

The Environment and Endangered Species/Species Extinctions 51

predation and competition can force populations to succumb. The magnitude of the effects of these causes of extinction depends on the size and degree of genetic connectedness of populations. Four types of stochastic processes can be distinguished: 1) demographic uncertainty, which is only a hazard for relatively small populations (numbering tens or hundreds of individuals); 2) environmental uncertainty, which is due to unpredictable changes in weather, food supply, disease, and the populations of competitors, predators, or parasites; 3) natural catastrophes such as floods, fires, or droughts; and 4) genetic uncertainty, random changes in genetic makeup to which several factors contribute.[14]

Agents of decline of species have been classified under four main headings: 1) overkill, 2) habitat destruction and fragmentation, 3) impact of introduced species, and 4) chains of extinction. These agents of decline have been described as the "the evil quartet."[15] Of these factors, introduced species and habitat destruction are responsible for most known extinctions and threatened species of vertebrates. Thirty-nine percent of species have become extinct through introductions and 36 percent through habitat loss. Hunting and deliberate extermination have also contributed significantly (23 percent of extinctions with known causes).

Overkill results from hunting at a rate above the maximum sustained yield. The most susceptible species are those with low intrinsic rates of increase (i.e., large mammals such as whales, elephants, and rhinos) because of their limited ability to recover quickly. Although such species usually have a high standing biomass when unharvested, they have a low maximum sustained yield, which is easily exceeded. They are even more vulnerable if they are valued either as food or as an easily marketable commodity. An example of this comes from Equatorial Guinea, where meat extraction from the bush was proportional to the purchasing power of the urban markets. The volume of bushmeat serving Malabo, the administrative capital in Bioko, was 70 percent greater than that being sold through Bata, in Rio Muni, though there was little difference in population size between the two centers (52,000 and 55,000 respectively).[16] Actual harvests far exceeded potential harvests on the island and have led to a drop in numbers of many species.[17]

Habitat destruction and fragmentation: Although habitats may be modified, degraded, or eliminated, they are more commonly fragmented. A large tract is often converted piecemeal to another land use. This practice is widespread throughout the world. Loss of habitat by a given proportion does not increase the vulnerability of a species, nor does it decrease the number of its members by that same proportion (except in the particular case of habitat cleared from the edge inward). Frequently, modification produces a patchwork pattern as it erodes the tract of habitat from inside and changes microclimates. Initially the areas occupied by the new land use form islands, which later multiply and enlarge until the new land use provides the continuous phase and the original habitat the discontinuous one. The vulnerability of species then increases disproportionately. A clear example of this comes from a study on the Atlantic rain forest fragments remaining in Pernambuco, Brazil.[18] The study's analyses of size, shape, and distribution of the forest fragments showed that a large proportion of these (48 percent) were less than 10 hectares, and only 7 percent were >100 hectares. A recent review of the literature on the effects of fragmentation provides empirical evidence from different parts of the world that in nearly all cases there is a local loss of species.[19] Isolated fragments suffer reduction in species richness with time after excision from continuous forest, and small fragments often have fewer species recorded for the same effort of observation than large fragments or areas of continuous forest. The mechanisms of fragmentation-related extinction include the deleterious effects of human disturbance during and after deforestation, the reduction of population sizes, the reduction of immigration rates, forest edge effects, changes in community structure, and the immigration of exotic species.

Impact of introduced species: Some species that have been introduced intentionally or unintentionally become invasive. These plants or animals often adversely affect the habitats and bioregions they invade economically, environmentally, and/or ecologically. In a number of well-recorded cases, these alien invasives have proceeded to exterminate native species by competing with them, preying upon them, or destroying their habitat. A review by Atkinson[20] has demonstrated that twenty-two species and subspecies of reptiles and amphibians have disappeared worldwide as a direct result of alien animals. In New Zealand alone, nine species of reptiles and amphibians and twenty-three bird species have become extinct since AD 1,000 through introductions. On the Pacific island of Guam the result of the introduction of one single alien species, the brown snake, has caused the extinction of most avifauna.[21]

The invasion of ecosystems by non-native species has occurred most significantly on islands, where indigenous species have often evolved in the absence of strong competition, herbivory, parasitism, or predation. As a result, introduced species thrive in those optimal insular ecosystems affecting their plant food, competitors, or animal prey. As islands are characterized by a high rate of endemism, the impacted populations often correspond to local subspecies or even unique species. A small number of mammal species are responsible for most of the damage to invaded insular ecosystems: rats, cats, goats, rabbits, pigs, and a few others.[22] The effect of alien invasive species may be simple or very complex, especially since a large array of invasive species, mammals and others, can be present simultaneously and interact among themselves as well as with the indigenous species. In most cases, introduced species generally have a strong impact, and they often are responsible for the impoverishment of the local flora and fauna.

The impact of introduced predators on islands has been clearly shown by the work carried out by Case and Bolger.[23] They analyzed lizard population sizes on tropical Pacific islands with and without mongooses. There was a significant depression of diurnal lizard populations, increasing the risk of extinction. Similarly, in the case of the black rat, the preeminent invasive species introduced to Pacific islands, there was a highly significant correlation between the arrival of the rat and the decline and extinction of five endemic genera of *Pomarea* monarch flycatchers.[24] The extinction of monarch populations after colonization by black rats (*Rattus rattus*) tended to take longer on larger islands than on smaller ones, and on islands without black rats, monarchs persist even where forests have been reduced by more than 75 percent. In contrast, there was no relationship between presence of Polynesian rats and monarch distribution.

Chains of extinction: The extinction of one species may bring about the demise of another; this is known as secondary extinction. Diamond illustrates this chain of events with the example of the near extinction of a plant (genus *Hibiscadelphus*) resulting in the disappearance of several of the Hawaiian honey creepers, its pollinators.[25] There are several instances of predators and scavengers dying out following the disappearance of the species that represented their source of food (keystone species). Another example of the loss of species interactions is the effect of the loss of pollinators on populations of the endangered Japanese primrose. Modeling studies demonstrated that pollinator extinctions (bumblebees, butterflies, and other insects) can lower the genetic variability of the flower populations through the loss of heterostyly (different positioning of stigmas and anthers).[26]

Vulnerability of Species to Extinction

There is considerable evidence that the number of species in an isolated habitat will decrease over time. The crucial issue for conservationists is whether it is possible to predict those species that are most at risk of extinction using knowledge of their biology and ecology. Some taxa are more affected than are others, resulting in groups of species with differential extinctions of populations. Survival capability and extinction vulnerability can vary by geological time periods, though the probability of a species becoming extinct is independent of the age of its clade. As explained by Raup, "the victims and survivors are not random samples of the pre-extinction biotas."[27] Some interrelated features make species susceptible to extinction. Several ecological or life history traits have been proposed as factors determining an animal species' sensitivity to extinction. Some species are more susceptible to rapid extinction through fragmentation of tropical rain forest than others. Laurance tested the efficacy of seven ecological traits (body size, longevity, fecundity, trophic level, dietary specialization, natural abundance in rain forest, and abundance in the surrounding habitat matrix) for predicting responses of sixteen nonflying mammals in tropical rain forests in Queensland.[28] The proneness of mammals to extinction appears to be inversely related to tolerance toward conditions in the prevailing matrix vegetation of the fragmented landscape. The ability of species to use modified habitats surrounding fragments appeared to be an overriding determinant of extinction proneness. Laurance argued that "matrix abundant" is a good predictor of extinction proneness, primarily because species that occur in the matrix should be most effective at dispersing between fragments. These mammals could either recolonize fragments following extinction or have lower rates of extinction because small populations in fragments were bolstered by the demographic and genetic contributions of immigrants. Furthermore, species that tolerate or exploit modified habitats may be adapted for ecological changes in fragments, particularly edge effects. For example, the arboreal folivores most abundant in modified habitats fed upon secondary trees that proliferated along corridors and fragment margins. Hence, in addition to being effective dispersers, these folivores probably responded positively to edge conditions in fragments.

Of course the characteristics of extinction-prone species are not independent, but tend to group together into categories of characteristics. For example, species of certain body sizes are similar in terms of their population densities, gestation length, birthrates, and other life history features. Furthermore, populations of species that have already undergone changes in their genetic variation will have a greater tendency to become extinct when a new disease, predator, or some other change occurs in the environment. For example, some authors have indicated that the lack of genetic variability in the cheetah (*Acinonyx jubatus*) could have been a contributing factor to the lack of disease resistance in this species.[29] Given this, the most susceptible species would be a large predator with narrow habitat tolerance, a long gestation period, and few young per litter, that is hunted for a natural product and/or sport, subject to inefficient game management, has a restricted distribution yet travels across international boundaries, is intolerant to humans, and has nonadaptive behaviors: something like a polar bear.[30]

The Meaning of *Rare Species*

The protection of rare species is an important focus of conservation efforts. Definition of categories of rarity must be based on geographic distribution, habitat specificity, and local population size. Distribution of a species can be quantified by the number of sites in which it is present, or on a biogeographic scale by the area of the distributional range. *Abundance* refers to the local population density, the number of individuals found in a given site. By virtue of their scarcity or restricted range, rare species are more prone to extinction. This fact is supported by theory and empirical data. Several demographic models show that the probability of extinction of a local population declines as its size increases, and field studies have supported these models. From a genetics viewpoint, low densities mean higher probabilities of depleting genetic variation and lower chances of long-term survival. However, rarity is not the only factor promoting extinction (see above). Karr, for example, found little support for the idea that initial rarity was the cause of the majority of bird extinctions in Barro Colorado Island, Panama.[31]

A classification of species according to their rarity based on geographic distribution, habitat specificity, and local population size was proposed by Rabinowitz for plants in the British Isles, one of the biologically best-known areas in the world.[32] In a later study, Rabinowitz et al. distinguished eight species categories but rejected the idea that these three ways of being rare were correlated (at least for British plants).[33] However, the magnitude of correlation between distribution and abundance is of prime importance, as suggested by Arita et al.[34] If rarity is seen as a combination of area of distribution (AD) and local density (LD), the number of species in each of the four possible categories is determined by the association between the variables.

Arita et al. tested the correlation between AD and LD of species using available data on Neotropical mammals.[35] They found a negative correlation between AD and LD for all species pooled, but coefficients varied from positive to negative in subsets defined by taxonomic and trophic characteristics. When classified into four main categories (using the medians of area of distribution and abundance), the slight negative correlation between AD and LD produced an apparently higher frequency of restricted/abundant and widespread/rare species than the other types. From the observed correlations, larger animals tended to fall in the lower right corner of the graph (widespread/low local density species); the upper right and lower left corners having animals of intermediate sizes, the exact position being affected by other factors. The conclusion was that because AD/LD correlation vanishes when body size is taken into account, rarity can be better classified by comparing only species of similar size. However, such a dichotomous classification can still be used even if the correlation between the variables is different from zero.[36] Species in the restricted and locally rare category are of major concern in terms of conservation, because they are likely to be at risk of extinction from demographic or genetic causes and an absence of other secondary sources.

Extinctions in Recent History

Most recent extinctions have been associated with European expansion in the fifteenth and sixteenth centuries (see Figure 3.1). However, in some parts of the world some species are known to have become extinct before the arrival of the Europeans. For example, the Polynesians who colonized the Hawaiian Islands in the fourth, fifth, and sixth centuries may have been responsible for the loss of around 50 of the 100 or so species of endemic land birds in the period between their arrival and that of the Europeans.

The number of species lost during the past 500 years documented by the International Union for Conservation of Nature (IUCN) is around 785

The Environment and Endangered Species/Species Extinctions 57

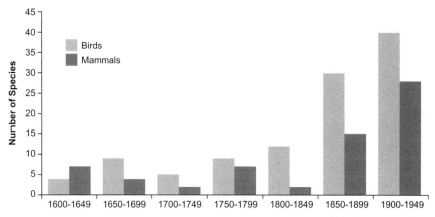

Figure 3.1 Number of birds and mammals listed extinct for the period 1600-1949. (Data from IUCN database. Available at: <http://www.iucnredlist.org/>. Accessed: October 1, 2010.)

worldwide. Many other extinctions, not included in this number, have likely occurred, but they have not yet been documented adequately enough to be listed as extinct. The list of extinct species is therefore likely to be an underestimate. Those listed provide the most detailed evidence on extinction available. An analysis of the IUCN database by Sax and Gaines revealed several emerging patterns for terrestrial vertebrate and plant species.[37] First, the pattern that emerges is a preponderant loss of island forms. This holds true generally when all causes of extinction are pooled and specifically when only extinctions that exotic species are believed to have contributed to are considered. Second, terrestrial vertebrates have disproportionately become extinct compared with plants, both in absolute terms and relative to the taxonomic richness of their respective groups. Third, the presumed causes of these extinctions are not evenly distributed among types of species interactions. Predation has been a far more important species interaction in causing extinctions than competition. Indeed, predation alone, in the absence of other factors like habitat destruction or pollution, is listed as being responsible for the extinction of >30 percent of vertebrate species. In contrast, competition is never listed as being the sole factor responsible for species extinction. Further, predation is listed as one of several contributing factors in >40 percent of terrestrial vertebrate extinctions, whereas competition is listed as a contributing factor in <10 percent of terrestrial vertebrate extinctions. This means that predation acting alone, or in concert with other factors, is believed to have contributed to the extinction of close to 80 percent of all terrestrial vertebrate species, whereas competition has contributed to <10 percent of these extinctions. These patterns suggest that terrestrial vertebrates are much more likely to become extinct from predation than competition.

There are difficulties in documenting extinctions in the past. Often the precise mechanisms for any individual extinction are difficult to confidently determine, because extinctions are often caused by multiple factors (species invasions, habitat destruction, human exploitation, pollution, and infectious disease) and most "documented" extinctions actually involve some speculation about the factors responsible. In addition, disagreement over species concepts and phylogenetic classifications of individual species may alter numbers.

Generally, if a taxon is not located for fifty years, it is considered extinct. It is impossible to eradicate an element of bias because it is not easy to state unequivocally that a species is no longer present when animals are known to persist unrecorded despite intensive efforts to locate them. The number of recorded extinctions could easily be an underestimate. Species that have never been described may have become extinct in historic times, but scientific taxonomy only began in the mid-eighteenth century, and our knowledge of the more diverse tropics is just emerging. According to the Millennium Ecosystem Assessment,[38] the rate of known extinctions of species in the past century is roughly 50–500 times greater than the extinction rate calculated from the fossil record of 0.1–1 extinctions per 1,000 species per 1,000 years (see Figure 3.2). The rate is up to 1,000 times higher than the background extinction rates if possibly extinct species are included.

"Distant past" refers to average extinction rates as estimated from the fossil record, whereas "recent past" refers to extinction rates calculated from known extinctions of species (lower estimate) or known extinction plus "possibly extinct" species (upper bound) (see Figure 3.2). "Future" extinctions in Figure 3.2 are model-derived estimates using a variety of techniques, including species-area models, rates at which species are shifting to increasingly more threatened categories, extinction probabilities associated with the IUCN categories of threat, and impacts of projected habitat loss on species loss with energy consumption. The time frame and species groups involved differ among the "future" estimates, but in general refer to either future loss of species based on the threat that exists today or current and future loss of species as a result of habitat changes taking place

The Environment and Endangered Species/Species Extinctions

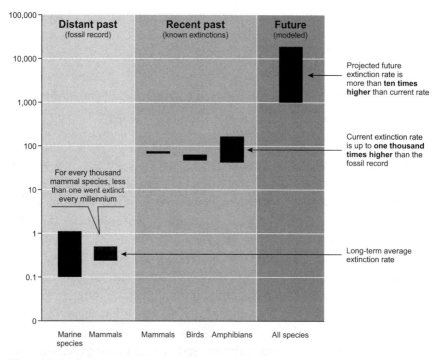

Figure 3.2 Species extinction rates expressed as per thousand species per millennium. (Redrawn from Fig. 1, Ecosystems and Human Well–being: Biodiversity Synthesis Millennium Ecosystem Assessment. Available at: <http://www.maweb.org/en/Synthesis.aspx>. Accessed: October 1, 2010.)

between roughly 1970 and 2050. Estimates based on the fossil record are low certainty; lower-bound estimates for known extinctions are high certainty, and upper-bound estimates are medium certainty; and lower-bound estimates for modeled extinctions are low certainty, and upper-bound estimates are speculative.

Present-day Extinction Rates

Estimates of how many species are being lost through habitat destruction are hampered by our ignorance of the total number of species (see above) and their distribution, as well as the patterns of habitat loss. In addition, our ignorance of the effects of deforestation on species also limits our ability to predict extinctions due to habitat loss and degradation. Currently, the estimates for species loss vary enormously, primarily because

- authors choose different baseline estimates of the number of species in the world (between 10 and 30 million);
- the proportion of species that reside in tropical forests (between 25 and 70 percent);
- the shape of the relationship between habitat loss and extinction; and
- the rate and extent of habitat loss (between 0.5 and 2 percent annual global deforestation).

All these sources of uncertainty in extinction estimates can lead to extremely large errors. Nonetheless, it is indisputable that a large extinction event is currently occurring. Despite the different ways in which these estimates were calculated, at a minimum, a substantial number of the world's species are likely to become extinct in the next decade. The current reduction of diversity is believed to be destined to approach that of the great catastrophes at the end of the Paleozoic and Mesozoic eras: in other words, the most extreme for 65 million years. The ultimate result is impossible to predict, but not something with which humanity will want to gamble.

A most significant point on the interpretation of extinction rates revolves around the definition of what is meant by "becoming extinct," as different authors attach distinct meanings to this. Using data for the comparatively well-studied birds and mammals (around 13,000 species), extinction rates for this century have been estimated at one species per year. Were this rate maintained, it would correspond to the average "species lifespan" of around 104 years; about two to three orders of magnitude shorter than estimated from the fossil record.[39] In contrast, the inferred extinction rate is almost two orders of magnitude longer than the impending extinction times (200–400 years). These extinction times have taken into account concepts such as habitat loss and species–area relations or have been derived from extrapolations of rates of species "climbing the ladder" of IUCN-defined categories of threat, from "vulnerable" to "endangered" to "extinct."[40] Others have used assessments of species-by-species extinction probability distributions as functions of time to calculate the expected times when half of the species in each of ten vertebrate taxa will be extinct.[41]

Although all these theoretical estimates are beset by many uncertainties, the discrepancy largely disappears if the notion of species committed to extinction is seen in a correct context; that is, that both species–area and "ladder-climbing" estimates are projected numbers of species. For example, Simberloff's estimate of 1,350 species of birds committed to extinction by the year 2015[42] cannot be misinterpreted as the predicted number to become extinct between 1986 and 2015 (an average of forty-five extinctions per year), since it would be implausibly high. A more empirical approach is to interpret "committed" to extinction as referring to any species whose populations in the wild are no longer viable and will inevitably become extinct, unless major conservation actions reverse current trends (by habitat restoration, elimination of introduced predators, captive breeding and reintroduction, etc.).

Why Conserve Biodiversity?

There are two good reasons for conserving biological diversity. The first is moral: it is right to do so. The second is practical: biological diversity supports human survival, notably through health, food, and industry.[43] The fundamental social, ethical, cultural, and economic values of biological resources have been recognized in most human disciplines, from religion to science. Given these multiple values, it is not surprising that most cultures accept the importance of conservation. Despite this, and in order to compete for the attention of government decision makers worldwide, policies regarding the protection of biological diversity must embrace economic values. Some conservationists take the view that biological resources are beyond quantification because they provide the biotic raw materials for every type of economic endeavor.

There is a multiplicity of ways of assessing the value of biological resources. Economists in particular have tried several approaches, but problems arise when applying a common formula to the myriad resources used by humans. A forest's value in terms of logs would be measured differently from its value for recreational purposes or watershed protection. Five main approaches are used to determine the value of biological resources:[44]

- Consumptive value: the value of natural products, e.g., firewood, fodder, and game meat. These are consumed directly and do not pass through a market.
- Productive value: the value of products that are commercially harvested, such as game sold in a market, timber, and medicinal plants.

- Nonconsumptive value: the indirect value of ecosystem functions, such as watershed protection, photosynthesis, regulation of climate, and soil production.
- Option value: the intangible value of keeping options open for the future.
- Existence value: the value attached to the ethical feelings of existence.

In 1992 numerous countries signed the United Nations Convention on Biological Diversity (CBD) in recognition of the present and future value of biological diversity and its significant reduction around the world. Signed by 150 government leaders at the 1992 Rio Earth Summit, the CBD is dedicated to promoting sustainable development. Conceived as a practical tool for translating the principles of Agenda 21, a program run by the United Nations related to sustainable development, into reality, the CBD recognizes that biological diversity is about more than plants, animals, and microorganisms and their ecosystems: it is about people and our need for food security, medicines, fresh air and water, shelter, and a clean and healthy environment in which to live. Signatories have legal obligations to develop national strategies and action plans for biodiversity conservation. The intention is for the CBD to be a powerful catalyst drawing together existing efforts to protect biological diversity and to provide strategic direction to the whole global effort in this area. The CBD thus provides a framework for worldwide action to conserve (and use biodiversity sustainably). It addresses the full range of biological diversity at genetic, species, and ecosystem levels in all environments, both within and outside protected areas. The CBD contains guiding concepts, such as the precautionary principle, and gives each country the responsibility for the conservation and sustainable use of its biological resources. Most important, the CBD provides for the needs of developing countries to enable them to implement its measures, including the provision of new and additional financial resources and appropriate access to relevant technologies.

In April 2002 at the sixth Conference of Parties (COP) to the CBD, governments committed themselves "to achieve by 2010 a significant reduction of the current rate of biodiversity loss at the global, regional and national level as a contribution to poverty alleviation and to the benefit of all life on Earth" (Decision VI/26). This "2010 Biodiversity Target" was later endorsed at the World Summit on Sustainable Development (WSSD)

The Environment and Endangered Species/Species Extinctions 63

and has been included in Millennium Development Goal 7 (MDG 7) under the "reducing biodiversity loss" target.

The 2010 Biodiversity Indicators Partnership (BIP) contributed to the 2010 Millennium Development Goals Report, to assess international efforts in reducing extreme poverty. The report highlights results from sixty indicators selected to measure progress toward the eight MDGs. The 2010 BIP has developed and contributed information for two of the indicators under MDG 7b: "to reduce biodiversity loss, achieving, by 2010, a significant reduction in the rate of loss."

The report features the BIP's "proportion of terrestrial and marine areas protected" (coverage of protected areas and overlays with biodiversity) indicator, created in partnership with the United Nations Environment Programme-World Conservation Monitoring Centre (UNEP-WCMC). It shows that although the proportion of protected areas has increased to 12 percent of the earth's land area and almost 1 percent of its sea area, more than two-thirds of areas designated as critical for conservation remain unprotected or only partially protected.

Also included are the findings of the BIP's "proportion of species threatened with extinction" (Red List Index) indicator, developed in partnership with Birdlife International, IUCN, and the Zoological Society of London (ZSL). These findings are equally alarming and show that the number of species facing extinction is increasing, with mammals in developing countries particularly at risk. Both BIP indicators presented in the report show that the world has failed to meet the 2010 biodiversity target.

Conclusion

There is prevailing evidence that the world's biodiversity is facing unprecedented threats from anthropogenic factors. Conservation is not just about avoiding extinctions, but also about restoring or recovering habitats and landscapes. The claim of those in favor of addressing the bigger environmental issues, or crises, is that by conserving ecosystems, species status will also benefit in the long term. Indeed, habitats must meet the needs of species that depend on them in order to ensure populations reach secure levels, while preventing other species from reaching such a perilous situation. Often, single-species conservation programs can be used as a means for protecting larger numbers of species and natural habitat using targeted taxa as representatives or symbols for habitat and sympatric fauna and flora. The latter concept revolves around the description of a single species that can be used to represent others, either from an ecological perspective, for example, keystone, or umbrella species, or by playing a public relations role, for example, flagship species.

The value of focusing on measurable objectives in conservation is receiving increased attention, due in part to a demand for accountability. Identifying and securing biodiversity priority areas should form the framework upon which to build other conservation actions. This process, known as systematic conservation planning, employs the analyses of numerical data on the distribution of biodiversity. Such quantitative methods can be applied to measures of biodiversity value, including phylogenetic diversity, species, higher taxa, vegetation or land classes, and most other biodiversity surrogates.

There is no single, objective way to measure the importance of the various components of biodiversity, and it is equally difficult to choose priorities for managing them. Since the 1980s there has been much emphasis in developing strategies to best allocate resources for species and habitat conservation on a global scale. However, the need for integrating more academic interests, such as how many species there are worldwide or their global distribution, must go hand-in-hand with more practical ways of ensuring the long-term viability of life on Earth.

Notes

1. E. O. Wilson, *Biodiversity* (Washington, DC: National Academy Press, 1988).
2. N. E. Stork, Insect diversity: Facts, fiction and speculation, *Biological Journal of the Linnean Society* 35 (1988): 321–337.
3. T. L. Erwin, Tropical forests: Their richness in Coleoptera and other arthropod species, *Coleopterist's Bulletin* 36 (1982): 74–75.
4. K. J. Gaston, The magnitude of global insect species richness, *Conservation Biology* 5 (1991): 283–296.
5. R. M. May, *Stability and complexity in model ecosystems* (Princeton, NJ: Princeton University Press, 1973).
6. R. M. May, How many species inhabit the Earth? *Scientific American* 267 (1992): 42–48.

7. G. Ceballos and J. H. Brown, Global patterns of mammalian diversity, endemism, and endangerment, *Conservation Biology* 9 (1995): 559–568.
8. W. W. Thomas and A. M. de Carvalho, *Estudio fitossociologico de Serra Grande, Uruçuca, Bahia, Brasil. XLIV Congresso Nacional de Botânica, São Luis, 24–30 de Janeiro de 1993, Resumos,* 1: 224 (Sociedade Botânica do Brasil, Universidade Federal de Maranhão, 1993).
9. J. Terborgh, S. K. Robinson, T. A. Parker III, et al., Structure and organization of an Amazonian forest bird community, *Ecological Monographs* 60 (1990): 213–238.
10. G. C. Stevens, The latitudinal gradient in geographical range: How so many species coexist in the tropics, *American Naturalist* 133 (1989): 947–949.
11. J. A. Estes, D. O. Duggins, and G. B. Rathbun, The ecology of extinctions in kelp forest communities, *Conservation Biology* 3 (1989): 252–264.
12. A. S. Romer, Time series and trends in animal evolution, in *Genetics, paleontology and evolution*, edited by G. L. Jepson, E. Mayr, and G. G. Simpson (Princeton, NJ: Princeton University Press, 1949), 103–121.
13. M. L. Rosenweig, *Species diversity in space and time* (Cambridge: Cambridge University Press, 1997).
14. M. L. Shaffer, Minimum viable populations: Coping with uncertainty, in *Viable populations for conservation*, edited by M. E. Soulé (Cambridge: Cambridge University Press, 1987), 69–86.
15. J. M. Diamond, Overview of recent extinctions, in *Conservation for the twenty-first century*, edited by D. Western and M. Pearl (New York: Oxford University Press, 1989), 34–37.
16. J. E. Fa, J. Juste, J. Perez del Val, and J. Castroviejo, Impact of market hunting on mammal species in Equatorial Guinea, *Conservation Biology* 9 (1995): 1107–1115.
17. L. Albrechtsen, D. W. Macdonald, P. J. Johnson, R. Castelo, and J. E. Fa, Faunal loss from bushmeat hunting: Empirical evidence and policy implications in Bioko Island, *Environmental Science & Policy* 10 (2007): 654–667.
18. P. Ranta, T. Blom, J. Niemelä, et al., The fragmented Atlantic rain forest of Brazil: Size, shape and distribution of forest fragments, *Biodiversity and Conservation* 7 (1998): 385–403.

19. I. M. Turner, Species loss in fragments of tropical rain forest: A review of the evidence, *Journal of Applied Ecology* 33 (1996): 200–209.
20. I. A. Atkinson, Introduced animals and extinctions, in *Conservation for the Twenty-first Century*, edited by D. Western and M. Pearl (New York: Oxford University Press, 1989), 54–69.
21. S. L. Pimm, The snake that ate Guam, *Trends in Ecology and Evolution* 2 (1987): 293–295.
22. F. Courchamp, J-L. Chapuis, and M. Pascal, Mammal invaders on islands: Impact, control and control impact, *Biological Reviews* 78 (2003): 347–383.
23. T. J. Case and D. T. Bolger, The role of introduced species in shaping the distribution of island reptiles, *Evolutionary Ecology* 5 (1991): 272–290.
24. J-C. Thibault, J-L. Martin, A. Penloup, and J-Y. Meyer, Understanding the decline and extinction of monarchs (Aves) in Polynesian Islands, *Biological Conservation* 108 (2002): 161–174.
25. J. M. Diamond, Overview of recent extinctions, in *Conservation for the twenty-first century*, edited by D. Western and M. Pearl (New York: Oxford University Press, 1989), 34–37.
26. I. Washitani, Predicted genetic consequences of strong fertility selection due to pollinator loss in an isolated population of *Primula sieboldii*, *Conservation Biology* 10 (1996): 59–64.
27. D. M. Raup, Biological extinctions in earth history, *Nature* 317 (1986): 384–385.
28. W. F. Laurance, Ecological correlates of extinction proneness in Australian tropical rain forest mammals, *Conservation Biology* 5 (1991): 79–89.
29. S. J. O'Brien and J. F. Evermann, Interactive influence of infectious disease and genetic diversity in natural populations, *Trends in Ecology and Evolution* 3 (1988): 254–259.
30. D. W. Ehrenfeld, The management of diversity: A conservation paradox, in *Ecology, economics, ethics: The broken circle*, edited by F. H. Borman and S. R. Kellert (New Haven, CT: Yale University Press, 1991), 26–39.
31. J. R. Karr, Population variability and extinction in the avifauna of a tropical land bridge island, *Ecology* 163 (1982): 1975–1978.

32. D. Rabinowitz, Seven forms of rarity, in *The biological aspects of rare plant conservation*, edited by H Synge (Chichester, UK: John Wiley & Sons, 1981), 205–217.
33. D. Rabinowitz, S. Cairns, and T. Dillon, Seven forms of rarity and their frequency in the flora of the British Isles, in *Conservation biology: The science of scarcity and diversity*, edited by M Soulé (Chichester UK: Wiley, 1986), 182–204.
34. H. Arita, J. G. Robinson, and K. H. Redford, Rarity in neotropical forest mammals and its ecological correlates, *Conservation Biology* 4 (1990): 181–192.
35. Arita et al., Rarity in neotropical forest mammals.
36. Rabinowitz et al., Seven forms of rarity.
37. D. F. Sax and S. D. Gaines, Species invasions and extinction: The future of native biodiversity on islands, *Proceedings of the National Academy of Sciences* 105 (2008): 11490–11497 (supplement 1).
38. Millennium Ecosystem Assessment, *Ecosystems and human well-being: Synthesis* (Washington, DC: Island Press, 2005).
39. D. M. Raup, Large-body impact and extinction in the Phanerozoic: An interpretation, *Paleobiology* 18 (1992): 80–88.
40. F. D. M. Smith, R. M. May, R. Pellew, et al., Estimating extinction rates, *Nature* 364 (1993): 494–496.
41. G. M. Mace, Classifying threatened species: Means and ends, *Philosophical transactions of the Royal Society of London* 344 (1994): 91–97.
42. D. Simberloff, Do species–area curves predict extinction in fragmented forest? in *Tropical deforestation and species extinction*, edited by T. C. Whitmore and J. A. Sayer (London: Chapman & Hall, 1992), 75–89.
43. J. A. McNeely, K. R. Miller, W. V. Reid, et al., *Conserving the world's biological diversity* (Gland, Switzerland; Washington, DC: IUCN, 1990).
44. McNeely et al., *Conserving the world's biological diversity*.

4

Environmental Noise and Health

Irene van Kamp, Wolfgang Babisch, and A. L. Brown

Introduction

Chronic exposure to noise in residential as well as work (and some recreational) situations can lead to a range of health effects. These are usually subdivided into well-being effects, such as annoyance and sleep disturbance, and clinical effects, such as hearing damage and cardiovascular diseases.[1–5] This chapter considers only the effects of environmental noise exposures, not those of occupational noise exposure.

Figure 4.1 shows the potential mechanisms by which noise can lead to health problems. The model is based on a publication of the Netherlands Health Council[6] and is one of the prevailing approaches to noise and health based on a cognitive stimulus-response model. The model assumes that most effects are a consequence of the appraisal of sound as noise. It is generally assumed that stress responses play an important role in the process by which environmental noise leads to health effects. However, sound can also directly lead to physiological responses due to interactions of the acoustic nerve with other parts of the central nervous system. This is particularly relevant during sleep. Noise exposure is associated with annoyance, sleep disturbance and activity disturbance, and stress responses. These effects are at the base of so-called instantaneous effects such as blood pressure increases and increased secretion of cortisol, responses considered to be risk factors for cardiovascular diseases and mental pathology. Responses are partly dependent on the noise characteristics of frequency, intensity, duration, and meaning, and partly on

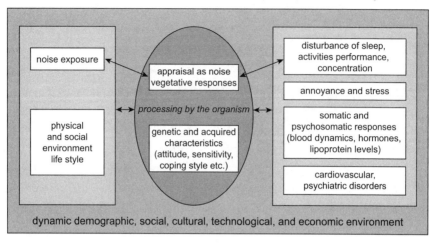

Figure 4.1 Conceptual framework for environmental noise and health. (Reprinted with permission from Health Council of the Netherlands [HCN]. Public health impact of large airports. HCN Report No. 1999/14E, The Hague, 2 September, 1999. http://www.gr.nl/pdf.php?ID=19&p=1)

nonacoustical aspects such as context, attitude, expectations, fear, noise sensitivity, and coping strategies.

Research in the past decades has shown that there is sufficient evidence for the development of hearing damage (at high noise levels generally experienced in occupational and some recreational settings, and above those experienced through exposure to environmental noise) and for development (at environmental levels of exposure) of the chronic effects of annoyance and sleep disturbance,[1,2] as well as mixed evidence for the development of hypertension and ischemic heart disease.[3–5] There is also growing evidence for effects on cognition and reading performance in children.[7–9] There is only weak evidence for immune system effects and no evidence for a direct relation with mental health.[1,10]

This chapter summarizes the evidence on environmental noise-related health effects, including effects of low-frequency noise and combined exposures. New approaches are also discussed in view of their potential to enhance our understanding of the differential health effects of exposure to sound.

Environmental Noise

It is useful to identify the sources and indicate the extent of exposure to environmental noise. Today more than half of the world's population lives in urbanized areas. Environmental noise is an ever-increasing problem as a result of this continuing urbanization, and the growth in movement by both surface and air transport modes in and near urban areas. Nighttime exposure is also expected to increase significantly due to trends toward a twenty-four-hour economy in urban areas. The most important sources of environmental noise are transport related: road, aircraft, and rail. There are various estimates of the exposure of populations to these sources, but the most comprehensive estimates are those emerging in Europe as a result of noise mapping pursuant to the Environmental Noise Directive (END).[11] The mapping has predominantly been of exposure to the day-evening-night level metric, L_{den} (see note 11 for a definition of this and other noise metrics). Although there has been considerable progress toward harmonization of noise exposure metrics,[11] there is still ongoing debate regarding the appropriate noise metrics to utilize to assess different health effects, particularly sleep disturbance.[12]

One estimate of the extent of exposure of dwellings in Europe to the most ubiquitous source, road traffic noise,[13] is shown in Figure 4.2. The data in this figure should be treated somewhat cautiously, as they represent an aggregate outcome from noise mapping across Europe, and there are

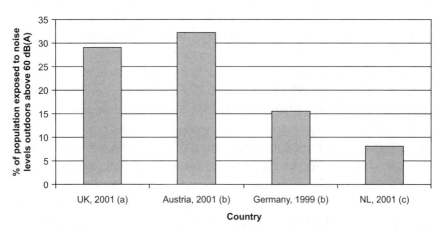

Figure 4.2 The percentage of the population (or, more correctly, the percentage of the population of urban dwelling units) exposed to road traffic noise levels in 5 dB bands of L_{den} in Europe. (Reprinted with permission from Vernon, J. et al. [2010]. Task 3 Impact Assessment and Proposal of Action Plan. A report prepared by RPA Ltd for DG Environment under Service contract No. 070307/2008/510980/SER/C3.)

large differences, which still need to be resolved, between countries in terms of assessment methods and completeness of data sets.

Though these high levels of exposure are for Europe, and comparable data are generally not available from elsewhere, it is unlikely that exposures in urban areas in the developing world, and in the world's mega-cities, are less. In fact, in many developing countries, noise levels and noise exposures are higher than those in the developed countries.[14]

Noise and Health

Annoyance

It is now widely accepted that annoyance is an endpoint of environmental noise that can be taken as a basis for evaluating the noise impact on exposed populations. For example, the Environmental Noise Directive (END) in Europe[11] recommends evaluating environmental noise exposures on the basis of estimated noise annoyance, in addition to evaluation on the basis of estimated sleep disturbance. In his review Gjestland concludes that only a few new annoyance studies have been performed in the past years and that most recent studies have been performed in Asia.[15] The ISO standard for annoyance has facilitated the comparison of annoyance data across studies.[16] Factors such as noise source, exposure level, and time of day of exposure only partly determine individual annoyance responses. Many nonacoustical factors, such as the extent of interference experienced, ability to cope, expectations, fear associated with the noise source, noise sensitivity, anger, and beliefs about whether noise could be reduced by those responsible, influence annoyance responses.[17] Generalized exposure-effect associations have been established for the effects of different noise sources on annoyance responses.[18, 19] Although the exposure-response relation for road traffic noise seems rather stable, there is growing evidence that the generalized curve for air traffic noise is no longer tenable and in need of an update.[20-22] Only a few studies have addressed annoyance responses of children. The Road Traffic and Aircraft Noise Exposure and Children's Cognition and Health (RANCH) study determined that children's annoyance can be measured reliably and validly by using a questionnaire. Exposure-response curves can be derived but are shaped somewhat differently than those found in adults, with higher levels of severe annoyance at the low end of the noise scale and lower levels at the high end.[23]

Sleep Disturbance

A distinction can be made between self-reported sleep disturbance—which can be interpreted in much the same way as nighttime annoyance—on the one hand, and physiological sleep disturbance, with insomnia-like symptoms or consequences, on the other. The former can be considered a well-being effect, the latter an ill-health effect. There is sufficient evidence that nighttime transport noise leads to acute sleep effects such as physiological response, arousal, awakening, sleep stage changes, and the amount of total sleep.[17, 24–26] It also leads to aftereffects such as self-reported sleep disturbance, reduced performance in the daytime, and cognitive effects. However, it is still unclear what the long-term health effects at the cognitive, physiological, emotional, and behavioral (performance) levels may be of these instantaneous and short-term effects of noise on sleep. According to the World Health Organization (WHO), there is nevertheless consensus about the biological plausibility that short-term sleep disturbances form a long-term health risk.[17] There is sufficient evidence that chronic sleep disturbance is related to self-reported overall sleep disturbance, insomnia-like symptoms, as well as increased medication use. For cardiovascular disease (CVD)-type effects and depression (and other diseases), no such relationships can be established based on current evidence. However, nighttime noise exposure, particularly, is considered to be a risk factor for CVD.[26] Recently a relationship was established between disturbed sleep and risk of type 2 diabetes.[27]

Regarding the appropriate noise metric to be used to assess sleep disturbance, measures of single noise events were, in past studies, primarily for exposure at the sleeper's ear, while energy-averaged measures were used to describe the outdoor façade exposure level of the sleeper's dwelling. From current evidence, it is clear that different indicators of sleep disturbance are related to different noise metrics and a combination of energy-averaged measures during the night (such as L_{night}), and measures of individual noise events, may be preferable.[13] An association with energy-averaged measures for the night period has been established for subjective sleep disturbance, motility, and awakenings, whereas single-event levels and the number of events (combined with levels) are more predictive of instantaneous and short-term effects of arousal, cardiovascular responses, and sleep-stage shifts. The additive value of noise event information is currently being analyzed in The Netherlands based on a large sleep study around Schiphol airport.[28]

In 2009 the Night Noise Guidelines were published.[17] Despite the many uncertainties, the document has suggested standard noise metrics for sleep disturbance, the effects to consider, the exposure-response relationships to apply, and the threshold levels to be used in preparing nighttime noise policies. Threshold levels, L_{night}, outdoors have been proposed, ranging from

- 30 dB(A) for no effect; to
- 30–40 dB(A) for some effects, but within acceptable limits, except for vulnerable groups; to
- 40–50 dB(A), where the effects are considerably increased and, for vulnerable groups, one could speak of severe effects; to
- Over 55 dB(A), where one could speak of a serious public health problem, with potential cardiovascular risk.

However, there is some concern about the extent to which these limits are applicable in any particular country—including in Asia, where studies are scarce. The WHO recognizes that the proposed limit values might not be realistically achieved in many places in the world where noise levels are already considerably higher than these and has therefore formulated interim target values of 55 dB (L_{night}, outdoor) and 40dB (L_{night}, outdoor), respectively.

Physiological and Cardiovascular Effects

Recent studies into the effects of noise on endocrine reactions such as cortisol and catecholamine show variable results that are hard to interpret.[29] Davies and van Kamp[4] have suggested that several factors that influence the variability seen in endocrine response to noise stimulation, including timing or measurement, type of stressor, controllability, individual response characteristic, and individual psychiatric sequelae, should be considered in future studies.

Several reviews have suggested that noise exposure is associated with blood pressure changes and ischemic heart disease.[3, 4, 30–32] The biological plausibility of the hypothesis of the effects of noise on the cardiovascular system is high and assumes that noise acts as a stress factor and as such has the potential of directly and indirectly causing disease.[31] The associations are weak for blood pressure changes and hypertension and somewhat stronger for ischemic heart disease.[33] For hypertension as well as ischemic heart disease, proposed threshold values[6, 34] range between L_{Aeq} levels of 65 to 70 dB environmental noise exposure outdoors. Recent meta-analyses partly confirm the conclusion of the WHO that the results of the studies investigating the association between road and aircraft noise exposure and cardiovascular disease converged.[34] No definite conclusions can be drawn regarding the exposure–effect relations or the possible threshold values. However, preliminary exposure–response curves are available that can be used for a quantitative risk assessment.[4] One restriction is that cardiovascular effects have only been investigated in limited population groups (middle-aged men). Since some studies regarding hypertension were cross-sectional, no final inferences can be made about causality and long-term effects. Besides study design, it is also relevant whether exposure and outcome have been assessed objectively. However, after adjustment for reporting bias, it is unlikely that ill health or noise sensitivity leads to noise exposure, so in most cases there is rather the risk of underestimating the effect. The meta-analyses on ischemic heart disease refer to prospective cohort and case-control studies, in which new and objectively diagnosed cases were detected over the study period and exposure was objectively assessed independently of disease incidence and therefore enabled more definite inferences.[29, 30]

Since 2000 a number of high-quality studies have addressed many of the past concerns of study design, power, analytical approach, exposure assessment, and outcome classification.[5] Results of studies of noise and blood pressure among children show inconsistent results across noise sources and studies.[5, 35] Key issues for future studies identified are these inconsistent effects in children as far as cardiovascular effects are concerned, vulnerable groups, and gender differences.[5] Exposure assessment should also be further improved, and improved harmonization in the measurement of cardiovascular endpoints is warranted (e.g., diagnosed versus self-reported hypertension). Systematic adjustment for confounding factors is also appropriate, and access of an individual to a quiet side within a noise-exposed dwelling may be an important effect modifier.

Mental Health

Recent reviews on noise effects and mental health[1, 10] concluded that there is no direct association between environmental noise and mental health, in both adults and children. Noise annoyance is consistently found to be an important mediator. Evidence for an effect of noise on psychological health suggests that, for both adults and children, noise is probably not associated with serious psychological ill-health but may affect quality of life and well-being. Conclusions from cross-sectional evidence should be treated with caution because poor mental health might go together with a negative evaluation of the environment or a larger susceptibility to noise in general. In future studies, a clearer distinction should be made between (diagnosed) mental health effects, medically unexplained symptoms, self-reported health, and well-being/quality of life. Also, a more contextual approach to this field is appropriate, with attention to both vulnerable groups and vulnerable locations, and the beneficial effect of the availability of areas with high acoustic quality for both.

Cognitive Effects

Studies into the cognitive effects of noise have been performed primarily in schoolchildren. During the last thirty years, a limited number of studies investigated the effects of long-term exposure to air, rail, and road traffic noise among primary schoolchildren. Cognitive effects were found on (comprehension) reading, attention, problem solving, and memory.[1, 7–9] The evidence for an association between noise and cognitive functioning was strongest for exposure to noise from air traffic. Performance on the more complex tasks was mainly affected.

The large-scale RANCH study, which compared the effects of road traffic and aircraft noise on children's cognitive performance in the Netherlands, Spain, and the United Kingdom, found a linear exposure–effect relationship between chronic aircraft noise exposure and impaired reading comprehension and recognition memory, after accounting for a range of socioeconomic and confounding factors.[7] No associations were observed between chronic road traffic noise exposure and cognition. Neither aircraft noise nor road traffic noise affected attention or working memory. A 5dBA increase in aircraft noise exposure (on an energy-averaged noise metric over sixteen hours of the day) was associated with a two-month delay in reading age in the UK and a one-month delay in The Netherlands.[8] This association remained after adjustment for aircraft noise annoyance and cognitive abilities, including episodic memory, working memory, and attention. It is not yet fully clear what the longer-term effects are on cognitive functioning. Preliminary results of a follow-up of the RANCH study on the UK sample have shown that the effects are persistent over a six-year period.[36] Most children still lived in the noisy area. Also, within the framework of the RANCH study, the neurobehavioral effects of road traffic and aircraft noise exposure in 553 primary schoolchildren living around Schiphol Amsterdam Airport was investigated making use of an automated test.[9] Effects of *school* noise exposure were observed in the more complex parts of the switching attention test: children attending schools with higher road or aircraft noise levels made more mistakes. Several mechanisms have been described in the literature to explain these findings; they include direct effects, teacher and pupil frustration, learned helplessness, and impaired attention. It has also been suggested that learning deficits are mediated through lack of attention, which is used as a coping mechanism to deal with unwanted sounds.[1]

Health Effects of Low-frequency Noise

Low-frequency noise (LFN) is sound with a long wave length and is usually defined as noise under a frequency of 100 Hz. Noise with a frequency less than 20 Hz is referred to as *infrasound*. Due to its characteristics, LFN can propagate over long distances, and the direction of an LFN source is often hard to determine. LFN can relate to a broad range of sources, and it is often very difficult to specifically identify the particular source. Potential sources are transport (rail, road, and air), navigation, industry, wind turbines, and diesel motors, as well as freight traffic, including rail marshaling activities. There are often several sources of LFN, such as air conditioners, ventilation systems, and refrigerators associated with dwellings. People who are disturbed by LFN often describe LFN as a *hum* or experience it as a pressure (in the head) and vibrations in the body.

Important effects of LFN described in the literature are annoyance, loss of concentration, and sleep disturbance.[37, 38] Also reported are health symptoms such as hearing loss; vertigo; balance problems; and physiological effects on breathing, heart rate, blood pressure, and cortisol levels. Occupational exposure to high levels of infra noise have been reported, but not for daily levels of exposure. Results from experiments into the effect on hearing loss, vertigo, and balance are inconclusive due to differences in design, frequencies, and sources, making comparison difficult. The number of studies into the physiological effects of LFN is limited. Results are often based on a single study, and the noise frequencies vary strongly between studies. Most of the effects cannot be distinguished from the physiological effects associated with non-low-frequency noise. An exception is the so-called vibro-acoustic disease,[39] a disorder characterized by a combination of neurological, respiratory, and cardiovascular symptoms. This disorder is not generally accepted and is very difficult to diagnose. Available results are exclusively derived from occupational studies and animal studies. Finally, it is not clear as yet from the available evidence which aspect(s) of low-frequency noise cause the physiological effects.

New Approaches

Burden of Disease of Environmental Noise

Health impact assessment (HIA) is increasingly used in the development of environmental and public health policies and regulations. It commonly involves the identification of environmental hazards and the quantification of the expected burden of disease (BoD). This (environmental) disease burden can be expressed in a variety of ways. The WHO and others increasingly use the DALY: disability adjusted life years, encapsulating the number of healthy life years. In view of the limited space in this chapter, we refer to Knol[40] and Knol and Staatsen[41] for a more detailed description and examples of integrated measures (see Figure 4.3). The long-term effects of transport noise are now reasonably well described for several health outcomes, and exposure–response relations have been established for them.[1-5] Based on these, a report is currently being finalized by the WHO on the noise-related burden of disease.[42]

After selecting a set of endpoints with sufficient evidence for a relationship with the risk factors under study, the expected environmental disease burden in a population can be quantified by combining population-density data with concentration distributions of a relevant exposure indicator (e.g.,

Environmental Noise and Health 79

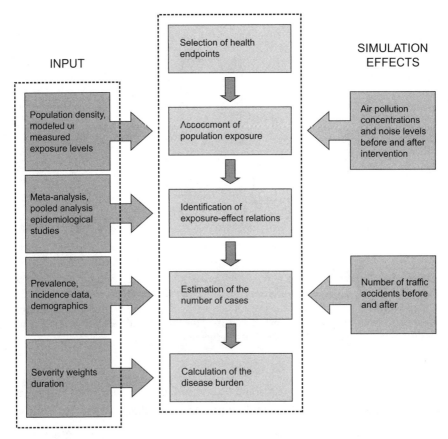

Figure 4.3 Model for calculating disease burden attributable to transport, showing sequential HIA steps, required input data (left column), and simulation effects (right column). (Reprinted with permission from Knol AB. Health and the environment: Assessing the impacts, addressing the uncertainties. [dissertation] University of Utrecht, Netherlands, 2010.)

concentration), information on exposure–response relationships, and estimates of the duration and severity of diseases.

Knol and Staatsen[41] applied this approach to calculate trends in BoD in The Netherlands. The (trends) in disease burden, related to exposure to PM_{10}, noise, radon, UV radiation, and damp houses, were expressed in DALYs and compared at the national level. Noise effects were selected based on existing evidence; they included severe annoyance, severe sleep disturbance, and mortality due to noise-induced hypertension. The results are presented in Figure 4.4. The study concluded that roughly 2 to 5 percent of the

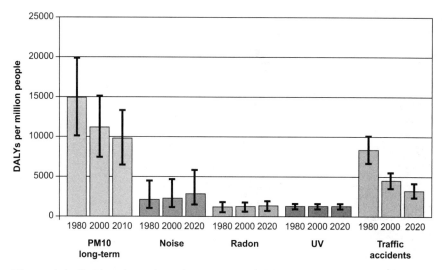

Figure 4.4 Estimated trends in the environmental burden of disease in the Netherlands, expressed in Disability-Adjusted Life Years (DALYs) with 95% confidence bands and grouped by environmental exposure and year. (Reprinted with permission from Knol AB. Health and the environment: Assessing the impacts, addressing the uncertainties. [dissertation] University of Utrecht, Netherlands, 2010.)

total disease burden can be attributed to the effect of (short-term) exposure to air pollution, noise, radon, total natural UV, and dampness in houses for the year 2000. Including the more uncertain effects of long-term PM_{10} exposure, this percentage was estimated to increase up to 13 percent, assuming no threshold for effects. The calculation for the year 2020 showed that the disease burden associated with noise exposure will probably increase up to a level where it is similar to the disease burden attributable to traffic accidents.

Combined Exposures

Recently, more integrated approaches have come forward, in which the health effects of combined noise sources or the combined effect of air pollution and noise are studied. As Stansfeld concluded in 2009,[43] the joint effects of noise and air pollution are increasingly being examined with a need for greater consideration of moderating factors in noise research.

There are indications that exposure to air pollution is associated with effects on the cardiovascular system.[44,45] Since people in urban areas often are exposed to both air pollution and noise, the effects on the cardiovascular system could be attributed to both exposure types. Epidemiological data linking cardiovascular disease prevalences with traffic-related air pollution and transportation noise are scarce, but recently several large studies have addressed the combined effects of noise and air pollution. Kluizenaar studied more than 18,000 people.[46] Noise exposure (road traffic) and air pollution were estimated using the URBIS model. The primary exposure indicators were the distance of the dwelling to the road (within 200 m), PM_{10}, and L_{den}, and hypertension was the main outcome. It was concluded that exposure to road traffic noise may be associated with hypertension in subjects who are between forty-five and fifty-five years old—considered to be a vulnerable group. Associations seem to be stronger at higher noise levels. The adjusted risk due to noise increases after adjustment for PM_{10} in this age group.

Results of the so-called cancer cohort,[45] which contains 120,000 people, was aimed at testing the hypothesis that the association between air pollution due to road traffic with mortality (CVD) is mediated or moderated by noise exposure. Results showed that cardiovascular mortality was associated with traffic intensity. The mortality was highest in areas with noise exposures over 65 dB. However, after adjustment for black smoke and traffic intensity, the effect of noise was reduced. The association between black smoke and mortality did not change after adjustment for noise. Since traffic intensity is also an important indicator of noise levels, the change in the exposure–response relationship between noise and CV mortality is disputable.

The Hypertension and Exposure to Noise near Airports (Hyena) study, an EU-funded multicenter study, looked into the combined effects of aircraft noise and road traffic noise on cardiovascular disease, in particular hypertension.[47] The number of participants was 4,800, aged forty-five to seventy, living around six European airports. Results indicated that long-term exposure is associated with excess risk of hypertension, primarily from nighttime aircraft noise, moderated by daytime road traffic exposures as well as several coping behaviors, such as closing windows.[48]

A Swedish study looked into the combined effects of long-term exposure to road traffic noise and air pollution and myocardial infarction (MI).[49] Results suggest a long-term effect of noise on increased risk for MI. Effect modification by air pollution was not strong. Most recently a large cohort study in Switzerland revealed an association between the exposure to air traffic noise and incidence of myocardial infarcts, while no effect was found from PM_{10} exposure.[50]

There is also increasing attention to the possible cognitive effects of air pollution. In 2008 the first epidemiological study investigating the effects of air pollution on children's cognitive functioning was presented.[51] The long-term concentration of black carbon particles from mobile sources was associated with decreases in cognitive test scores among 202 primary schoolchildren living in Boston. It is hypothesized that particles move to the brain tissue, where they may cause oxidative stress and inflammatory reactions. Since children in urban areas often are exposed to several environmental exposures simultaneously, it is possible that the associations found in this study could also be attributed to traffic-related air pollution and not to road traffic and aircraft noise exposure. Conversely, the effects found in the studies investigating the relationship between air pollution and cognitive functioning could also be attributed to noise exposure. More research is necessary to disentangle the effects of traffic-related air pollution and noise exposure. Currently, secondary analysis on the RANCH data,[9] matched with air pollution data for the Dutch sample, is addressing this topic.

Soundscape Approaches, Meaning, and Context

Another trend is the soundscape approach.[52, 53] This originally was oriented not toward health but toward the meaning of sound in the environment and planning involving the protection and creation of varied soundscapes. However, the concept could potentially link with health through the restorative function of people's experiences of areas of high acoustic quality, and through its emphasis on context and meaning. This approach is still in its infancy, and evidence of the beneficial health effects of areas of high acoustic quality is still lacking, as is the mechanism by which restoration might occur.

In all the noise and health studies discussed above, environmental noise was considered as a pollutant, a somewhat unavoidable waste product and an aversive stimulus that leads to negative responses. This stimulus-response approach is at the base of most noise research and policy in which the emphasis is on threshold levels, norms, and interventions aimed at reducing levels. However, this almost-exclusive attention to physical noise metrics is starting to shift toward attempting to understand meanings

TABLE 4.1
Noise Control Approach versus the Soundscape Approach

Noise Control	Soundscape
Concerns sounds of discomfort	Concerns sounds of preference
Integrates sounds from all sources	Differentiates between sound sources (wanted vs. unwanted)
Manages by reducing levels	Manages by wanted sounds masking unwanted sounds
Sound as a waste	Sound as a resource

Source: Adapted from A. L. Brown, Soundscapes and environmental noise management, *Noise Control Engineering Journal* 58 (5) (2010): 493–500.

and the role of context in people's perceptions of, and reaction to, acoustic environments. This soundscape approach considers the acoustical environment more broadly as a resource, not merely as a waste product (see Table 4.1). It shifts the focus from the physically measured levels of exposure to the meaning of the sound heard—through making a distinction between wanted and unwanted sounds—and the role of context, often geographically defined, in shaping human perception and experience of the acoustical environment.

Research into the soundscape of "quiet areas" has shown that people prefer sounds of water, nature, and humans above mechanical sounds.[52, 54] (See Figure 4.5.)

There is some indication that certain areas contribute to restoration. Often it is assumed that this is related to aspects of quiet and green, but evidence is still limited, and insight into the mechanisms needs further attention. Most studies address the restorative effects of natural recreational areas outside the urban environment.[55–57] The question is whether natural areas within, and in the vicinity of, urban areas contribute to psychophysiological and mental restoration after stress as well. Does restoration require the absence of urban noise? Besides potential immediate restorative effects, there may be long-term effects of access to environmental amenities in the immediate living environment. Dutch cross-sectional studies found that residents in green neighborhoods report better general health.[58, 59] Do natural environments (micro/macro) positively influence long-term general health and well-being, and which environmental aspects

Figure 4.5 Example of an acoustic environment—a New York City park—with potential restorative outcomes. (Photo by A.L. Brown.)

are important? The soundscape approach is aimed at location-specific acoustical quality and its immediate effects on people's perceptions and well-being. The long-term effects of access to high acoustic quality areas need to be studied in more depth.

The soundscape approach is still highly acoustical. The meaning people give to sounds and noise, or to an exposure context, strongly influences their reactions and accompanying health effects. Human noise can be considered as a product of behavior, which in turn is a consequence of human needs. As a result of these needs, people produce sounds and expose themselves to sounds, partly with a purpose and partly as an unintended by-product of their activities. The meanings of these sounds can be negative and positive; they are partly generic, partly culturally defined; and they change over time. The noise control approach has been focused exclusively on the negative aspects and meanings of noise. But in order to understand the driving forces behind noise, it is important to study its positive aspects as well. The shift from decibels to meanings and context offers important cues for future research and policy.

In the context of noise and health, it is important to understand the pathway by which these meanings lead to well-being and health via restoration. Against this background, a provisional conceptual working model has been developed (see Figure 4.6). An extensive literature review[60] showed that people have a range of motives to produce sound: a need for sensation and excitement, control of identity, control over the aesthetics of a noise environment (soundscape), a sense of belonging, maintaining working rhythms and motivation, expression of revolt/rebellion, power, perceived company or accompanied solitude, in and exclusion of groups (commercial or safety motives), and defense/imposing behavior. The model takes the societal and personal needs that lead to sound and noise as a point of departure. These acoustical stimuli result in physiological changes and inner reactions, which can be positive as well as negative: emotion; pleasure; a sense of power; or feelings of anxiety, stress, and annoyance. These physiological and emotional reactions are dependent not only on levels of sound/noise but also on meanings. Context, personal factors such as noise sensitivity, and the degree to which someone is capable to restore also play a role. This process finally influences behavior as well as health and well-being, which in their turn affect motives, reactions, and restoration. The arrows in the model tentatively indicate how these

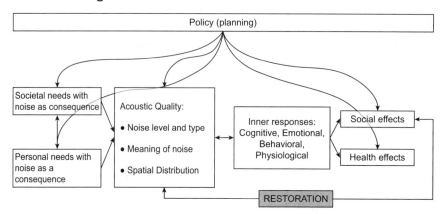

Figure 4.6 Provisional framework on societal aspects of noise. (Reprinted with permission from J. Devilee, E. Maris, I. van Kamp, The societal meaning of sound and noise: a new perspective. Invited paper Internoise, 2010, Lisbon.)

processes are interrelated, while taking planning and geographical distribution of acoustic quality into account.

Further, the function of an area, and the time dimension, should be considered in future studies. Current policy tends to focus almost exclusively on the residential situation. This is justified to the extent that people spend, on average, sixteen hours at home per day and, for most people, a quiet home is crucial and seen as the most important place to relax and restore from daily pressures. But when other activities and functions are considered, the current residence-only approach may be too limited. Current policy is aimed only at chronic exposures and long-term effects.

At all levels a sense of control on the one hand and a sense of helplessness on the other play a key role. From the stress literature there is ample evidence that learned helplessness is related to a range of health problems. There is also evidence that learned helplessness is linked to socioeconomic status. It has also been shown that noise plays a part in this, but its exact influence is difficult to determine. This is due partly to a lack of studies and partly to an accumulation of psychosocial and physical stressors, which makes it difficult to pinpoint the specific role of noise. Thinking in terms of acoustical quality sheds a different light on the spatial and demographic distribution of noise. The local orientation of the soundscape approach seems quite fit to study the acoustical quality in deprived areas and groups as well as at sensitive locations (e.g., school environments) and thus has the potential to contribute to an understanding of the field of social health inequalities. It could be fruitful to study these types of mechanisms in more depth.

Conclusion

There is now well-documented evidence of distinct health effects of noise. Research in the past decades has shown that there is sufficient evidence of the development of hearing damage at high noise levels generally experienced in occupational and some recreational settings, but above those experienced through exposure to environmental noise. There is also sufficient evidence of the development of the chronic effects of annoyance and sleep disturbance at environmental levels of noise exposure (exposure predominantly to transport noise) and mixed evidence of the development of hypertension and ischemic heart disease. There is growing evidence of effects of environmental noise on cognition and reading performance in children, but no or weak evidence of immune system effects, and no evidence of a direct relationship with mental health.

Exposure–response relationships are available for some of these health outcomes, and together with increasingly better estimates of population exposures to environmental noise, a (tentative) start has been made to calculate the environmental-noise related burden of disease in DALYs.

Issues for future noise and health research include the tenability of generalized exposure–response relationships regarding annoyance with aircraft noise, the long-term effects of sleep disturbance including the risk of diabetes type 2, harmonization of measures regarding CVD, conceptual issues in relation to mental health outcomes, and the robust but limited evidence on cognitive effects of noise on children and its long-term consequences. Detailed studies on the added value of additional noise metrics based on maximum levels, number of events, and duration should be pursued with respect to sleep disturbance.

The health effects of low-frequency noise are still highly anecdotal, while the number of sources of low-frequency noise might be increasing especially in view of energy-saving measures. Insight into the combined effects of noise and air pollution is growing, although not yet conclusive.

Work being done in Europe and Canada should shed new light on this topic.

Current innovation in noise and health research includes a more integrated approach to environmental stressors and a more contextual approach to acoustical quality (the soundscape approach)—the latter possibly related to restorative effects of high-quality acoustic environments. Both potentially offer important insights. A model was presented that uses human motives to produce and receive sounds as a point of departure. Apart from noise levels, the meaning and distribution of sounds (geographic as well as demographic) are key concepts as well as the effects on social behavior and health. A sense of control likely plays a key role. The challenge for future research is to include this sense of control, and its counterpart, a sense of helplessness, together with motives and meanings, alongside the individual's exposure to the acoustic environment.

Notes

1. C. Clark and S. A. Stansfeld, The effect of transportation noise on health and cognitive development: A review of recent evidence, *International Journal of Comparative Psychology* 20 (2007): 145–158.
2. European Commission Working Group on Health and Socio-Economic Aspects of Noise 2004, *Position paper on dose-effect relationships for night time noise,* http://ec.europa.eu/environment/noise/pdf/position paper.pdf.
3. W. Babisch, Transportation noise and cardiovascular risk: Updated review and synthesis of epidemiological studies indicate that the evidence has increased, *Noise & Health* 8 (2006): 1–27.
4. W. Babisch, Road traffic noise and cardiovascular risk, *Noise & Health* 10 (38) (2008): 27–33.
5. H. Davies and Irene van Kamp, Environmental noise and cardiovascular disease: Five year review and future directions, in *Proceedings of the 9th Congress of the International Commission on the Biological Effects of Noise (ICBEN)*, Mashantucket, July 21–25, 2008 (on CD), edited by B. Griefahn.
6. Health Council of the Netherlands (HCN), *Public health impact of large airports*, HCN Report No. 1999/14E, The Hague, September 2, 1999, http://www.gr.nl/pdf.php?ID=19&p=1.

7. S. A. Satisfied, B. Berglund, C. Clark, et al., Aircraft and road traffic noise and children's cognition and health: A cross-national study, *Lancet* 365 (2005): 1942–1949.
8. C. Clark, R. Martin, E. van Kempen, T. Alfred, J. Head, and H. W. Davies, et al., Exposure-effect relations between aircraft and road traffic noise exposure at school and reading comprehension: The RANCH project, *Noise & Health* 8 (2006). 30, 58.
9. E. van Kempen, I. van Kamp, E. Lebret, et al., Does transportation noise cause neurobehavioral effects in primary school children? A cross-sectional study, *Environmental Health* 9 (2010), http://www.ehjournal.net/content/9/1/25.
10. I. van Kamp and H. Davies, Environmental noise and mental health: Five year review and future directions, in *Proceedings of the 9th Congress of the International Commission on the Biological Effects of Noise (ICBEN)*, Mashantucket, July 21–25, 2008 (on CD), edited by B. Griefahn.
11. Directive 2002/49/EC of the European Parliament and of the Council of 25 June 2002 relating to the assessment and management of environmental noise, *Final report on noise*, 2002.
12. S. A. Janssen, M. Basner, B. Griefahn, and H. M. E. Miedema, Environmental noise and sleep disturbance, Background paper for WHO, working group on risk assessment of environmental noise, 2011. See also L. Fritschi, A. L. Brown, R. Kim, D. Schwela, and S. Kephalopoulos, S. (eds.), *Health risk assessment of environmental noise in Europe* (Bonn: WHO, Regional Office for Europe, 2011).
13. J. Vernon, et al., *Task 3 impact assessment and proposal of action plan A*, 2010 (report prepared by RPA Ltd for DG Environment under service contract no. 070307/2008/510980/SER/C3).
14. D. Schwela, Noise policies in Southeast Asia, in *Proceedings of Internoise, 2006*, Hawaii.
15. T. Gjestland, Research on community response to noise—in the last five years, in *Proceedings of the 9th Congress of the International Commission on the Biological Effects of Noise (ICBEN)*, Mashantucket, July 21–25, 2008 (on CD), edited by B. Griefahn.
16. ISO, *Acoustics: Assessment of noise annoyance by means of social and socio-acoustic surveys*, ISO/TS 15666:2003 (2003).

17. WHO, *Night noise guidelines (NNGL) for Europe* (Bonn: EU Centre for Environment and Health, Bonn Office Grant Agreement 2003309 Between the European Commission, DG Sanco, and the World Health Organization, Regional Office for Europe, 2009).
18. E. E. M. M. van Kempen, B. Staatsen, and I. van Kamp, *Selection and evaluation of exposure-effect relationships for health impact assessment in the field of noise and health*, RIVM report 630400001/2005, http://rivm.openrepository.com/rivm/bitstream/10029/7412/1/630400001.pdf.
19. F. B. Bernard and I. H. Flindell, *Estimating dose-response relationships between noise exposure and human health impacts in the UK* (technical report, 2009).
20. S. A. Janssen, H. Vos, and H. M. E. Miedema, The role of study characteristics in changes in aircraft noise annoyance over time, in *Proceedings Euronoise* (Edinburgh, Scotland, 2009).
21. E. E. M. M. van Kempen and I. van Kamp, *Annoyance from air traffic noise: Possible trends in exposure-response relationships*, Report 01/2005 MGO EvK Reference 00265/2005 (RIVM, Netherlands, 2005).
22. W. Babisch, D. Houthuijs, G. Pershagen, E. Cadum, K. Katsouyanni, M. Velonakis, M.-L. Dudley, H.-D. Marohn, W. Swart, O. Breugelmans, G. Bluhm, J. Selander, F. Vigna-Taglianti, S. Pisani, A. Haralabidis, K. Dimakopoulou, I. Zachos, L. Järup, and HYENA Consortium, Annoyance due to aircraft noise has increased over the years—Results of the HYENA study, *Environment International* 35 (8) (November 2009): 1169–1176.
23. E. E. M. M. van Kempen, I. van Kamp, R. K. Stellato Mats, et al., Annoyance reactions to air- and road traffic noise of schoolchildren: The RANCH study, *Journal of the Acoustical Society of America* 125 (2) (February 2009).
24. M. Basner and B. Griefahn, Environmental noise and sleep disturbance (background paper for WHO, Working Group on Risk Assessment of Environmental Noise, 2009).
25. A. Muzet, Environmental noise, sleep and health (review), *Sleep Medicine Reviews* 11 (2) (2007): 135–142.
26. Health Council of the Netherlands, *Effects of noise on sleep and health*, Publication no. 2004/14 (The Hague: Health Council of the Netherlands, 2004).

27. E. Donga, M. van Dijk, J. G. van Dijk, et al., A single night of partial sleep deprivation induces insulin resistance in multiple metabolic pathways in healthy subjects, *Journal of Clinical Endocrinology & Metabolism* 95 (6) (2010): 2963–2968.
28. W. Passchier-Vermeer, H. Vos, J. H. M. Steenbekkers, et al., *Sleep disturbance and aircraft noise. Exposure-effect relationships*, Report no. 2002.027. 2002 (Leiden, TNO-PG, 2002).
29. W. Babisch, Stress hormones in the research of cardiovascular disease: Effects of noise, *Noise & Health* 5 (18) (2003): 1–11.
30. W. Babisch, Traffic noise and cardiovascular disease: Epidemiological review and synthesis, *Noise & Health* 2 (8) (2000): 9–32.
31. E. E. M. M. van Kempen, H. Kruize, et al., The association between noise exposure and blood-pressure and ischemic heart disease: A meta-analysis, *Environmental Health Perspectives* 110 (2002): 307–317.
32. W. Babisch and I. van Kamp, Exposure-response relationship of the association between aircraft noise and the risk of hypertension, *Noise & Health* 11 (44) (2009): 161–168.
33. Th. Bodin, M. Albin, J. Ardö, et al., Road traffic noise and hypertension: Results from a cross-sectional public health survey in southern Sweden, *Environmental Health* 8 (2009): 38.
34. WHO, *Guidelines for community noise*, edited by B. Berglund, T. Lindvall, D. H. Schwela, and K. T. Goh (Geneva: World Health Organisation, Guideline Document, 2000).
35. E. van Kempen, I. van Kamp, P. Fischer, et al., Noise exposure and children's blood pressure and heart rate: The RANCH project. *Occupational and Environmental Medicine* 63 (2006): 632–639.
36. C. Clark, S. Stansfeld, and J. Head, The long-term effects of aircraft noise exposure on children's cognition: Findings from the UK RANCH follow-up study, in *Proceedings Euronoise* (Edinburgh, Scotland, 2009).
37. T. Koeman and R. van Poll, *Low frequency noise, sources, health effects and guidelines* (in Dutch) (RIVM, 2008).
38. W. K. Persson, J. Bengtsson, A. Agge, and M. Bjorkman, A descriptive cross-sectional study of annoyance from low frequency noise installations in an urban environment, *Noise & Health* 5 (2003): 35–46.
39. N. A. A. Castelo Branco and M. Alves-Pereira, Vibroacoustic disease, *Noise & Health* 6 (23) (2004): 3–20.

40. A. B. Knol, Health and the environment: Assessing the impacts, addressing the uncertainties (dissertation, University of Utrecht, Netherlands, 2010).
41. A. B. Knol and B. A. M. Staatsen, *Trends in the environmental burden of disease in the Netherlands, 1980–2020*, RIVM report no. 500029001(Bilthoven: National Institute for Public Health and the Environment, 2005).
42. L. Fritschi, L. Brown, R. Kim, et al., *Health risk assessment of environmental noise in Europe*.
43. S. A. Stansfeld, New directions in noise and health research, in *Proceedings Euronoise* (Edinburgh, Scotland, 2009).
44. R. D. Brook, Cardiovascular effects of air pollution, *Clinical Science* 115 (2008): 175–187.
45. R. Beelen, G. Hoek, D. Houthuijs, et al., The joint association of air pollution and noise from road traffic with cardiovascular mortality in a cohort study, *Occupational and Environmental Medicine* 66 (2009): 243–250.
46. Y. De Kluizenaar, R. T. Gansevoort, H. M. E. Miedema, and P. E. de Jong, Hypertension and road traffic noise exposure, *Journal of Occupational and Environmental Medicine* 49 (5) (2007): 484–492.
47. L. Jarup, W. Babisch, D. Houthuijs, et al., Hypertension and exposure to noise near airports: The HYENA study, *Environmental Health Perspectives* 116 (3) (2008): 329–333.
48. W. Babisch, D. Houthuijs, G. Pershagen, et al., Hypertension and exposure to noise near airports: Results of the HYENA study, in *Proceedings of the 9th Congress of the International Commission on the Biological Effects of Noise (ICBEN)*, Mashantucket, July 21–25, 2008 (on CD), edited by B. Griefahn.
49. J. Selander, M. E. Nilsson, G. Bluhm, et al., Long-term exposure to road traffic noise and myocardial infarction, *Epidemiology* 20 (2) (2009): 272–279.
50. A. Huss, A. Spoerri, M. Egger, and M. Röösli, Aircraft noise, air pollution, and mortality from myocardial infarction. *Epidemiology* 21 (6) (2010): 829–836.
51. S. F. Suglia, A. Gryparis, R. O. Wright, et al., Association of black carbon with cognition among children in a prospective birth cohort study, *American Journal of Epidemiology* 167 (2008): 280–286.

52. S. R. Payne, W. J. Davies, and M. D. Adams, *Research into the practical and policy applications of soundscape concepts and techniques in urban areas* (London: Welsh Assembly Goverment/Department of the Environment/The Scottish Government/Department for Environment, Food and Rural Affairs, 2009).
53. A. L. Brown, Soundscapes and environmental noise management, *Noise Control Engineering Journal* 58 (5) (2010): 493 500.
54. M. Zhang and J. Kang, Towards the evaluation, description, and creation of soundscapes in urban open spaces, *Environment and Planning B: Planning and Design.* 34 (1) (2007): 68–86.
55. Health Council of the Netherlands and Council for Spatial Environment and Nature Research, *Natuur en gezondheid. Invloed van natuur op sociaal, psychisch en lichamelijk welbevinden*, 2004/09, A02a (Den Haag: Gezondheidsraad en RMNO, 2004) (in Dutch; English summary).
56. Health Council of the Netherlands, *Quiet areas and health*, Publication no. 2006/12 (The Hague: Health Council of the Netherlands, 2006) (in Dutch; English summary).
57. T. Hartig, G. W. Evans, L. D. Jamner, et al., Tracking restoration in natural and urban field settings, *Journal of Environmental Psychology* 23 (2003): 109–123.
58. R. A. Verheij, J. Maas, P. Groenewegen, et al., Green space, urbanity, and health: How strong is the relation? *Journal of Epidemiology and Community Health* 60 (2006): 587–592.
59. J. Maas, R. A. Verheij, P. P. Groenewegen, et al., Green space, urbanity and health: How strong is the relation? *Journal of Epidemiology and Community Health* 60 (2006): 587–592.
60. J. Devilee, E. Maris, and I. van Kamp, The societal meaning of sound and noise: A new perspective (paper presented at Internoise 2010, Lisbon).

Acknowledgments

The contributions of Jeroen Devilee, Eveline Maris, and Elise van Kempen to parts of the new work summarized in this chapter are gratefully acknowledged. We would also like to thank Danny Houthuijs for his valuable comments on a previous version of parts of this chapter (published in the proceedings of ICA 2010).

5

Environmental Epigenetics

Mihalis I. Panayiotidis, Aglaia Pappa, and Dominique Ziech

Introduction

The roots of epigenetics can be found in Aristotle's theory of epigenesis, which emphasized developmental changes to be gradual and qualitative. In the 1940s epigenetics was addressed by Conrad Waddington, who referred to the field as "factors that influence how a genetic predisposition can ultimately play out in a biological or clinical outcome."[1] At that time, a gene's role in development was unknown, and many developmental geneticists were striving to better understand how gene(s) influence an organism's development. Waddington's epigenetic approach was actually quite different from what traditional developmental genetics were concerned with at that time. Though both approaches were attempting to clarify the relationship between genetics and phenotypic variation, Waddington was more concerned with the genetics underlying phenotypic variation(s). His approach was met with little enthusiasm; nevertheless he continued to defend his hypothesis by acknowledging that most genetic variations (such as mutations) have little or no effect on the resulting phenotype. In addition, a consensus within the scientific community regarding epigenetics was still lacking; it was even suggested by some scientists that the term be abandoned. Despite these obstacles, Waddington persevered and defined epigenetics as a process involving genes switching "on" and "off" in distinct patterns in order to make specialized cells in the body. Waddington argued that patterns of gene expression, not genes themselves, are what define each cell type.[1]

In the 1990s epigenetics was defined as the sum of all genetic and nongenetic factors acting on cells to selectively control gene expression and increase phenotypic complexity during development. In 1996 Arthur Riggs and colleagues, in their book, *Epigenetic Mechanisms of Gene Regulation*, further defined epigenetics as "the study of mitotically and/or meiotically heritable changes in gene function that cannot be explained by changes in DNA sequence."[1] In the twentieth century, the molecular basis of epigenetics was studied in a variety of organisms, and since then the field has been integrated into biology and intersects with developmental biology, evolution, genetics, and ecology. In today's molecular era, the term encompasses all heritable information within cells that are not the DNA sequence itself.[1]

As one can see, the concept of epigenetics has changed greatly. In fact, the epigenetic approach has now replaced the methods utilized by many classical developmental genetics, and many genetic events have been redefined as epigenetic ones. For example, it is now well accepted that the variations among liver, skin, and kidney cells are of epigenetic, not genetic, origin. In addition, the ability of scientists to restore differentiated somatic cells back to a pluripotent state (i.e., induced progenitor stem cells) does not change the DNA in these cells, but rather induces alterations in their epigenetic program.[2,3] Moreover, epigenetics is hypothesized to be a key process in the manifestation of various adult diseases such as neurodevelopmental disorders,[4] cardiovascular disease,[5] type 2 diabetes,[6] obesity,[7] Alzheimer's disease,[8] autism,[9] asthma,[10] and bipolar disorder.[11] Finally, today's epigenetic research is focused on elucidating the association between environmental factors (diet, stress, prenatal nutrition, environmental and chemical exposures, etc.) and the epigenetic ability to imprint genes passed from one generation to the next.[12]

Epigenetics

Background

As mentioned previously, *epigenetics* is defined as the study of mechanisms or pathways that initiate and maintain heritable patterns of gene expression without changing the DNA sequence itself and instead lead to modifications in histone and DNA methylation patterns that can turn genes "on" and/or "off." As a result, epigenetics is a key player in the development and differentiation of various cell types in an organism.[3] Essentially, the genes that a tissue does not need are turned "off" and are not expressed (also referred to as gene silencing), whereas the genes that a tissue needs are turned "on" and expressed. For example, the genes needed to make hair cells are present within other cell types (e.g., liver cells), but they have been silenced, or turned "off." It is important to understand that all our cells share the same DNA and therefore contain the entire genome, which includes some 30,000 genes.[13] In other words, all our cells are genetically identical, and it is epigenetics that contributes to allowing cells to be essentially different.

Mechanisms

Epigenetics literally means "on" or "over" the genetic information encoded in DNA, and it results in chemical changes to the "epigenome." In parallel to the term *genome* (which defines the complete set of genetic information contained in the DNA of an organism), the *epigenome* generally refers to the complete set of characteristics of epigenetic pathways in an organism that are overlaid on top of the genome. In other words, think of the genome as a computer's hardware, and the epigenome as the software that "tells" the computer what to do. Similarly, we can think of the transmission of genetic information as constituted by the chemical letters of the DNA sequence and epigenetic transmission as residing in the fonts of these letters and in the punctuation. A perfect example is the following sentences: "I love running," and "I **love** running!!!" Though the words are the same, the second sentence is a much stronger statement. Epigenetic modification of genes plays a major role in how strongly a gene is turned "on." The epigenome is a complex that consists of DNA methylation and the chromatin and essentially forms a second code overlaid on top of the DNA sequence code of the genome. Though each organism has a single genome, the same individual has multiple epigenomes, which may differ by cell and tissue type and may also change over the lifetime of an organism. Along with the genome, the epigenome is duplicated in daughter cells during cell growth and division. In many instances, such as cellular differentiation, the epigenome may change to establish a new pattern of gene expression and may even vary between any two cells, even of the same type.[13]

On the other hand, lifestyle habits (smoking, alcohol intake, diet, etc.) can directly alter the epigenome, correlating with changes in epigenetic markers and a potential vulnerability to the development of disease—a key difference between genetics and epigenetics.[12, 14] Finally, researchers have identified four types of epigenetic pathways: (a) histone modifications, (b) DNA methylation, (c) nucleosome remodeling, and (d) noncoding RNA-mediated pathways. Alterations to any of these epigenetic pathways lead to variations in chromatin structure, which in turn result in genome modulations. Such epigenetic modulations are known to intertwine with each other and ultimately lead to altered gene regulation and expression patterns. It is very likely that other pathways beyond the four mentioned above will be discovered in the future.[15] To date, however, histone modifications and DNA methylation are the most studied epigenetic pathways and are the focus of this review.

Histone Modifications

The chromatin constitutes a structural complex of histone proteins and DNA. *Histones* are small globular proteins that are known to wrap around DNA and participate in the regulation of gene expression. Histones are located in a region called the histone tail and store epigenetic information through myriad chemical post-translational modifications such as acetylation, phosphorylation, and methylation, to name a few. Some modifications are reversible and dynamic and are often associated with the inducible expression of individual genes.[16] Chemical modifications in histone proteins are proposed to form a "histone code," which in turn is seen as an epigenetic extension of the DNA coding potential. In essence, when histones are chemically modified, gene expression is directly altered by genes becoming inactivated or activated. The functional consequence of histone modification(s) depends on the type of residue—for instance, lysine (K) or arginine—and also on the specific site within the same residue where the modification occurs (i.e., on K9, K4, or K20). For example, when the K9 residue is methylated, genes are inactivated or silenced. On the contrary, when the K4 residue is methylated, genes are activated. Furthermore, when specific histone residues are acetylated, genes become transcriptionally active, and when they are deacetylated, gene expression is repressed. Essentially, when the histone tail is chemically altered, the corresponding

DNA Methylation

DNA methylation is an essential component in the development of mammals, but despite twenty-five years of work, researchers still do not know exactly how this is so. What is known, however, is that methylation changes the interactions between proteins and DNA, which result in alterations to the chromatin structure and thus cause either decreased or increased transcriptional rates. More specifically, DNA methylation is a biochemical modification that attaches directly to the DNA itself. This biochemical modification consists of methyl (CH_3) groups that are covalently added to the fifth carbon of the cytosine ring (one of the five nucleotides in the nucleic acids of DNA and RNA), forming a residue called 5-methyl cytosine. The addition of methyl groups to the 5-position of cytosines alters the structure of the major groove in DNA to which the DNA proteins bind. The widespread methylation of the C5 position of cytosine residues in DNA is one of the most fundamental "epigenetic markers" and can be copied after DNA synthesis, resulting in heritable changes in chromatin structure.[16-19] Epigenetic research has shown that mammals appear to have taken advantage of such methylation changes to provide a heritable mechanism for altering DNA-protein interactions in inducing gene silencing.[18]

Along the linear DNA chain are sites where a cytosine molecule is linked (via a phosphate group) to a guanine molecule (another one of the five nucleotides in the nucleic acids of DNA and RNA), thus forming what we know as CpG sites. Regions of DNA that have a high density of CpG sites are, in turn, referred to as CpG islands. DNA methylation occurs predominantly within CpG islands, whereby either too much (hypermethylation) or too little (hypomethylation) occurs. The removal of methyl groups leads to a hypomethylation state, whereas the addition of methyl groups leads to a hypermethylation one. It has been estimated that as much as 80 percent of all CpG sites in the mammalian genome are methylated.[16, 20] Methylation of CpG islands in promoter regions of genes (DNA segments where the initiation of a gene's transcription is initiated) leads to the binding of methylated CpG binding proteins (MBDs), which consequently block the initiation of transcription and thus repress transcriptional activation.[18] Finally, DNA methylation, in conjunction with post-translational histone modifications, is known to be involved in the regulation of chromatin, which leads to either a transcriptional activation or repression (gene silencing).[16, 20]

Epigenetics, Disease, and the Environment

Up until recently, many diseases were thought of solely as genetic; however, it is now recognized that numerous diseases have in fact more of an epigenetic than genetic origin.[13, 16] Interestingly, the first example of a human disease to be identified with such an epigenetic origin was cancer.[13] More specifically, in 1983 widespread loss of DNA methylation (hypomethylation) was observed in colorectal cancers—the first epigenetic alteration to be implicated in human cancers.[2, 21] Only recently, epigenetic research has suggested that during the development of a neoplasm, the degree of hypomethylation of genomic DNA increases as the lesion progresses from a benign proliferation of cells to an invasive cancer. In fact, a malignant cell can have 20 to 60 percent less genomic methylation than its normal counterpart.[13] Hypermethylation, on the other hand, has also been identified as a major contributor in cancer development.[18] In fact, it has been suggested that CpG island hypermethylation, specifically within the promoter regions of tumor-suppressor genes, is a major event that can potentially serve as a diagnostic tool.

Recently, research has estimated that 100 to 400 hypermethylated CpG islands occur within the promoter region of a given tumor type. It is important to note, however, that we still do not understand how CpG islands become hypermethylated in some types of cancer and not in others. Nevertheless, hypermethylation of the CpG island promoter regions is known to affect genes involved in processes such as cell cycle, DNA repair, apoptosis, and angiogenesis—all of which are known to have a key role in cancer development. For example, when DNA repair genes (such as *hMLH1*, *BRAC1*, *MGMT*) are hypermethylated, they become inactivated, and thus the repair of genetic mistakes is hindered, thereby opening the way to cellular neoplastic transformation. In addition, the subsequent hypermethylation of CpG islands in the promoter regions of tumor-suppressor genes in cancer cells is associated with specific histone markers, namely deacetylation of histones H3 and H4, loss of H3K4 trimethylation, and gain of H3K9 methylation and H3K27 trimethylation. Such DNA methylation and histone modification patterns have potential clinical use; they are hypothesized to serve as diagnostic tools, prognostic factors, and even predictors of responses to treatments. For example, the hypermethylation of the *GSTP1* (glutathione S-transferase) gene is found in 80 to 90 percent of patients with prostate cancer and is not found in the corresponding benign tissue. Therefore, on such basis, the detection of *GSTP1* methylation could help to distinguish between the malignant and benign state, thus allowing for an epigenetic therapy to have a clinical application.[22]

Other rare diseases (such as immunodeficiency, centromeric instability, facial anomaly syndrome, and a common kind of intellectual disability in young girls) are known to be at least partially caused by changes in the methylation machinery (via either hyper- or hypomethylation-induced epigenetic changes in gene expression levels). In fact, recent epigenetic research has suggested that in many of these diseases, DNA methylation plays a major role in the proper development of an organism after birth. That is, DNA methylation is essential for the survival of mammals, and impairment of DNA methylation results in a substantial increased risk for disease, including cancer.[18]

As mentioned previously, environmental toxins, such as heavy metals and pesticides, are known to disrupt DNA methylation and chromatin patterns. The European library of commercial compounds, called REACH, has recently identified a list of about 200,000 chemicals in order to determine potential toxicities. This outreach was initiated by the growing number of recent studies showing strong correlations between environmental toxicity and disease manifestation.[23] For example, recent data have suggested that *in utero* exposure to vinclozolin, a commonly used pesticide, might be dormant in the fetus and yet be activated ten, twenty, or even fifty years later and consequently cause cancer.[24] Other data have shown that when pregnant rats were exposed to high doses of pesticides, their offspring (next three generations) developed high rates of infertility.[25, 26] A separate study showed that exposure to estrogenic and anti-androgenic toxins decreased male fertility and could be inheritable among multiple generations.[27] Similarly, environmental exposures to metals, such as tin and cadmium, were also shown to have significant health effects. For example, in a study led by Bruce Blumberg at the University of California, Irvine, a child who was prenatally exposed to tin demonstrated a propensity to become overweight throughout his or her life. Blumberg explained these findings by proposing that prenatal exposure to tin may be strongly implicated in the increased production and systemic circulation of chemicals, which he called "obesogens." Although the exact relationship between tin exposure and obesogens remains to be clarified, Blumberg's findings are a perfect example of just how potent the relationship between the environment and a person's epigenetic programming can be.[28]

Another example of how detrimental exposure to environmental carcinogens can be (in terms of damaging the epigenome) is the well-established carcinogen tobacco smoke. Evidence suggests that tobacco smoke is strongly linked to alterations in epigenetic patterns. More specifically, tobacco smoke has been shown to disrupt DNA methylation patterns and lead to the demethylation of metastatic genes in lung cancer cells. In 2006 a groundbreaking paper published in the *European Journal of Human Genetics* presented the major finding that the sons of men who smoked during pre-puberty were at a higher risk for obesity and other health problems in adulthood. In particular, of the 14,024 fathers studied, 166 stated that they had started smoking before the age of eleven—just as their bodies were preparing to enter puberty. The researchers claimed that the period around puberty is a valid developmental stage to detect epigenetic changes, as boys are genetically isolated before puberty because they cannot form sperm. That is, if the environment is going to imprint epigenetic marks on genes in the Y chromosome, what better time to do it than when sperm starts to form? When the researchers examined the sons of those 166 early smokers, it turned out that the boys had significantly higher body mass indexes than other boys by the age of nine, suggesting that the sons of men who smoked during pre-puberty were at a higher risk for obesity.[29]

Diet and nutritional states have also been strongly implicated in having negative health consequences for the epigenome and overall human health. In the 1980s Dr. Lars Olov Bygren suggested that "powerful" environmental conditions (e.g., starvation) could leave an imprint on the genetic material in eggs and sperm, which can short-circuit evolution and pass along new traits in a single generation. A study by Bygren published in 2001 in *Acta Biotheoretica* suggested that children who went from normal eating habits to gluttony in a single winter season produced sons and grandsons who lived far shorter lives.[19] The same results were observed in female lineages, meaning that daughters and granddaughters of girls who had gone from normal to gluttonous diets also lived shorter lives. Simply put, the data suggest that over a single winter season of overeating, a youngster could initiate a biological chain of events that would lead grandchildren to die decades earlier. The explanation for this relationship lies in epigenetics, as it provides evidence that lifestyle choices, such as eating too much, can change the epigenome in ways that cause obesity genes to be strongly expressed while longevity genes are expressed too weakly.[19]

Yet another pivotal study displaying a strong correlation among diet, epigenetic modifications, and an organism's development utilized the agouti strain of mice. The authors of this study established two genetically identical strains of mice: one with brown fur and an average weight of about 32 grams, in which the agouti gene was hypermethylated and thus silenced, and the other with yellow fur and a weight of around 63 grams (obese), in which the gene was hypomethylated and thus activated. In order for the agouti gene to be hypermethylated, the first strain was fed a diet supplemented with rich sources of methyl groups (vitamin B12, folic acid, choline, and betaine) in comparison with the second strain. Data showed that the observed obesity in the second strain of mice was due to the hypomethylation-induced expression of the agouti gene, which in addition was shown to be more susceptible to diabetes and cancer.[13]

Furthermore, a recent study has shown that when female mice were exposed to a diet modified to contain 50 mg/kg of BPA (Bisphenol A, an industrial chemical used in polycarbonate plastic production and epoxy resin linings in metal-based food and beverage cans), the offspring were also exposed to BPA, *in utero* as well as during lactation, leading to a higher proportion of yellow-coated than brown (control) mice. However, when the offspring were supplemented with methyl donors, they returned to methylation levels and coat color comparable to the control ones, suggesting that BPA causes a hypomethylation-induced effect on the epigenome of the offspring.[30] In fact, in two separate recent studies, exposures of mice to *in utero* high (5mg/kg) or *gestation* low (20µg/kg) doses of BPA were documented to cause hypomethylation of the *Hoxa10* gene in the uterus (a gene involved in the regulation of gene expression, morphogenesis, and differentiation)[31] and altered the epigenome in the forebrain of the offspring, respectively.[32] Finally, in another study, *NotI* loci (involved in brain development) were also shown to be hypomethylated after exposure to BPA at 3.0 µg per gram of powder food.[33]

Behavioral "changes" between mother and infant (such as in conditions of chronic and unpredictable maternal separation) have also been observed to cause changes in the epigenome. For example, a recent study has shown that infants born under conditions of such behavioral changes exhibited depressive-like behaviors and altered behavioral responses when in aversive environments. Interestingly, data strongly implicated alterations in DNA methylation patterns, in both directions (meaning either hypo- or hypermethylated), in the promoter region of several candidate genes (such as the genes coding for methyl CpG-binding protein 2 [MeCP2], serotonin receptor 1A, monoamine oxidase A [MAOA], cannabinoid receptor-1 [CB1], and corticotrophin releasing factor receptor 2 [CRFR]).[34]

Epigenetic markers have also been identified in individuals afflicted with post-traumatic stress disorder (PTSD). A recent study has analyzed 14,000 genes from blood samples extracted from PTSD patients and concluded that these individuals had six to seven times the number of genes involved in such phenomena as inflammatory and immune responses, sensory perception of sound, and response to xenobiotic stimuli as people without PTSD.[35]

Epigenetic Therapies

The study of epigenetics allows for the understanding of the dynamic relationship between genes and the environment.[36] Given that epigenetics is at the heart of phenotypic variation in health and disease, it seems likely that understanding and manipulating the epigenome holds great potential for preventing and treating human disease. Epigenetics offers an important window into understanding how the environment interacts with disease and how modulating these interactions could potentially improve human health.[21] The International Human Epigenome Consortium (IHEC)—a sequel to the Human Genome Project—was launched in February 2010 in Paris, France. The consortium aims to map 1,000 reference epigenomes in an attempt to allow researchers to create a robust and trustworthy epigenome template(s) in order to help establish a baseline for the creation of epigenetic therapies.

Thus far, the U.S. Food and Drug Administration (FDA) has approved only a handful of epigenetic therapies. The first epigenetic drug was approved by the FDA in 2004 and was used to treat patients with myelodysplastic syndrome. Research showed that individuals who took the drug *azacitidine* had reduced symptoms of the disease and thus resulted in a better quality of life.[37] Azacitidine's usage was first proposed by Lewis Silvermann at Mount Sinai School of Medicine in New York. Silvermann found that azacitidine could be used at lower doses than called for by traditional chemotherapeutic protocols and had the potential to reverse genes being silenced into actively expressed ones.[38] The drug has proven effective as such; it is now marketed as *Vidaza* and is commonly used in patients with leukemia. It is important to note that one of the advantages of epigenetic therapy lies in its reduced toxicity. More specifically, in standard trials a drug is administered at the highest possible dosage that will not harm a patient. In addition, toxicity tests are normally performed during a drug's developmental stage in order to establish a baseline concentration—one that is both effective and nontoxic to the patient. In contrast, epigenetic therapy offers an advantage in minimizing toxicity and consequently the side effects that traditionally accompany conventional drug therapies.

One of the most exciting epigenetic therapies currently in development is targeted to stroke patients. Preliminary results suggest that inhibition of two key enzymes known to be involved in epigenetic alterations (a DNA methyltransferase and a histone deacetylase) is essential in improving the pathophysiology of stroke by promoting brain repair, functional reorganization, and learning and memory. More specifically, epigenetic alterations in DNA methylation, histone modification, and chromatin remodeling have all been identified as contributing factors in the development of stroke.[39] However, although such epigenetic therapy shows potential in the ability to repair injury to the central nervous system, data suggest that there are also therapeutic limitations. One such limitation involves the nonspecific activation of genes within normal cells and the potential for epigenetic modifications to revert back to their previous state. Although the latter may be corrected with continuous re-treatment, it still poses an unpredictable barrier for epigenetic therapy. Nevertheless, reversibility is indeed one of the most attractive advantages of epigenetic therapy—that is, the ability to reverse epigenetic alterations. More research is needed into controlling the potency of reversibility of epigenetic alterations. Ultimately the goal is to reverse only those epigenetic alterations that are known to be implicated in the disease state and to allow all other "non-associated disease" epigenetic patterns to remain as is.[40]

Conclusion

Epigenetics refers to heritable changes in gene function that cannot be explained by changes in DNA sequence. Today, researchers are focused on elucidating the association between environmental and lifestyle factors and the epigenetic ability to imprint genes passed from one generation to the next. To this end, environmental toxins and carcinogens (e.g., pesticides, heavy metals, and tobacco smoke), diet and nutritional states, behavioral changes between mother and infant (e.g., conditions of chronic and unpredictable maternal separation), and post-traumatic stress disorder have all been shown to cause epigenetically induced gene alterations that ultimately result in a disease state. Thus epigenetic alterations can provide a means for understanding how the environment interacts with disease and how modulating these interactions could potentially improve human health. This, in turn, is of paramount importance because understanding and manipulating the epigenome holds great potential for preventing and treating human disease.

Notes

1. E. Jablonka and M. J. Lamb, The changing concept of epigenetics, *Annals of the New York Academy of Sciences* 981 (2002): 82–96.
2. A. P. Feinberg, Epigenetics at the epicenter of modern medicine, *Journal of the AMA* 299 (11) (2008): 1345–1350.
3. A. Bird, Perceptions of epigenetics, *Nature* 447 (7143) (2007): 396–398.
4. A. L. Gropman and M. L. Batshaw, Epigenetics, copy number variation, and other molecular mechanisms underlying neurodevelopmental disabilities: New insights and diagnostic approaches, *Journal of Developmental and Behavioral Pediatrics* 31 (7) (2010): 582–591.
5. J. M. Ordovas and C. E. Smith, Epigenetics and cardiovascular disease, *National Review of Cardiology* 7 (9) (2010): 510–519.
6. L. M. Villeneuve and R. Natarajan, The role of epigenetics in the pathology of diabetic complications, *American Journal of Physiology—Renal Physiology* 299 (1) (2010): F14–25.

7. J. Campion, F. I. Milagro, and J. A. Martinez, Individuality and epigenetics in obesity, *Obesity Review* 10 (4) (2009): 383–392.
8. F. Coppede, One-carbon metabolism and Alzheimer's disease: Focus on epigenetics, *Current Genomics* 11 (4) (2010): 246–260.
9. D. Grafodatskaya, B. Chung, P. Szatmari, and R. Weksberg, Autism spectrum disorders and epigenetics, *Journal of the American Academy of Child and Adolescent Psychiatry* 49 (8) (2010): 794–809.
10. S. M. Ho, Environmental epigenetics of asthma: An update, *Journal of Allergy and Clinical Immunology* 126 (3) (2010): 453–465.
11. A. Petronis, Epigenetics and bipolar disorder: New opportunities and challenges, *American Journal of Medical Genetics Part C, Seminars in Medical Genetics* 123C (1) (2003): 65–75.
12. D. Ziech, R. Franco, A. Pappa, et al., The role of epigenetics in environmental and occupational carcinogenesis, *Chemical-Biological Interactions* 188 (2) (2010): 340–349.
13. C. A. Cooney, Epigenetics—DNA-based mirror of our environment? *Disease Markers* 23 (1–2) (2007): 121–137.
14. J. C. Mathers, G. Strathdee, and C. L. Relton, Induction of epigenetic alterations by dietary and other environmental factors, *Advanced Genetics* 71 (2010): 3–39.
15. C. Bonisch, S. M. Nieratschker, N. K. Orfanos, and S. B. Hake, Chromatin proteomics and epigenetic regulatory circuits, *Expert Review of Proteomics* 5 (1) (2008): 105–119.
16. I. M. Adcock, P. Ford, K. Ito, and P. J. Barnes, Epigenetics and airways disease, *Respiratory Research* 7 (1) (2006): 21.
17. G. Riddihough and E. Pennisi, The evolution of epigenetics, *Science* 293 (5532) (2001): 1063–1067.
18. P. A. Jones and D. Takai, The role of DNA methylation in mammalian epigenetics, *Science* 293 (5532) (2001): 1068–1070.
19. L. O. Bygren, G. Kaati, and S. Edvinsson, Longevity determined by paternal ancestors' nutrition during their slow growth period, *Acta Biotheoretica* 49 (1) (2001): 53–59.
20. R. Franco, O. Schoneveld, A. G. Georgakilas, and M. I. Panayiotidis, Oxidative stress, DNA methylation and carcinogenesis, *Cancer Letters* 266 (1) (2008): 6–11.
21. A. P. Feinberg, Phenotypic plasticity and the epigenetics of human disease, *Nature* 447 (7143) (2007): 433–440.

22. M. Esteller, Epigenetics in cancer, *New England Journal of Medicine* 358 (2008): 1148–1159.
23. P. Matthiessen and I. Johnson, Implications of research on endocrine disruption for the environmental risk assessment, regulation and monitoring of chemicals in the European Union, *Environmental Pollution* 146 (1) (2007): 9–18.
24. C. Guerrero-Bosagna, M. Settles, B. Lucker, and M. K. Skinner, Epigenetic transgenerational actions of vinclozolin on promoter regions of the sperm epigenome, *PLoS One* 5 (9) (2010): e13100.
25. M. D. Anway, A. S. Cupp, M. Uzumcu, and M. K. Skinner, Epigenetic transgenerational actions of endocrine disruptors and male fertility, *Science* 308 (5727) (2005): 1466–1469.
26. M. K. Skinner and C. Guerrero-Bosagna, Environmental signals and transgenerational epigenetics, *Epigenomics* 1 (1) (2009): 111–117.
27. M. D. Anway and M. K. Skinner, Epigenetic transgenerational actions of endocrine disruptors, *Endocrinology* 147 (6) (2006): S43–49.
28. S. Kirchner, T. Kieu, C. Chow, et al., Prenatal exposure to the environmental obesogen tributyltin predisposes multipotent stem cells to become adipocytes, *Molecular Endocrinology* 24 (3) (2010): 526–539.
29. M. E. Pembrey, L. O. Bygren, G. Kaati, et al., Sex-specific, male-line transgenerational responses in humans, *European Journal of Human Genetics* 14 (2) (2006): 159–166.
30. A. J. Bernal and R. L. Jirtle, Epigenomic disruption: The effects of early developmental exposures, *Birth Defects Research Part A, Clinical and Molecular Teratology* 88 (10) (2010): 938–944.
31. J. G. Bromer, Y. Zhou, M. B. Taylor, L. Doherty, and H. S. Taylor, Bisphenol-A exposure in utero leads to epigenetic alterations in the developmental programming of uterine estrogen response, *FASEB Journal* 24 (7) (2010): 2271–2280.
32. T. Yaoi, K. Itok, K. Nakamura, et al., Genome-wide analysis of epigenomic alterations in fetal mouse forebrain after exposure to low doses of bisphenol A, *Biochemical and Biophysical Research Communications* 376 (3) (2008): 563–567.
33. S. Tando, K. Itoh, T. Yaoi, et al., Effects of pre- and neonatal exposure to bisphenol A on murine brain development, *Brain Development* 29 (6) (2007): 352–356.

34. T. B. Franklin, H. Russig, I. C. Weiss, et al., Epigenetic transmission of the impact of early stress across generations, *Biological Psychiatry* 68 (5) (2010): 408–415.
35. M. Uddin, A. E. Aiello, D. E. Wildman, et al., Epigenetic and immune function profiles associated with posttraumatic stress disorder, *Proceedings of the National Academy of Science USA* 50 (2) (2010): 107 124.
36. A. P. Auger and C. J. Auger, Epigenetic turn ons and turn offs: Chromatin reorganization and brain differentiation, *Endocrinology* (2010), DOI:10.1210/en.2010-0793.
37. C. E. Vigil, T. Martin-Santos, and G. Garcia-Manero, Safety and efficacy of azacitidine in myelodysplastic syndromes, *Journal of Drug Design, Development and Therapy* 4 (2010): 221–229.
38. L. R. Silvermann, J. F. Holland, R. S. Weinberg, et al., Effects of treatment with 5-azacytidine on the in vivo and in vitro hematopoiesis in patients with myelodysplastic syndromes, *Leukemia* 7 (1) (1993): 21–29.
39. I. A. Qureshi and M. F. Mehler, Emerging role of epigenetics in stroke: Part 1, DNA methylation and chromatin modifications, *Archives of Neurology* 67 (11) (2010): 1316–1322.
40. Q. Lu, X. Qiu, N. Hu, et al., Epigenetics, disease, and therapeutic interventions, *Ageing Research Review* 5 (4) (2006): 449–467.

6

Climate Change and Health

Jenny Griffiths

Introduction

Climate change appears to be accelerating. It seems that there is little time left in which to stabilize greenhouse gas emissions sufficiently to reduce the catastrophic consequences of global warming.[1] This chapter summarizes the evidence demonstrating how delay in getting emissions under control quickly will put the lives and well-being of billions of people at increased risk in the next decades. The risk is substantially increased by global environmental threats, including land degradation, a high rate of extinction of species, and human population growth, which interact with climate change. But health and environmental professionals have a major opportunity to reap the health dividend from policies and practices, such as low-carbon diets and active travel, that both reduce greenhouse gas emissions and also improve health. Action at individual and community levels can make a real difference in directly reducing carbon emissions and making it safe for governments to take the necessary legislative or regulatory action. In so doing, legislators can be reminded that they will be saving lives.

The Impacts of Climate Change on Health

Overview

The most exhaustive and authoritative review to date of the evidence of the health impacts of climate change is the contribution of Working Group II to the Fourth Assessment Report of the Intergovernmental Panel on

Climate Change, published in 2007.[2] This expressed the level of confidence the authors had in the evidence on the direction and magnitude of selected health impacts. Only impacts that the Intergovernmental Panel on Climate Change had very high, high, or medium confidence in as occurring or going to occur are considered in this section. As the major impacts of climate change are largely projected for the future, the health effects are mainly expressed in the literature as risks and probabilities.

It can nonetheless be stated with some confidence that "[c]limate change is the biggest global health threat of the 21st century."[3] The fundamental determinants of health, the conditions that make life possible—good air quality, water, food, and social stability—are already affected by climate change in many parts of the world and will deteriorate inexorably without rapid and resolute global action. The impacts are and will be particularly felt by the most vulnerable in society—the young, the elderly, and the poor—and therefore exacerbate inequalities in health, particularly due to poverty.

A report from the Global Humanitarian Forum in 2009 estimated that climate change is *already* responsible for 300,000 deaths a year and that 325 million people are seriously affected.[4] The weather is now more unpredictable, increasing the frequency and severity of natural disasters such as heat waves and floods. In low-income countries in particular, natural disasters result in increases in fatal and nonfatal injuries, food- and waterborne disease, and vector-borne disease, and heighten the risk of malnutrition.[5]

Specific Health Impacts

There is a range of specific likely health impacts of climate change: deaths, injuries, and increases in specific diseases. This summary is drawn from Mala Rao's review of the evidence.[6]

Extreme weather events are likely to become more frequent and intense. Heat waves can cause substantial mortality and morbidity; for example, the European heat wave of 2003 was responsible for more than 35,000 extra deaths. Floods and tropical cyclones have the greatest impact in South Asia and Latin America and result in deaths from drowning, and unsafe or unhealthy living conditions. Low- and middle-income countries with poor sanitation and lack of safe water report risks of diarrheal diseases, cholera, cryptosporidiosis, and typhoid fever. In the United States,

Hurricane Katrina killed more than 1,000 people, uprooted 500,000 others, and caused more than $100 billion worth of damage.

Climate change is an important determinant of air quality, such as ground-level ozone and fine particulate matter. Poor air quality is associated with higher levels of respiratory disease (such as pneumonia, chronic obstructive pulmonary disease, and asthma) and cardiovascular disease.

Increased carbon dioxide and temperatures are altering the timing, production, and distribution of pollens. The prevalence of allergenic diseases such as asthma and rhinitis has increased in many countries.

There is a correlation between rising temperatures and the incidence of common forms of food poisoning such as salmonellosis. In the United Kingdom, additional food poisoning notifications of 4,000, 9,000, and 14,000 are projected for +1, +2, and +3 degrees Celcius of increase in temperature.[7] Warming of the seas may stimulate toxin-producing algal blooms that can contaminate shellfish. Mercury levels may build up in fish and increase the health risks of consumers.

Warming is resulting in chemical processes that deplete stratospheric ozone and, as a result, increase exposure to ultraviolet radiation. Adverse effects of excess exposure to ultraviolet radiation include skin cancers, sunburn, and cataracts.

Health Impacts in Lower-Income Countries

Growth in population, rapid urbanization, and land use changes interact with climate change to magnify its impacts.[8] The brunt of global warming will fall on South Asia, Africa, and the Eastern Mediterranean region. A World Health Organization (WHO) study[9] estimated that climate change caused the following loss of Disability-Adjusted Life Years per million population in the year 2000 in these areas: Africa region, 3071.5; Eastern Mediterranean, 1586.5; South-East Asia Region, 1703.5. A Disability-Adjusted Life Year is the sum of years of life lost due to premature death and years of life lived with disability.

Drought associated with climate change may result in higher food prices and food and water shortages, which could lead to severe undernutrition and greater vulnerability to infectious diseases.[10] There may be changes in the distribution, intensity, and seasonality of meningococcal meningitis and a greater burden of diarrheal disease in countries already lacking access to safe water. Crop failures can drive smallholders and farmers to suicide. Food and water shortages can result in conflict over natural resources and people becoming environmental refugees. Consequential population displacement and mass migrations have a range of associated mental and physical health risks.

About fifty countries are home to the world's poorest communities and at the greatest risk of violent conflict as a result of climate change reinforcing existing social, economic, and political instability. Prolonged drought and resulting crop failures in Africa and elsewhere are resulting in rural-urban migration and increasing the risk of transmission of vector-borne diseases. The war and resulting human tragedy in Darfur, Sudan, is recognized as the first example, the conflict having been triggered by an ecological crisis resulting at least in part from climate change.[11]

Major shifts in the patterns and distribution of vector-borne diseases are anticipated in association with climate change.[12] Globally, temperature increases of two to three degrees centigrade would increase the number of people who in climatic terms are at risk of malaria by around 3 to 5 percent, that is, several hundred million. It is projected that by 2085, 50 to 60 percent of the global population would be at risk of dengue fever (caused by a virus transmitted by the bite of an infected mosquito). Finally, some northward expansion of tick-borne disease (such as Lyme disease) is projected in parts of Europe.

The Natural Environment and Health

Global Environmental Threats Interacting with Climate Change

The natural environment is a key determinant of human health, demonstrated schematically in Figure 6.1, which summarizes the main influences on health. The impact of the human species on the natural environment over the last 10,000 years, including intensive agriculture, concentration in urban centers, and industrial production of goods requiring massive quantities of natural resources, has been too rapid to enable natural systems to adapt.[13]

The work of the United Nations Environment Program (UNEP) reminds us that climate change is taking place against a background of major global environmental threats, including a high rate of extinction of species and the challenge of feeding a growing human population. The UNEP

Figure 6.1 The determinants of health in our neighborhoods (H. Barton and M. Grant, 2006). A health map for the local human habitat. *The Journal of the Royal Society for the Promotion of Health*, 2006; 126 (6): 252–253. Used by permission.

report *Global Environment Outlook: Environment for Development* (GEO-4),[14] from which the information below is summarized, was published in 2007, some twenty years after the World Commission on Environment and Development (the Brundtland Commission) produced its seminal report, *Our Common Future*.[15]

Major areas of global environmental damage are, like climate change, at risk of passing the point of no return. The sixth major extinction of life on earth is underway—the last one was when the dinosaurs were wiped out—and this extinction is entirely anthropogenic. Current biodiversity changes are the fastest in human history. Species are becoming extinct a hundred times more quickly than the rate shown in the fossil record; it is estimated that we are losing three species *an hour.* Of the major vertebrate groups that have been assessed comprehensively, over 30 percent of amphibians, 23 percent of mammals, and 12 percent of birds are threatened.

In the oceans, there are "dead zones" deprived of oxygen. Fishing capacity is estimated at 250 percent more than is required to catch the oceans' sustainable production. Fish consumption more than tripled from 1961 to 2001, and fish stocks are close to exhaustion in some areas.

Loss of fertile land through degradation, especially in Africa, is a threat as serious as climate change and biodiversity loss. It affects up to a third of the world's people, through pollution, soil erosion, nutrient depletion, water scarcity, and salinity. A dwindling amount of fresh water is available for humans and other creatures to share. Climate change interacts with degraded environmental systems, causing further deterioration and threats to human health. So, for example, the increase in droughts, floods, and other extreme weather events exacerbates the loss of fertile land for food production.

The combination of climate change with these other environmental threats means that, as Adrian Lister has said, "We are destroying the context in which [the human species] arose and to which we are adapted."[16]

But the UNEP report reminds us that we can successfully tackle environmental issues if we have the will: 12 percent of the earth's surface is now within protected areas such as nature reserves, and governments and industry have reduced the production of ozone-damaging chemicals by 95 percent in response to the identification of holes in the ozone layer.

Nature, Green Space, Health, and Climate Change

Green spaces, particularly trees, can play a useful role in reducing levels of carbon dioxide and can also help to reduce the health risks from climate change. Green shade in hot weather can help to keep us cool, and green areas can reduce the "heat island" effect in built-up urban areas.[17]

Nature and green space also promote health and well-being. They improve air quality and reduce noise in urban areas. Enabling people to access green space, which promotes exercise and social interaction, can help to prevent obesity, cardiovascular disease, mental ill-health, antisocial behavior, and health inequalities.[17] A large-scale study in the UK of over 336,000 patient records showed significantly less health inequality between rich and poor groups in areas with more green space than between similar groups in areas with less green space. The association between income deprivation and mortality differed significantly across the groups. There was 25 percent lower all-cause mortality in areas with high concentrations of green space compared to areas with low concentrations. For circulatory disease, there was a one-third (29.7 percent) lower mortality in greener areas.[18] Of course there are fewer parks in deprived areas than in affluent areas. In England, the most affluent 20 percent of wards have five times the amount of general green space per person than the most deprived 10 percent of wards.[19]

What's Good for the Climate Is Good for Health

The "co-benefits" of access to nature and green space for health and for reducing the impact of climate change are one example of the evidence that, although climate change will inevitably kill and damage the health of large numbers of people, we can also be very positive: policies that reduce greenhouse gas emissions will result in large health gains, with sizeable reductions in many of the major killers, including heart disease, cancer, obesity, diabetes, road deaths, and air pollution. For example, consuming fewer animal products reduces carbon emissions while reducing heart disease and stroke; less use of the car and more walking and cycling do the same.[20] Low-carbon societies also offer the prospect of improved quality of life, focusing on the community, self-sufficiency, family, relationships, creativity, and contact with nature. All of these are demonstrably good for our health and well-being.

In high-income societies, each of the following accounts for about one-quarter of greenhouse gas emissions: energy (to heat, light, and run our homes and workplaces); motorized transport; food and drink (including production, transportation, and retail); and production and consumption of all other goods and services.[21] The first three are discussed in the following paragraphs.

Energy Use

One international collaborative study[22] reporting on household energy use in the UK estimated that a strategy combining improved insulation and ventilation control, switching from household fossil fuel use (gas, coal, oil) to electricity, and a reduction in average indoor temperature to 18 degrees centigrade would have appreciable benefits to health (estimated as 850 fewer Disability-Adjusted Life Years per million population per year), mainly due to improved indoor air quality and warmer indoor air temperatures in winter preventing deaths and disability from hypothermia. Savings of carbon dioxide were considerable (0.6 megaton of carbon dioxide per million population in one year).

Active Travel

Another collaborative study[23] estimated the impact of lower-emission motor vehicles and increased "active travel" (walking and cycling). The lower emissions from motor vehicles would reduce both greenhouse gas emissions and the health burden from urban outdoor air pollution. Increasing the distances walked and cycled would lead to large health benefits, including reductions in the prevalence of ischemic heart disease, cerebrovascular disease, depression, dementia, and diabetes. There is, however, a risk of more pedestrian and cycle injuries. Creation of safe urban environments for mass active travel requires prioritization of the needs of pedestrians and cyclists over those of motorists.

Agriculture and Food

A large proportion of agricultural greenhouse gas emissions is related to the production of foods from livestock, along with associated deforestation and other changes in land use. An international collaborative study[24] suggested that a combination of agricultural technological improvements, such as improved manure management, and a reduction in production of foods from animal sources (by about 30 percent by 2030), could reduce greenhouse gas emissions from food and agriculture by 50 percent from the 1990 level by 2030. Through a reduction in total calories, a lower intake of saturated fat, and reduced meat-related carcinogens, the health outcomes would include lower levels of obesity, type 2 diabetes, cardiovascular disease, and colorectal and some other cancers. The reduction on the burden of ischemic heart disease, in particular, was estimated to be about 15 percent in the UK.

Reducing greenhouse gas emissions would have the additional benefit of reducing air pollutants such as nitrogen oxides, sulfur dioxide, particulate matter, and mercury. These pollutants are linked with respiratory diseases, including chronic bronchitis.[25]

So we can cut carbon and, in doing so, improve health and reduce health inequalities, both within and between societies around the world.

Taking Action

The director-general of WHO has stated: "Climate change is a price that we are paying for short-sighted policies. The pursuit of economic wealth took precedence over protection of the planet's ecological health, and over the most vulnerable in society. Fundamentally we are all facing a choice about values: improving lives, protecting the weakest, and fairness."[26]

Visions of a Low-Carbon Healthy Society

If people are asked, "What is the most life-enhancing thing that a low-carbon society would offer?" they may respond with insights such as the following:[27]

"All the things that matter to me in life are low-carbon: families, friends, communities, the arts. I look forward to the rediscovery of those joys."

"Consuming more definitely does not make you happy. I will delight in the fact that less is more."

"I will grow my own vegetables and cook more."

"No more airport queues, no more traffic jams."

"Birdsong—and much more of it than now!"

One way of engaging people in any setting (work, home, or community) in tackling climate change and promoting health is to ask them to brainstorm, individually or in groups, their vision of what people's lives would be like in a healthy, low-emission society. This helps people to construct an optimistic picture of the future and to reflect on their present lives; they feel less anxious about the future being a terrifying prospect in which their standard of living will have plummeted. Following are some of the ideas that groups of people in England have proposed for a better, healthier, low-carbon future.[27]

- Quality of life will be seen as the new status symbol—people will boast about their *quality of life* rather than their *standard of living*. Quality of life is, of course, closely related to well-being.
- We will each have a personal carbon allowance, stored on a plastic card, which we will choose to use carefully for selected carbon-consuming activities (operating within national and international carbon cap and trade schemes).
- Environmentally derived natural resources will all be priced at a realistic level, that is, much higher than now, which will radically reduce consumption of material goods. We will have learned to take pride in doing more with less.
- There will be no gasoline-powered cars: all private vehicles will be renewably powered or zero emission; the motorways will be closed to cars; and fleets of buses will take people from city to city, as buses have the lowest carbon emissions of any motorized transport.[28]
- Road-building programs will be converted into a massive public transport expansion program.
- Urban residential areas will be closed to car traffic, and people will get around on foot, by bike, or by bus.
- Children will *all* go to school by bus, by bike, or on foot.
- All houses will be carbon-neutral, retrofitted with a package of insulation and technology—all suitable houses will have micro-generation, and there will be solar panels around local community sites.
- Fuel poverty (wherein people spend more than 10 percent of their income on heating and still suffer from damaging cold in winter, due to poor insulation and heating systems) and its associated morbidity and mortality will be a thing of the past.
- All houses will have visual displays giving real-time information on their carbon consumption.
- Aviation will become a once-in-a-lifetime luxury: perhaps a single lifetime trip on an airplane as a retirement present or a twenty-first birthday present. People will take a year or two off at an appropriate

time in their lives for "slow travel" around the world, using local public transport, walking, and cycling; several airports will close or their traffic will be substantially reduced.
- Holidays will mainly be taken locally, with a much stronger emphasis on knowing and conserving nature, arts and craft skills, and other forms of personal creativity.
- There will be much more emphasis on local activities and community interactive events.
- The obesity epidemic will be a thing of the past. People will be fitter, healthier, and happier.

Sustainable Development: A Framework for Achieving a Vision

The ideas summarized in the previous section enable the more sustainable use of natural resources. To achieve the best possible levels of health and well-being while reducing carbon emissions by the necessary 80–90 percent in high-income countries, we need to influence policy makers to adopt the framework of *sustainable* development, underpinned by "living within our means" as regards natural resources. The pursuit of carbon-emitting economic growth has not provided economic and social stability, which are key determinants of health, for much of the world's population. In high-income countries, consumerism and overconsumption are associated with loss of well-being, stress, and dissatisfaction.

Prosperity has to be redefined so that it does not depend on constant increases in economic growth. It should be defined by people's capability to flourish physically, psychologically, and socially (in other words, their health and well-being). Better health for all—and more happiness and fulfillment—means a reasonable quality of life for all within ecological limits, including a reduced level of greenhouse gases in the atmosphere.[29]

Sustainable development means ensuring social progress to meet the needs of everyone, with effective protection of the environment and prudent use of natural resources. It has a number of important characteristics.[29] It draws attention to the needs and claims of future generations; climate change is happening on our watch and will be our legacy. Sustainable development focuses on alternatives to today's economic growth to improve well-being and achieve long-term equity within and between societies around the world. It looks for opportunities for investing in synergistic measures or co-benefits that reduce environmental damage, promote social justice, and narrow health inequalities, particularly to improve life for the poorest people in the global population.

At the policy level, lasting prosperity to improve health and reduce carbon emissions in high-income countries by the 80–90 percent needed to ensure humanity's long-term future will have to be based on an economic strategy based on the principles of sustainable development, including equity.[30] The available paid work will have to be shared, with much more exchange of goods and services. Improving the work-life balance across society will improve quality of life for individuals and communities as well as contributing to macro-economic stability. Inequalities of income and wealth, which are closely linked to health inequalities, will have to be reduced.[31] Supporting these policies, the argument for fiscal reform or ecological tax reform, a shift in the burden of taxation from economic goods to ecological bads (e.g., pollution) has been accepted for at least a decade. Appropriate mechanisms must be found to move the basis of taxation from incomes to the consumption of natural resources and greenhouse gas emissions. Economic measurement frameworks therefore must be adjusted to account systematically for ecological and social factors.

Resilient communities will be more self-sufficient in their use of resources, with less movement of people outside of the local area for work, shopping, and leisure. Communities will enjoy devolved responsibility for spatial planning, resulting in ubiquitous shared public spaces. The dominant culture of consumerism, which has damaged people's well-being, will be modified through community debate, interaction, and local experience and replaced by programs that encourage people to focus on their core values of relationships, creativity, and helping others.[32]

Sustainable development has been described as the trinity of "green, well and fair":[33] light in use of natural resources including carbon, focused on health and well-being, and socially just.

Advocacy: The Health Arguments

Advocacy is action by individuals and groups to generate the legitimizing momentum to allow governments to feel politically safe in taking appropriate legislative and regulatory action. A critical mass of popular support is needed to create sufficient momentum so that change becomes self-sustaining. Small changes in public consensus can sometimes bring about swift changes in political consensus. Advocacy in the field of climate change and health can come from at least three different sources. First, campaigning pressure is valuable, for example, from health and environmental lobby groups and individuals who lobby politicians. Second, some determined activists are challenging the law and obtaining the support of the law courts. There are examples in the UK, and no doubt elsewhere, of juries concluding that nonviolent direct action, for example, trespass in airports and power stations, is legally justifiable because its purpose is to prevent climate change causing greater damage to property around the world.[34] Third, groups and communities use local action to advocate for change, such as the Transition Town movement, described below.

The health-based arguments are powerful in advocacy for action on climate change, for a number of reasons. First, as argued previously, much of what needs to be done to reduce the threat of climate change should be done anyway for health reasons; this is known as the "health co-benefits" argument. Second, the language of the health arguments is positive and promises better, longer lives (it is not the language of "giving up," as in "carbon rationing"). The health benefits are individual and yield personal benefits quite quickly, for example, from taking more exercise or having more contact with the natural environment, in contradistinction to the long-term carbon reduction arguments of benefiting one's grandchildren. Health professionals are trusted by the public; in the UK opinion polls suggest repeatedly that doctors are *the* professionals most trusted to tell the truth.[34] Health and environmental professionals form a large workforce, are well-organized through professional associations, and enjoy power and influence in most societies. And finally, health practitioners are value-driven, seeing health protection and saving the planet as issues of social justice.

The language developed to combat cigarette consumption is very similar to the language appropriate to climate change reduction. If you stop smoking by age thirty-five, your risk of harm and life expectancy return to normal within a few years, and a further benefit is that nobody else suffers the harmful effects of exposure to your smoke. If you drive your car less, fly less, and reduce your carbon emissions in other ways, you will help to reduce the effects of climate change. Both sets of changes will help you enjoy your grandchildren, and if you reduce your carbon footprint, your grandchildren may have a planet to enjoy.[34]

The battle against tobacco also provides lessons for the battle against climate change in respect to the nature of the evidence. Sustainable development and public health are based on the "precautionary principle": when there is a real risk to health or society, decisions to reduce the risk have to be taken on the best evidence available. There may not be certainty. Forty years ago, the tobacco industry hired experts to denigrate the growing scientific consensus. The same is happening with climate change today.

Effective advocacy requires health professionals to be organized in order to influence the decision-making process internationally (particularly the UN Framework Convention on Climate Change, which meets most years) and also nationally. Many professional associations are now connecting with the growing international health lobby for action on climate change.[35] Health and environmental practitioners can also be effective advocates by influencing the strategic planning systems and processes led by local government, helping to ensure that health is central to the design of their local communities. A good example is promoting accessibility for active travel, for walking and cycling, designing the built environment to make walking and cycling and use of public transport easier.[36]

Sustainable Communities

Many believe that action by communities, rather than by individuals (too little) or governments (too little, too late), is the most likely pathway to a more sustainable future.[37] To make the necessary radical reduction in carbon emissions, communities have to have available locally as many as possible of the essential elements needed to thrive: work, food, and social and cultural facilities. Localization is therefore a pivotal policy and is of course the opposite of the travel-hungry strategies for employment, consumption, and leisure pursued in the last half century in many countries. Resilience and self-sufficiency are also arguably very important if communities are to survive what are likely to be very challenging economic times in the next half century as the oil reserves run out.

Creating healthy, sustainable, low-carbon communities requires considerably more empowerment—influence and control over the social and economic strategies pursued locally—than people typically currently enjoy. In reviewing the evidence on community engagement, the National Institute for Health and Clinical Excellence in England[38] concluded that positive health outcomes result from equal partnerships between communities and decision makers. The more control local people have, the more well-being they will feel.

A community-based approach is also essential if behavior is to change toward lower-carbon living. Efforts to educate or persuade people to change their travel and consumption habits to be less carbon-emitting are likely to be unsuccessful unless cultural and social norms are addressed. People learn through groups and networks: "I will if you will." Jackson concluded from his work on motivating sustainable consumption that

> [a] key lesson from this review is the importance of community-based social change. Individual behaviors are shaped and constrained by social norms and expectations. Negotiating change is best pursued at the level of groups and communities. Social support is particularly vital in breaking habits and in devising new social norms.[39]

Transition Towns

The international Transition Towns movement is one of the best-known examples of communities beginning to shape their own sustainable and healthier futures. Started in 2005 by Rob Hopkins in Totnes in South West England,[40] at the time of writing there are about 320 "official" transition initiatives around the world, a number that has doubled in the last two years. The aim of the transition movement is to help people to reconnect with the living environment around them, and with their local communities, to wean them off their addiction to fossil fuels and create communities that are more self-supporting and resilient, and therefore healthier.

The transition movement's links with health are diverse: through health-enhancing projects such as growing fruit and vegetables, planting trees, and reskilling people so that they can be more self-sufficient in repair and maintenance, cookery, and other domestic skills, as well as through the benefits to health of reducing carbon emissions.

Low-Carbon Healthcare Systems

Healthcare systems themselves should be playing a major part in leading the way to a healthier, low-carbon society. The measurement of carbon emissions from different types of healthcare and the organization of lower-carbon healthcare delivery are still in their infancy. Work undertaken so far[41] does, however, give us some pointers for the development of healthcare systems over the next twenty years.

First, people need to be supported in taking responsibility for their own health. As budgets become more constrained, the burden of responsibility for health is likely to shift back toward individuals. The health literacy of the population needs to be improved, with a strong focus on prevention. Even where healthcare spending is declining, resources should be allocated to supporting people with their own health improvement. Health professionals, respected by their patients and clients, working in systems with massive size and reach in society, are in the best possible position to take a leadership role in showing that low-carbon lifestyles can be happier and healthier.

Once carbon is properly priced, transport and carbon-intensive healthcare technology in hospitals will become even more expensive. Information and communication technology in people's homes and local communities will therefore be very important in giving them access to specialized professional advice.

Healthcare organizations, as major employers, as substantial purchasers of goods and services—creating travel, producing waste, consuming energy, and commissioning buildings on a vast scale—can put their muscle and spending power to work for health improvement and sustainable development, thereby helping to improve health, reduce inequalities, and safeguard the environment for the benefit of whole communities.[42]

Where to Start: Steps Toward a Low-Carbon, Healthier Lifestyle

Table 6.1 summarizes the steps that can be taken toward a low-carbon, healthy life. We all need to start somewhere. If action at the policy or community level seems impossible at this moment, try one or more of the simple interventions listed in Table 6.1. It is easy to be an exemplar, a role model for others. Any of these actions will both improve health and well-being and make a personal contribution to tackling climate change. They can be undertaken by any individual or family, be scaled up for action at the population level, or form the basis for advice by health and environment practitioners.[43]

You can make a real difference. If you make one or two changes, and talk gently about them to others at home, at work, or in the community,

TABLE 6.1
Steps Toward a Low-Carbon, Healthy Lifestyle

Action	Benefits to Health and the Climate
Have at least one or two meat-free days each week.	The risk of cardiovascular disease and diabetes and some cancers is reduced; carbon emissions are reduced.
Walk or cycle every day.	Even one or two fewer car journeys each week can reduce carbon emissions from travel by 10 percent. Walking and cycling have proven benefits for mental and physical health.
When you do drive a motor vehicle, try to reduce your speed.	Both carbon emissions and road traffic accidents are directly related to speed.
Ensure the maximum possible insulation in your own home and encourage others to do the same.	Insulation keeps buildings warm in winter and cool in summer, therefore reducing the adverse health effects of both heat and cold, and saves carbon (and money) by reducing energy expenditure.
Have contact with nature and safe green space.	Walking and gardening in green space improves mental and physical health.
Join with others in improving the quality of local green space.	Your own and others' health will benefit, and you will help to preserve the living environment.

Source: Summarized from L. Stewart and A. Maryon-Davis, How you can make a real difference, in *The health practitioner's guide to climate change*, edited by J. Griffiths, M. Rao, F. Adshead, and A. Thorpe (London: Earthscan, 2009), http://www.earthscan.co.uk/?tabid=74742.

you can be sure that others will follow. They too will be demonstrating that what's good for the climate is good for health.

Conclusion

Climate change is the largest global health threat of the twenty-first century. It is killing people now. But it is also true that lower greenhouse gas emissions and the promotion of good health are inextricably linked. The "diseases of affluence," such as diabetes and heart disease, have almost overwhelmed healthcare systems in some countries. Health-enhancing, sustainable lives, radically reducing our demands on the earth's natural resources, valuing nature and green space (which are vital to human health and well-being), can result in huge health dividends in terms of lower levels of major killers such as cancer and heart disease, and also reduce health inequalities.

All great social movements (e.g., civil rights, women's liberation) and previous major health campaigns (e.g., family planning tobacco control) were started by committed enthusiasts who generated a spark for change.[44] Health and environmental professionals, respected and influential in society, can play a crucial role in raising and maintaining public debate on climate change and health, as well as leading and implementing action at all levels, from policy to community to individuals.

Notes

1. J. Hansen, M. Sato, P. Kharecha, et al., *Target atmospheric CO_2: Where should humanity aim?* (2008), http://avxiv.org/abs/0804.1126v3 (accessed July 9, 2010).
2. U. Confalonieri, B. Menne, R. Akhtar, et al., Human health, in *Climate change 2007—Impacts, adaptation and vulnerability: Contribution of Working Group II to the fourth assessment report of the Intergovernmental Panel on Climate Change*, edited by M. L. Parry, O. F. Canziani, J. P. Palutikof, et al. (Cambridge, UK: Cambridge University Press, 2007), 391–431.
3. A. Costello, M. Abbas, A. Allen, et al., Lancet and University College London Institute for Global Health Commission: Managing the health effects of climate change, *Lancet* 373 (2009): 1693–1733.
4. Global Humanitarian Forum, *The anatomy of a silent crisis: Climate change human impact report* (Geneva: Global Humanitarian Forum, 2009), http://www.eird.org/publicaciones/humanimpactreport.pdf (accessed December 3, 2010).
5. World Health Organization (WHO), *How much disease would climate change cause?* (2003), http://www.who.int/globalchange/environment/en/chapter7.pdf (accessed July 9, 2010).
6. M. Rao, Climate change is deadly: The health impacts of climate change, in *The health practitioner's guide to climate change*, edited by J. Griffiths, M. Rao, F. Adshead, and A. Thorpe (London: Earthscan, 2009), http://www.earthscan.co.uk/?tabid=74742.
7. S. Kovats (ed.), *Health effects of climate change in the UK, an update of the Department of Health report 2001/2002* (London: Department of Health/Health Protection Agency, 2008).

8. A. Haines, R. S. Kovats, D. Lendrum-Campbell, and C. Corvalan, Climate change and human health: Impacts, vulnerability, and mitigation, *Lancet* 367 (2006): 2101–2109.
9. WHO, *How much disease would climate change cause?*
10. Rao, Climate change is deadly.
11. D. Smith and J. Vivekananda, *A climate of conflict: The links between climate change, peace and war* (London: International Alert, 2007).
12. Rao, Climate change is deadly.
13. D. Stone, Health and the natural environment, in *The health practitioner's guide to climate change*, edited by J. Griffiths, M. Rao, F. Adshead, and A. Thorpe (London: Earthscan, 2009), http://www.earthscan.co.uk/?tabid=74742.
14. United Nations Environment Programme (UNEP), *Global environment outlook: Environment for development (GEO-4)* (2007), http://www.unep.org/geo/geo4/media/ (accessed July 9, 2010).
15. G. H. Brundtland, et al., *Our common future: Report of the World Commission on Environment and Development* (Oxford: Oxford University Press, 1987).
16. A. Lister, The biotic effects of climate change: Past, present and future (paper presented at the conference on climate change and its impact on health, January 29, 2008), http://events.rcplondon.ac.uk/archiveevent/0801climate.aspx (accessed July 9, 2010).
17. Faculty of Public Health, in partnership with Natural England, *Great outdoors: How our natural health service uses green space to improve wellbeing*, http://www.fph.org.uk/uploads/r_great_outdoors.pdf (accessed July 9, 2010).
18. R. Mitchell and F. Popham, Effect of exposure to natural environment on health inequalities: An observational population study, *Lancet* 372 (9650) (2008): 1655–1660.
19. Commission on Architecture and the Built Environment (CABE), *The green information gap* (2009), http://www.cabe.org.uk/publications/community-green (accessed July 8, 2010).
20. A. Haines, et al., Public health benefits of strategies to reduce greenhouse-gas emissions: Overview and implications for policy makers, *Lancet* (2009): DOI:10.1016/S0140-6736(09)61759-1.

21. Faculty of Public Health, *Sustaining a healthy future: Taking action on climate change*, special focus on the NHS (2009), http://www.fph.org.uk (accessed July 8, 2010).
22. P. Wilkinson, et al., Public health benefits of strategies to reduce greenhouse-gas emissions: Household energy, *Lancet* (2009): DOI:10.1016/S0140-6736(09)61713-X.
23. J. Woodcock, et al., Public health benefits of strategies to reduce greenhouse-gas emissions: Urban land transport, *Lancet* (2009): DOI:10.1016/S0140-6736(09)61714-1.
24. S. Friel et al., Public health benefits of strategies to reduce greenhouse-gas emissions: Food and agriculture, *Lancet* (2009): DOI:10.1016/S0140-6736(09)61753-0.
25. Health and Environment Alliance, *Air quality "co-benefits" should be considered in climate policies* (2010), http://www.env-health.org/a/3533 (accessed July 7, 2010).
26. M. Chan, Cutting carbon, improving health, *Lancet* (2009): DOI:10.1016/S0140-6736(09)61993-0.
27. The content of this subsection is derived from the author's discussions with groups of colleagues and citizens.
28. G. Monbiot, *Heat: How to stop the planet burning* (London: Allen Lane, Penguin Group, 2006).
29. T. Jackson, *Prosperity without growth? The transition to a sustainable economy* (London: Sustainable Development Commission, 2009), http://www.sd-commission.org.uk.
30. Sustainable Development Commission, *Sustainable development: The key to tackling health inequalities* (London: Sustainable Development Commission, 2010), http://www.sd-commission.org.uk.
31. The Marmot Review, *Fair society, healthy lives: A strategic review of health inequalities in England post-2010* (2010), http://www.marmotreview.org (accessed July 8, 2010).
32. Jackson, *Prosperity without growth?*
33. A. Coote and J. Franklin, *Green, well, fair: Three economies for social justice* (London: New Economics Foundation, 2008).
34. M. Gill and R. Stott, Leadership: How to influence national and international policy, in *The health practitioner's guide to climate change*, edited by J. Griffiths, M. Rao, F. Adshead, and A. Thorpe (London: Earthscan, 2009), http://www.earthscan.co.uk/?tabid=74742.

35. For more information on organized health advocacy, see the Climate and Health Council, http://www.climateandhealth.org.
36. H. Barton, M. Grant, and P. Insall, How to help to plan a healthy, sustainable, low-carbon community, in *The health practitioner's guide to climate change*, edited by J. Griffiths, M. Rao, F. Adshead, and A. Thorpe (London: Earthscan, 2009), http://www.earthscan.co.uk/?tabid=74742.
37. J. Griffiths, How to take action in the community, in *The health practitioner's guide to climate change*, edited by J. Griffiths, M. Rao, F. Adshead, and A. Thorpe (London: Earthscan, 2009), http://www.earthscan.co.uk/?tabid=74742.
38. National Institute of Health and Clinical Excellence (NICE), *Community engagement*, NICE public health guidance 9 (London: NICE, 2008).
39. T. Jackson, (2005) *Motivating sustainable consumption: A review of evidence on consumer behaviour and behavioural change*, A report to the Sustainable Development Research Network (2005), http://www.sd-research.org.uk/post.php?p=126 (accessed July 9, 2010).
40. For more information about Transition Towns, see http://www.transitionnetwork.org (accessed July 9, 2010).
41. NHS Sustainable Development Unit, with Forum for the Future, *Fit for the future: Scenarios for low-carbon healthcare 2030* (Cambridge, UK: NHS Sustainable Development Unit, 2009), http://www.sdu.nhs.uk.
42. A. Coote, ed., *Claiming the health dividend: Unlocking the benefits of NHS spending* (London: King's Fund, 2002).
43. J. Nurse, D. Basher, A. Bone, and W. Bird, An ecological approach to promoting population mental health and well-being—a response to the challenge of climate change, *Perspectives in Public Health* 130 (1) (2010): 27–33.
44. L. Stewart and A. Maryon-Davis, How you can make a real difference, in *The health practitioner's guide to climate change*, edited by J. Griffiths, M. Rao, F. Adshead, and A. Thorpe (London: Earthscan, 2009), http://www.earthscan.co.uk/?tabid=74742.

Acknowledgments

The author would like to thank all the editors and contributors to the book *The Health Practitioner's Guide to Climate Change* (Earthscan, 2009).

7

Oceanic Pollution

Stephen B. Weisberg and Karen E. Setty

Introduction

Human civilizations have long used the world's oceans as receptacles for waste. The phrase "the solution to pollution is dilution" epitomizes the common pollution management strategy used over past centuries, based on the assumption that human actions could not significantly impact such a large diluent as the ocean. In the last several decades, though, increases in scientific understanding of oceanic processes and capacity limits for pollution assimilation have led to recognition that land-based activities can profoundly affect the oceans. Dramatic increases in the size of human populations and associated waste are exerting more pressure on oceans than ever before. As scientific understanding of ocean systems has evolved, environmental strategies and programs to address pollutants are also becoming more sophisticated.

Ocean Pollution Sources and Effects

Ocean pollution is generally categorized as that coming from "point" or "nonpoint" sources. Point sources are easily identifiable conveyances, such as industrial pipes, municipal outfalls, and waste disposed from offshore platforms. Nonpoint sources, such as atmospheric deposition, terrestrial runoff, and mobile sources (e.g., marine vessels), come from a more diffuse area. Runoff from land is the most prominent nonpoint source of marine pollution and can contain a variety of pollutants, such as fertilizers and pesticides from agricultural fields, lawns, and golf courses; oil and grease from roads and parking lots; bacteria and nutrients from livestock waste; and toxic chemicals from abandoned mines.[1] On undeveloped land, water can be naturally filtered as it soaks into the ground. When land is developed for human uses, though, it is often covered by impervious surfaces, such as roofs, streets, driveways, sidewalks, and parking lots. Water falling on these areas runs off the surface instead of being used onsite. This change, known as hydromodification, releases a larger amount of runoff more quickly into the local drainage system (e.g., stream or storm drain) as compared to undeveloped land.

Effects of pollution can range from mild disturbance of normal physical, chemical, or biological functions to a widespread loss of aquatic life. Physical effects of marine pollution include boat-motor clogging or animal entrapment. Chemical effects include changes in ocean acidity or increases in levels of contaminants in seafood. Biological effects include disease or reproductive problems in marine organisms. The effects of single pollutant types are fairly well understood owing to controlled studies; however, marine ecosystems are typically exposed to combinations of different pollutants from different sources. This can lead to synergistic effects, in which the presence of one pollutant strengthens or exacerbates the influence of another. In addition to pollutants, marine ecosystems are also subject to an array of other stressors, such as habitat encroachment and harvesting of marine resources. In combination, these stressors can cause greater damage to an ecosystem than pollution alone.

Major Ocean Pollution Classes

Although they are diverse, ocean pollutants are often grouped into classes by similarity in their properties, sources, and effects (see Table 7.1). The extent and impacts of some pollutant classes are better understood than others. For example, toxic substances and fecal waste have been monitored and managed for decades, whereas nutrients and greenhouse gases are relatively new to the field of pollution management. The extent and impacts of other ocean pollutant classes, such as noise, heat, radioactive materials, and salty brine, are relatively limited or unclear at present, and thus are not presented in this chapter. Still, monitoring and research is ongoing at some level for virtually every type of ocean pollutant.

TABLE 7.1

Major Ocean Pollution Classes and Their Primary Sources and Effects

Pollutant Class	Sources	Effects
Nutrients	Municipal wastewater, runoff from agricultural and urban areas, industrial discharges, animal waste, atmospheric deposition	Eutrophication, harmful algal blooms, reduced water clarity, hypoxia, smothering and death of marine organisms
Toxic substances	Industrial and municipal wastewater; atmospheric deposition; runoff from roads, landfills, agricultural and urban areas; contaminated soil/sediment	Acute (i.e., death) and chronic (e.g., reproductive failure) effects on marine organisms, bioaccumulation through the food chain
Greenhouse gases	Combustion of fossil fuels, livestock operations, paddy field agriculture, landfills, refrigerants	Global climate change, rise in sea surface temperature, ocean acidification, coral reef bleaching, loss of sea ice
Fecal waste	Municipal sewage, combined sewer overflows, infrastructure leaks and spills, marine vessels, livestock, aquaculture, wildlife, pet waste	Spread of disease-causing agents via water ingestion, water contact, and/or seafood consumption
Invasive species	Marine vessels, ballast water, aquaculture operations, transport by travelers, household aquariums and gardens	Loss of native species, habitat destruction
Debris	Land-based litter, marine vessels	Death of marine organisms due to entanglement or ingestion, aesthetic impacts, adsorption and ingestion of toxic chemicals
Oil	Natural seeps, runoff from roads and parking lots, leaks and spills during petroleum extraction and transport	Death of marine organisms due to entrapment, hypothermia, dehydration, or toxicity (if ingested); habitat destruction; reduced light penetration

Nutrients

Nutrients such as nitrogen and phosphorous are vital to living organisms, but they can have deleterious effects on coastal ecosystems when present in excess. The process of eutrophication, or long-term nutrient overenrichment, begins when primary producers like algae and phytoplankton respond to the abundance of nutrients in the environment with explosive growth and reproduction. As these organisms die off, they sink to the bottom and decompose, which consumes the dissolved oxygen in the water. This creates localized low oxygen (hypoxic) or no oxygen (anoxic) "dead zones," where some aquatic organisms cannot survive. Increasingly frequent observations of algal blooms, fish kills, and dead zones in recent years have all been attributed to excessive nutrient pollution.

In addition to hypoxia, overabundance of algae can also cause mechanical damage to water intakes or boat motors and aesthetic issues like unpleasant odors and cloudy or discolored water. Algal blooms that physically cloud the water can block the sunlight needed by other aquatic life to survive. Algae that contain a reddish or brownish pigment, referred to as red tides, can affect tourism by discoloring the water. Some algae also produce toxins that can accumulate in the bodies of filter-feeding fish and shellfish as they process large volumes of affected water, which are then passed to humans via seafood consumption. Algal toxins have been known to sicken or kill marine mammals, fish, and birds and may also produce respiratory ailments in humans.

In the United States, nutrients delivered to the coastal ocean come primarily from diffuse nonpoint sources, such as runoff from urban and agricultural areas and aerial deposition from power plants and motor vehicles.[1] Nutrients exist in both organic and inorganic forms, naturally cycling through various media in the environment, including soil, water, air, and living organisms. With the expanded use of fossil fuels and chemical fertilizers in the last half of the twentieth century, human activity has significantly altered natural nutrient cycles.[2] Perhaps the single largest change stems from the widespread use of synthetic inorganic fertilizer in the 1950s, following its invention during World War I.[3] Globally, approximately 20 million metric tons (Mt) of phosphorus are mined each year for fertilizers; almost half of this returns to the ocean—an amount equivalent to roughly eight times the natural input.[4] Similarly, industrial fixation of atmospheric nitrogen to ammonia fertilizers creates about 80 Mt of reactive nitrogen annually, much of which eventually makes its way to inland and coastal waterways.[4]

The magnitude of nutrient loading alone does not necessarily predict the severity of eutrophication effects, because other factors can influence the ecosystem's reaction. In the United States, ocean eutrophication is most prevalent along the Atlantic and Gulf Coasts.[1] With heavily populated coastlines and intense agricultural practices, the amount of nitrogen added to these waters has increased about fivefold since preindustrial times and is expected to continue rising into the future.[5] Importantly, these areas are also more prone to temperature-induced stratification than marine systems along the U.S. West Coast, where eutrophication is far less common.[1] Stratification of the water column can play a role in the eutrophication process by blocking exchange of oxygenated surface waters with the oxygen-deprived bottom layer.

The problem of nutrient pollution is extensive; two-thirds of U.S. estuaries and bays are considered either moderately or severely degraded by eutrophication.[6] One of the most prominent examples is found in the Gulf of Mexico. First documented in the early 1970s, an extensive dead zone has formed every year during the summer near the mouth of the Mississippi River.[7] This dead zone is thought to be caused by water column stratification in combination with excess nutrients, particularly nitrogen, in runoff from the Mississippi-Atchafalaya River Basin.[2] A majority of the nutrient loading to the Gulf comes from distant agricultural areas along the upper Mississippi and Ohio Rivers, nearly 1,600 kilometers upstream. Other eutrophication hot spots in the United States are the Chesapeake Bay and the western basin of Long Island Sound.[2] Internationally, large-scale eutrophication has also occurred in the Baltic Sea, eastern North Sea, northern Adriatic Sea, northwestern Black Sea, and Japan's Seto Inland Sea. In total, the United Nations Environment Program has identified 150 known dead zones around the world, ranging from less than 1 up to 70,000 square kilometers in size.[8] This number is double the number in 1990, and many of the zones are expanding in size.

Currently there is no single, comprehensive national strategy to limit nutrient pollution in the United States; however, a number of local, state, regional, and federal programs are focused on protecting and restoring coastal habitats.[3] Approaches to mitigating nutrient pollution include reducing nutrient usage, controlling losses to the environment at the source, and sequestering or removing nutrients from ambient waters or wastewater.[2] In inland areas, a variety of programs are being implemented to control nutrient release (e.g., from agricultural fields and residential communities) using "best management practices." For example, restoration of vegetated buffers and wetlands around water bodies can help to trap nutrient-rich runoff and prevent it from reaching waterways. Though a more expensive approach, some coastal wastewater treatment plants have been upgraded to reduce the concentration of nutrients in discharges. On a larger scale, the U.S. Environmental Protection Agency (EPA) is working with a number of states to develop regulatory numeric nutrient criteria that will protect beneficial uses, as well as establishment of total maximum daily loads for managing critically impaired water bodies.[9]

Toxic Substances

The mention of toxic pollution brings to mind large barrels marked with a skull and crossbones, though it is often much less conspicuous. Toxic pollutants commonly found in marine environments include industrial organic chemicals (e.g., polychlorinated biphenyls [PCBs], tetrachlorobenzene), trace metals (e.g., cadmium, copper, lead, mercury), pesticides and their by-products (e.g., dichlorodiphenyltrichloroethane [DDT], tributyltin), and by-products of combustion and industrial processes (e.g., polycyclic aromatic hydrocarbons [PAHs], dioxins).[2] More than 100,000 different chemicals are currently registered for use in the United States, and approximately 30,000 chemicals are widely used (more than one metric ton/year) in international commerce.[10, 11] The vast majority of these are industrial chemicals, followed by food additives, cosmetics ingredients, pharmaceuticals, and pesticides.

Legacy contaminants are those that were discharged in the past and have persisted in the environment. Even though they are currently prohibited in the United States, legacy pollutants like DDT may continue to persist in coastal environments for decades to centuries. More recently discovered pollutants are referred to as contaminants of emerging concern (CECs), because there is relatively little information about their potential risks; for many, methods to detect and quantify their presence in the environment do not yet exist. Though most CECs are not regularly monitored in the environment, assessment and screening processes in the United States, Europe, and Canada have yielded lists prioritizing the constituents of highest concern.

A chemical is considered toxic if it causes death (acute toxicity) or impairment to the endocrine, immune, or reproductive systems (chronic toxicity). The toxic chemicals of greatest concern are widespread and persistent, accumulate in biological tissues, and/or induce effects at extremely low concentrations.[2] Acute toxicity is fairly well understood. Controlled experiments in the laboratory expose test organisms to measured quantities of chemicals over a set time period and reach an indisputable endpoint (i.e., death). However, chronic toxicity for marine organisms, in both laboratory tests and the environment, is subject to confounding factors such as differences in an organism's feeding habits, mobility, and exposure to chemical mixtures.

The effects of toxic substances introduced into oceans can be diverse and far-reaching. For example, although they have direct impacts on exposed organisms living in contaminated marine sediments, many toxic chemicals are lipophilic (fat soluble) and will also bioaccumulate, meaning that they concentrate in organisms farther up the food chain.[1] Top-level predators, including humans who consume seafood, eventually accumulate the highest chemical burden. As a result, federal, state, and local government agencies issue fish consumption advisories to prevent overexposure to toxic chemicals like mercury and PCBs, especially in vulnerable populations (e.g., young children, pregnant women).[12] Predatory marine fish, birds, and mammals are also at high risk. Brown pelican populations, for instance, were once endangered by excessive accumulation of DDT, which causes eggshell thinning and reproductive failure.[2]

Toxic contaminants reach marine environments through a variety of mechanisms, including aerial deposition, runoff, and point source discharge. These substances tend to sorb to organic particles in soils and sediments. They often become concentrated in ecosystem transitional areas like bays, marinas, estuaries, and outfall zones that have historically received large amounts of land-based discharge or runoff, and where natural processes such as particle coagulation, agglomeration, and settling occur. Through disruption and resuspension, contaminated sediments can actually become a source of pollution to the overlying marine waters. Examples include DDT off the Palos Verdes coast in Southern California and PCBs in San Francisco and San Diego Bay.[2]

Emissions of toxic pollutants date as far back as the domestication of fire, when heavy metals released by burning firewood altered cave environments.[13] Wide-scale mining and metal-working activities during the Roman Empire, the operation of large furnaces with tall smokestacks starting in the sixteenth century, and the Industrial Revolution in the seventeenth and eighteenth centuries greatly extended the reach and impact of toxic substances. In the twentieth century, many new chemicals (e.g., PCBs) were developed in an effort to improve quality of life. Emissions of some chemicals peaked during the 1970s and have begun to decline again with the implementation of environmental controls.[13] For example, the phase-out of leaded gasoline in North America and Europe during the 1970s brought a significant decrease in lead emissions.

Like nutrients, solutions to toxic pollution range from reducing chemical usage and release to isolating or removing them from marine ecosystems. Since 1972, the regulation of point contaminant sources under the Clean Water Act has drastically reduced loads of known toxic substances to inland and coastal water bodies in the United States. Measures to control potentially toxic chemicals vary among other regions of the world. The Stockholm Convention, a global agreement to eliminate or restrict the use of persistent organic pollutants, came into force in 2004 to work toward addressing the worst chemical offenders, also known as the "dirty dozen."[14] These twelve priority contaminants include DDT, PCBs, chlordane, heptachlor, and toxaphene. Nine additional chemicals, including polybrominated diphenyl ethers (PBDEs), which are used as flame retardants, were added to the convention in 2009.

Sediment dredging and capping are two of the few effective and affordable approaches available for removing or isolating toxic substances in marine ecosystems. Still, they have some considerable drawbacks. Sediment dredging can reintroduce buried contaminants into the water column and requires the removed sediment to be incinerated or placed in a landfill. Sediment caps provide physical barriers between the contaminants and local marine life, but questions remain about their overall effectiveness, stability, and permanence relative to cost. Natural remediation (leaving the sediment undisturbed and allowing chemical decomposition or additional sediment deposition to reduce contaminant concentrations) is also used, but requires a longer period to produce results.

Like sediment cleanup, regulatory bans may not be feasible or effective in all cases. Banning a toxic substance does little unless it is replaced with a less-toxic or nontoxic alternative. In an alternative approach, the "Green Chemistry" movement seeks to reduce the risk posed by chemical contaminants by rethinking the alternatives and developing more benign modes of synthesis, processing, and use.[15] Examples are replacing plastic in disposable flatware with biodegradable potato starch and using less toxic chemicals for bleaching paper. Though this movement is gaining momentum, chemical-intensive and polluting practices remain the norm in many cases. Silver and gold mining in South and Central America, for example, often use mercury as part of the process for separating these elements, resulting in extensive contamination of waterways.[13]

Greenhouse Gases

Global climate change is a common topic in the news today, referring to the phenomenon of human-induced alteration of global climate patterns due to over-enrichment of heat-trapping gases in the atmosphere. Greenhouse gases permit survival of life on Earth, and many are naturally present in the atmosphere. As in the case of nutrients, though, too much of a seemingly harmless substance can alter natural geochemical cycles and devastate ecosystems. The most common greenhouse gases are carbon dioxide (CO_2), methane (CH_4), nitrous oxide (N_2O), and refrigerants like chlorofluorocarbons (CFCs).

Though most commonly thought of as an atmospheric pollutant, greenhouse gases also have the potential to harm ocean ecosystems through ocean acidification and warming. Oceans absorb CO_2 from the atmosphere, forming a weak carbonic acid when the molecule combines with water. Due to drastic increases in atmospheric CO_2, the acidity of the world's oceans is being subtly altered. Marine organisms require a specific range of environmental conditions to survive, and even a slight change can impair their life functions or cause death. In particular, increased acidity makes development of calcareous shells and skeletons more difficult, and their dissolution easier.[16] Though acidification of marine waters can be readily observed in a controlled lab setting, the extent of the potential effects on ocean ecosystems is not well understood, owing to a scarcity of research.

In addition to acidification effects, greenhouse gases induce climate pattern alterations that raise sea surface temperatures in certain regions of the world. Marine organisms may not be able to function or survive in waters outside their normal temperature range. Corals are particularly at risk, as they expel the algae that allow them to survive with just a 1–3°C increase in sea surface temperature. This phenomenon is called coral reef bleaching. Other climate change effects on the ocean include sea level rise, altered migration cycles, and sea ice melt, which reduces and fragments habitat for polar species.

The largest source of greenhouse gases is combustion of fossil fuels, followed by land use change (e.g., conversion of a forested area to agricultural fields).[17] Other significant sources include livestock, paddy field farming, and landfills. Greenhouse gas emissions increased dramatically during the twentieth century. Average annual emissions of CO_2 from fossil fuel combustion went up by more than 10 percent in just one decade, from the 1990s to the 2000s.[17] As a result of higher emissions, average concentrations of CO_2 in the atmosphere have risen from about 280 parts per million (ppm) in preindustrial years to 379 ppm as of 2005 (see Figure 7.1).[17] The complex balance of the carbon cycle is affected not only by increased CO_2 emissions, but also by changes in absorption and transformation of CO_2. Globally, oceans absorb approximately 24.4 Mt of CO_2 each day, which helps to mitigate rising atmospheric levels.[16] CO_2 is also sequestered by trees and other plant life in the world's forests and rainforests.

Addressing greenhouse gases will require a global approach. Several major actions have already been taken to slow, stop, or reverse the trends in greenhouse gas emissions. In the 1992 Monaco declaration, a group of 155 scientists urged policy makers to take immediate action to address ocean acidification, based on the available science and its potential to severely affect marine organisms, food webs, biodiversity, and fisheries. In addition, an international treaty called The United Nations Framework Convention on Climate Change was produced during the Earth Summit in Rio de Janeiro, Brazil, in 1992. Its objective is to stabilize atmospheric greenhouse gas concentrations at a level that would prevent dangerous interference with climate forcing. The Kyoto Protocol is an updated agreement, adopted in 1997 and entered into force in 2005, which set binding targets for developed nations to reduce greenhouse gas emissions. In 2009 the EPA elected to classify CO_2 as a pollutant, paving the way for its regulation in the United States.

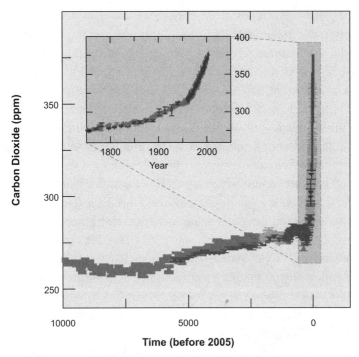

Figure 7.1 Change in atmospheric CO_2 concentrations over the last 10,000 years and since 1750 (inset) based on ice core and atmospheric samples. (Adapted from Solomon S, Qin D, Manning M, et al. Climate Change 2007: The Physical Science Basis. Contribution of Working Group I to the Fourth Assessment Report of the Intergovernmental Panel on Climate Change. Cambridge, United Kingdom and New York, NY: Cambridge University Press, 2007.)

Fecal Waste

Fecal waste can pollute marine waters through the addition of both nutrients and pathogens, the effects of which include eutrophication and spread of disease to marine organisms and humans. Because nutrient pollution was discussed separately, this section focuses solely on the impacts of pathogens from fecal pollution. People who have contact with or ingest fecal-contaminated water may develop a range of acute health problems, such as rashes or gastrointestinal illness, which occasionally lead to death. This is especially an issue where water contact activities, like bathing and clothes washing, are common. Beaches along U.S. coasts are monitored to warn visitors when high levels of bacteria are present, indicating an elevated health risk. Exposure to pathogens can also occur through the consumption of contaminated seafood. In this way, it impacts both inland and coastal populations. Treating diseases spread by fecal pollution can be costly, and economic impacts from lost tourism and work productivity may also be extensive.

Sources of fecal waste include humans, livestock, aquaculture, pets, and wildlife. In the United States, animal feedlots produce about 500 Mt of manure annually, more than three times the amount of sanitary waste produced by humans.[1] Nonpoint sources of fecal pollution, like pet waste left on streets, can be washed into waterways following rain events. Point sources to the ocean include sewage discharges, leaks, spills, and concentrated animal feeding operations. In developing countries, an estimated 90 percent of all wastewater is discharged directly into rivers, lakes, or oceans without any form of treatment.[18] Even in developed nations like the United States, fecal pollution commonly occurs from sewage spills, aging infrastructure, leaking septic tanks, and combined sewer overflows (CSOs). CSOs are an outdated method of preventing excess loading to wastewater treatment plants during periods of heavy rain by allowing sewage to flow into storm drains, which discharge without treatment into local water bodies. Globally, an estimated 245,000 square kilometers of coastal ecosystems are affected by fecal pollution.[18] It is most common along heavily inhabited coastlines and in developing nations (see Figure 7.2), and migration toward popular but densely populated coastlines exacerbates this issue. Currently, twenty-one of the world's thirty-three "mega-cities" are located on the coast, and more than half of the U.S. population lives within 100 kilometers of the coast.[18]

One strategy to address fecal pollution is to improve wastewater treatment and sanitation practices. This may take the form of costly construction or upgrade projects to install modern wastewater treatment facilities. Alternatively, there are many other technologies that can be applied to a given region at a lower cost. These include composting toilets, UV disinfection, constructed wetlands, and reuse of wastewater for crop irrigation. Recent efforts to increase global access to wastewater sanitation have been moderately successful, increasing global coverage from 55 to 60 percent from 1990 to 2000, though much work remains.[19] In addition, routine and consistent monitoring of coastal waters is necessary to understand health risks and protect beachgoers.[19] The 1999 Annapolis Protocol was an effort proposed by the EPA and the World Health Organization to build an integrated

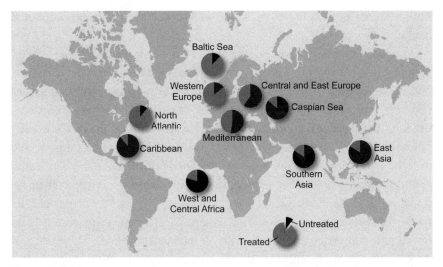

Figure 7.2 Ratio of treated to untreated wastewater discharged in 10 world regions. (Adapted from United Nations Environment Programme, UN-HABITAT, GRID-Arendal. Corcoran E, Nellemann C, Baker E, Bos R, Osborn D, Savelli H [eds]. Sick Water? The central role of wastewater management in sustainable development. A Rapid Response Assessment. Arendal, Norway: Birkeland Trykkeri, 2010.)

global protocol for monitoring recreational bathing waters. Another important task is reducing nonpoint sources of fecal pollution. The EPA has taken a step forward by electing in 2002 to treat large concentrated animal feeding operations as point sources under the Clean Water Act.[20]

Invasive Species

Though harmless in their natural habitat, invasive species have the potential to drastically alter ecosystems when introduced into areas outside their native one. Whether transported accidentally or purposefully, invasive species can out-compete native species, alter habitat, change food webs, and introduce new diseases.[1] They may drastically reduce the numbers of certain native species or cause them to become extinct, resulting in loss of genetic biodiversity. Without the natural balance of community diversity, ecosystem services like water purification, recreation, and food and pharmaceutical production may collapse. This imparts an economic burden, particularly for coastal communities. The introduction of invasive species into U.S. oceans and the Great Lakes has been estimated to cost the country millions to billions of dollars each year in terms of economic and ecological damage.[20]

Aquatic invasive species can be transported within a ship's ballast water, attached to the hulls of marine vessels, with travelers or their belongings, and as pets or plants for gardens or aquariums. Aquaculture involving nonnative or genetically modified organisms, intended to augment food production, also presents a potential problem if the cultured organisms escape into the wild. It is estimated that close to one million Atlantic salmon have escaped from farm pens along the west coast of North America.[1] Concerns about invasive species have increased, as globalization has allowed easier transport of people and goods around the world. Invasive species can be found on every continent, including Antarctica. The rate of marine species introduction has risen exponentially over the last 200 years.[1] There are 374 documented invasive species in U.S. waters, 150 of which arrived after 1970. More than 175 non-native species live in San Francisco Bay alone.[1] Unlike some other forms of pollution that degrade over time, invasive species often persist, increase in numbers, and spread into larger areas.[20]

Addressing invasive species will require management policies that prevent their transport and introduction and control their numbers once introduced.[20] International ballast water management standards and effective enforcement are necessary, as ballast water is a prime method of aquatic invasive species transport. Public education can also reduce instances of accidental species introduction. In the United States, the Nonindigenous Aquatic Nuisance Prevention and Control Act, as amended in 1996 by the National Invasive Species Act, is the primary legislation regarding aquatic invasive species and ballast water management.[20] It established an Aquatic Nuisance Species Task Force, responsible for facilitating coordination among federal, regional, and state agencies, as well as promoting research, prevention, control, monitoring, and information dissemination programs. Still, many coastal states in the United States have yet to develop an approved management plan. In response to the first international meeting on this topic in 1996, a partnership called the Global Invasive Species Program was formed to minimize the spread and impact of invasive species.

Debris

Marine debris, such as cigarette butts, plastic bottles, and used food containers, is possibly the most visible type of ocean pollutant and the easiest

Oceanic Pollution 147

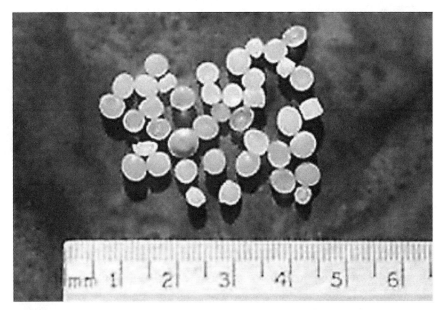

Figure 7.3 Small pre-production plastic pellets. (Photo credit: Shelly Moore, Southern California Coastal Water Research Project.)

to separate from water. Even so, marine debris breaks into smaller pieces over time and is redistributed by ocean currents into remote areas far from the source.[21] As plastic debris breaks down, especially in the case of pieces smaller than five millimeters, it is often mistaken for food and ingested by fish. One primary source of debris in this size class is preproduction plastic pellets, which are shipped to manufacturers for use in creating final products and commonly lost during transport (see Figure 7.3). Along with the physical effects of ingestion (e.g., blockage or filling of the gut), plastic debris can also absorb and transfer toxic organic chemicals.[20] Larger debris items (e.g., plastic six-pack soda rings, fishing line, and discarded fishing nets) are notorious for entangling marine animals and can lead to death by interfering with life activities, including the ability to eat, breathe, swim, and avoid predators. Aside from marine organisms, items with sharp edges can pose a risk to swimmers and divers. Debris also impacts aesthetics and may damage coastal tourism revenues.

Debris is produced as a by-product of modern societies. Starting in the mid-twentieth century, new petroleum-based materials were produced that could resist biological and oxidative decay. These products have many benefits during their useable lives, but upon disposal can persist in the environment for decades to centuries. A study conducted in the United States from 2001 to 2006 found that plastic straws, plastic bottles, plastic bags, metal beverage cans, and balloons were the most abundant types of marine debris.[22] Approximately 80 percent of marine debris comes from onshore sources that are either intentionally dumped or unintentionally washed or blown off the land, while the remaining 20 percent comes from offshore sources like ships and platforms.[20] Land-based litter can be washed or blown into waterways and storm drains, which in coastal areas usually drain untreated into the ocean. Once it enters the ocean, debris can sink, float, wash up on the shore, or travel with ocean currents far out to sea.

The best approach to lessening marine debris pollution is to reduce, reuse, or recycle materials at the source. Controls and fines can also be implemented to reduce littering and accidental release of materials from trash receptacles, trucks, boats, and landfills. The last mode of defense against marine debris is manually removing it from the environment (e.g., through river or beach cleanups). A number of initiatives have been implemented from the local to international levels to reduce marine debris. Notably, the annual International Coastal Cleanup campaign (organized by The Ocean Conservancy with funding from the EPA) has removed over 40,000 metric tons (t) of debris from more than 209,000 kilometers of shoreline.[20] Divers participating in this campaign also target underwater trash, and at the same time, data are collected to continually monitor marine debris prevalence and sources.

Oil

Like marine debris, oil is a highly visible ocean pollutant that can harm or kill wildlife. Because most oil products are less dense than seawater, they float along the ocean surface and tend to adhere to the bodies of marine animals. Oil-soaked plumage makes birds less buoyant, impairs flight ability, and increases their vulnerability to temperature fluctuations. Oil in the fur of marine mammals reduces its insulating effects, leading to hypothermia. Most animals coated in oil do not survive unless they are physically cleaned by humans. On top of these mechanical effects, crude oil and refined petroleum products contain varying types and concentrations of toxins. While attempting to clean themselves, animals often ingest oil, which can cause kidney damage, impair liver function, and irritate the digestive tract. Ongoing exposure to smaller amounts of oil can impair feeding, growth, development, and reproduction of marine organisms and increase susceptibility to disease. Oil also reduces light penetration into marine waters, which can stunt the photosynthetic activities of algae and plants. Humans exposed to marine oil pollution may be affected by skin irritation and respiratory ailments.

Large-scale accidental oil spills are a highly recognized source of pollution in marine environments; however, land-based runoff is an equally significant source. The equivalent of the 1989 *Exxon Valdez* oil spill, about 42 million liters of oil, runs off land into coastal waters approximately every eight months.[1] Oil can be released from cars, lawnmowers, boats, jet skis, and airplanes. It runs off of paved roads and parking lots and is funneled into waterways via storm drains. Natural seeps account for about 60 percent of oil inputs to oceans in North America, and 45 percent worldwide, while the remainder derives from human activities, including petroleum use, transport, extraction, exploration, and production.[23] The Deepwater Horizon oil spill is the most recent of several major accidental releases (over 100,000 t) since the 1960s. Caused by a drilling rig explosion in April 2010, it released millions of liters of oil per day into the Gulf of Mexico for more than eighty days.

Oil spills are both costly and difficult to address. Available cleanup methods include booms, skimmers, absorbent materials, gelling agents, fire, dispersant chemicals, and bioremediation with bacteria. Legislation and management measures, like the 1990 U.S. Oil Pollution Act and International Convention on Oil Pollution Preparedness, Response and Cooperation, have been enacted to reduce accidental spills from petroleum production and transport operations both in the United States and abroad. Programs to address nonpoint sources of oil are not as universal, but many mechanisms to minimize oily runoff are available. These include advertising campaigns to educate the public about dumping oil in storm drains and low-impact development measures that prevent oil-tainted stormwater from rushing offsite.

Ocean Pollution Legislation

As knowledge of marine pollution sources and effects has improved, better legislation has been enacted in many countries to limit society's impact on the oceans. Visibility of ocean pollution rose during the 1950s and peaked in the early 1970s. Initial concerns related to identifiable point sources, specifically large discharge pipes containing municipal and industrial waste. From the mid-1970s to the 1990s, pollution from ships became a major concern following several high-profile oil spills. Since some of the biggest pollution sources were targeted in the past few decades, a greater percentage of the remaining pollution inputs can now be traced back to less conspicuous nonpoint sources. Thus, attention in recent years has shifted to addressing these. Even more recently, concerns have arisen over the continuous production of new chemicals and detection of unregulated compounds in aquatic environments. As a result, environmental scientists and managers are now paying attention to a wide variety of pollution release and transport mechanisms, ranging from cosmetic additives to carbon dioxide.

In general, stronger pollution legislation has been enacted in developed countries (e.g., in North America and Europe) as compared to developing countries (e.g., in Africa and south Asia). Ocean policy in the United States has amalgamated over the years as an array of laws have been issued, often in response to an environmental crisis.[1] These include the Clean Water Act, Coastal Zone Management Act, Marine Mammal Protection Act, and Ocean Dumping Act, enacted in 1972; the Endangered Species Act of 1973; and the 1990 Ocean Pollution Act. Altogether, there are over 140 laws in the United States pertaining to oceans and coasts.[24] The Clean Water Act has been successful at addressing point sources and some nonpoint sources of pollution. In Southern California, for example, inputs of many pollutants have been reduced by 90 percent or more with better wastewater treatment, leading to the recovery of fish communities, kelp beds, and certain seabird populations.[1] Further, advances and upgrades in coastal wastewater treatment over the past few decades have kept up with growth in coastal populations and increased loading of wastes.[2]

At an international level, marine pollution likewise became a concern midway through the twentieth century.[25] Following the Torrey Canyon oil spill in 1967, the International Maritime Organization convened an international conference to prepare an agreement for controlling ship-based pollution. The International Convention for the Prevention of Pollution from Ships (MARPOL) was adopted in 1973 and 1978 and amended many times through 2009.[25] It covers ship-based pollution from oil, chemicals, harmful substances in packaged form, sewage, garbage, and air pollution. Later, the globally recognized United Nations Convention on the Law of the Sea (UNCLOS), adopted in 1982 and brought into force in 1994, took steps toward addressing land-based pollution sources.[20] It contains several articles pertaining to coastal protection and conservation, covering coastal drilling, mining, ocean dumping, and atmospheric pollution. Parties to UNCLOS must take steps to protect fragile ecosystems and endangered species, as well as notify others when a threat to the marine environment is imminent.

Conclusion

Even though legislation and management programs to protect marine ecosystems, such as the U.S. Clean Water Act, have been largely effective at curbing pollution, ocean conditions have still deteriorated in many ways since they came into force. It is estimated that each year coastal waters still receive a total of about 22 trillion liters of wastewater, 41,700–59,900 t of toxic organic chemicals, and 69,100 t of toxic metals.[26] The contribution of nonpoint sources of pollution to coastal waters has not been significantly reduced over the last thirty years, except where the manufacture or use of a contaminant (e.g., lead in paint and gasoline) has been banned or dramatically changed.[2]

History shows that management activities to address ocean pollution usually start with the largest contributors. Pollution emissions have been cut in half by implementing point source regulations, but remaining cuts will be more difficult as pollution management is subject to the law of diminishing returns. In addition, regulation of certain types of pollutants like invasive species or CECs cannot be addressed in the same ways that have been successful for other pollutants like lead and DDT. With this in mind, a strong scientific understanding of pollution issues, along with creative and cost-effective management policies, will be needed.

Another prime need is universal governance. Regulations applied by a single nation will have only a limited effect on the interconnected oceans. Participation in international conventions and adherence to their guidelines is typically voluntary, and vast disparities exist in the pervasiveness and degree of municipal and industrial wastewater treatment among developed and developing nations. In addition, contaminants that have been banned in developed countries are still produced and utilized in many developing countries. At a national level, marine pollution sources are not constrained to coastal areas, and there is a need for integrated ocean policies that recognize the roles of multiple stakeholders near the coast and far inland.

Notes

1. Pew Oceans Commission, *America's living oceans: Charting a course for sea change* (Arlington, VA: Pew Oceans Commission, 2003).
2. D. F. Boesch, R. H. Burroughs, J. E. Baker, et al., *Marine pollution in the United States: Significant accomplishments, future challenges* (Arlington, VA: Pew Oceans Commission, 2001).
3. National Research Council, *Clean coastal waters: Understanding and reducing the effects of nutrient pollution* (Washington, DC: National Academy Press, 2000).
4. J. Rockström, W. Steffen, K. Noone, et al., Planetary boundaries: Exploring the safe operating space for humanity, *Ecology and Society* 14 (2) (2009): 32.
5. R. W. Howarth, D. Anderson, J. Cloern, et al., Nutrient pollution of coastal rivers, bays, and seas, *Issues in Ecology* 7 (2000): 1–15.
6. S. B. Bricker, C. G. Clement, D. E. Pirhalla, et al., *National estuarine eutrophication assessment: Effects of nutrient enrichment in the nation's estuaries* (Silver Spring, MD: National Oceanic and Atmospheric Administration, 1999).
7. R. E. Turner, N. N. Rabalais, and D. Justic, Gulf of Mexico hypoxia: Alternate states and a legacy, *Environmental Science and Technology* 42 (2008): 2323–2327.
8. United Nations, *Coastal area pollution: The role of cities* (Nairobi, Kenya: United Nations Environment Programme, United Nations Human Settlement Programme, 2005).
9. *Water Quality criteria for nitrogen and phosphorus pollution* (Washington, DC: U.S. Environmental Protection Agency, 2010), http://www.epa.gov/waterscience/criteria/nutrient/ (accessed May 20, 2010).
10. *Managing contaminants of emerging concern in California: Developing processes for prioritizing, monitoring, and determining thresholds of concern* (Fountain Valley, CA: California Ocean Science Trust, National Water Research Institute, San Francisco Estuary Institute, Southern California Coastal Water Research Project, Urban Water Research Center of the University of California Irvine, 2009).

11. D. C. Muir and P. H. Howard, Are there other persistent organic pollutants? A challenge for environmental chemists, *Environmental Science and Technology* 40 (23) (2006): 7157–7166.
12. *Fish advisories* (Washington, DC: U.S. Environmental Protection Agency, 2010), http://www.epa.gov/waterscience/fish/ (accessed June 1, 2010).
13. J. O. Nriagu, A history of global metal pollution, *Science* 272 (5259) (1996): 223–224.
14. Stockholm convention on persistent organic pollutants (POPs) (Geneva, Switzerland: Stockholm Convention Secretariat, 2008), http://chm.pops.int/Home/tabid/36/language/en-US/Default.aspx (accessed May 26, 2010).
15. *Green chemistry* (Washington, DC: U.S. Environmental Protection Agency, 2010), http://www.epa.gov/greenchemistry/ (accessed June 1, 2010).
16. J. P. Gattuso, L. Hanson, and Epoca Consortium, European project on ocean acidification (EPOCA): Objectives, products, and scientific highlights, *Oceanography* 22 (4) (2009): 190–201.
17. S. Solomon, D. Qin, M. Manning, et al., *Climate change 2007: The physical science basis; Contribution of Working Group I to the fourth assessment report of the Intergovernmental Panel on Climate Change* (Cambridge, UK, and New York: Cambridge University Press, 2007).
18. United Nations Environment Programme, UN-HABITAT, GRID-Arendal, in *Sick water? The central role of wastewater management in sustainable development: A rapid response assessment*, edited by E. Corcoran, C. Nellemann, E. Baker, et al. (Arendal, Norway: Birkeland Trykkeri, 2010).
19. R. E. Bowen, A. Frankic, and M. E. Davis, Human development and resource use in the coastal zone: Influences on human health, *Oceanography* 19 (2) (2006): 62–71.
20. U.S. Commission on Ocean Policy, *An ocean blueprint for the 21st century: Final report* (Washington, DC: U.S. Commission on Ocean Policy, 2004).
21. C. J. Moore, S. L. Moore, M. K. Leecaster, et al., A comparison of plastic and plankton in the North Pacific central gyre, *Maritime Pollution Bulletin* 42 (12) (2001): 1297–1300.

22. S. B. Sheavly, *National marine debris monitoring program: Final program report, data analysis and summary* (Washington, DC: Ocean Conservancy, 2007).
23. National Research Council, *Oil in the sea III: Inputs, fates, and effects* (Washington, DC: National Academy Press, 2003).
24. Sea Grant Law Center, University of Mississippi, *Governing the oceans* (Washington, DC: National Oceanic and Atmospheric Administration, U.S. Department of Commerce, 2002), www.oceancommission.gov/documents/gov_oceans/gov_oceans.html (accessed May 18, 2010).
25. International Maritime Organization, *International Convention for the Prevention of Pollution from Ships, 1973, as modified by the protocol of 1978 relating thereto (MARPOL)* (International Maritime Organization, 2002), http://www.imo.org/conventions/contents.asp?doc_id=678&topic_id=258 (accessed May 20, 2010).
26. United Nations Development Programme, *Conserving biodiversity, sustaining livelihoods: Experiences from GEF-UNDP biological diversity projects* (New York: United Nations Development Programme Bureau for Development Policy, 2002).

Abbreviations Used

Chlorofluorocarbon (CFC)
Combined sewer overflow (CSO)
Contaminant of emerging concern (CEC)
Dichlorodiphenyltrichloroethane (DDT)
International Convention for the Prevention of Pollution from Ships (MARPOL)
Metric tons (t)
Million metric tons (Mt)
Polybrominated diphenyl ether (PBDE)
Polychlorinated biphenyl (PCB)
Polycyclic aromatic hydrocarbon (PAH)
Ultraviolet (UV)
United Nations Convention on the Law of the Sea (UNCLOS)
United States Environmental Protection Agency (EPA)

8

Fundamentals of Environmental Epidemiology

Marc A. Strassburg

And I wish to give an account of the other kinds of waters, namely, of such as are wholesome and such as are unwholesome, and what bad and what good effects may be derived from water; for water contributes much towards health.

—Hippocrates (340 B.C.)[1]

Introduction

Epidemiology is often defined as the use of the scientific method with its proper reasoning and logic to study and develop theories about why a wide variety of factors and their distributions relate to various outcomes of health or ill-health in the population. On a more basic level, epidemiology is a science that keeps us honest, or as Rudyard Kipling wrote in his story "The Elephant's Child":[2]

I KEEP six honest serving-men
(They taught me all I knew);
Their names are What and Why and When
And How and Where and Who.

The word *epidemiology* comes from the Latin or Greek: *Epi* indicates among us, *demos* means people, and *logos* or logy indicates science, scientific study, or theory. Environmental epidemiology is the branch of public health that deals with environmental conditions/factors and hazards that may pose a risk to human health (i.e., the study of those exposures usually not in the control of the individual). For example, although smoking poses a severe hazard to one's health, it is the exposure to secondhand smoke in particular that falls clearly in the domain of environmental health. Overall, environmental epidemiology identifies and quantifies exposures to environmental contaminants, conducts risk assessments and risk communication, provides surveillance for adverse health effects, and provides health-based guidance on levels of exposure to such contaminants. Much of the knowledge base of environmental health is dependent on groups of multidisciplinary scientists, including epidemiologists. Epidemiologic research in this field is aimed at clarifying associations between environmental factors (exposures) and disease risk in humans. Unfortunately much of the evidence is based on observational human studies or those carried out on animals or in the laboratory. All suffer from issues inherent when there is an absence of human experimental studies, most notably difficulties in generalization and external validity. The domain of environmental epidemiology traditionally includes air pollution, water pollution, and occupational exposures to physical and chemical agents. More recently areas of focus have expanded and, for example, range from evaluating the increasing toxic levels of arsenic in major waterways, to the study of emerging infectious diseases, to attempting to measure the impacts of the thinning of the ozone layer and secular climatic changes on the global environment and human health. In the field of environmental epidemiology, many advances have been made in the area of occupational health, where exposures are large and can be well documented. However, measuring lower levels of exposures in the general population poses considerable challenges. Thus, new approaches are being developed to derive impacts of the environmental burden of disease on the population. In addition to biogenetic advances, one analytic example is the World Health Organization's (WHO) use of global estimates of deaths and disability adjusted life years (DALY). Such calculations help to illustrate what the WHO estimates to be as much as 24 percent of global disease being caused by environmental exposures[3] (see Figures 8.1 and 8.2).

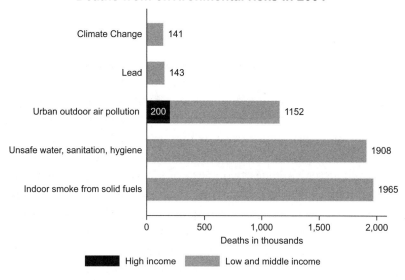

Figure 8.1 Deaths from environmental risks in 2004. (http://www.who.int/quantifying_ehimpacts/global/envrf2004/en/index.html; accessed on October 20, 2010)

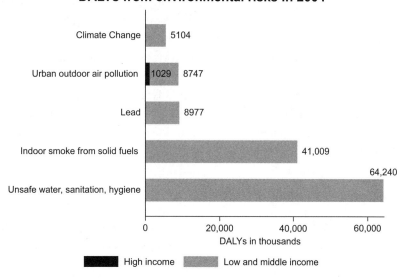

Figure 8.2 DALYs from environmental risks in 2004. (Preventing disease through healthy environments: Towards an estimate of the environmental burden of disease. A. Prüss-Üstün and C. Corvalá. WHO ISBN 92 4 159382 [2006] http://www.who.int/entity/quantifying_ehimpacts/publications/preventingdisease.pdf. Used by permission of WHO.)

The Concept of Individual Risk versus Population Rates

Simply stated, *risk* is the likelihood of an effect occurring in an individual, based on a certain type of exposure or dose. This can best be illustrated by the common dose–response models demonstrated by Sir Richard Doll and others in their landmark studies confirming the increasing risk of lung cancer in smokers as they increased the number of daily cigarettes consumed—the resulting measure is called the relative risk (e.g., someone who smokes two packs of cigarettes a day might be ten times as likely to develop lung cancer as those who don't smoke, etc.).[4] Thus, such a dose–effect or dose–response relationship is basically how a biological organism reacts when exposed at different levels to a toxic substance. If the response shifts as the substance quantitatively shifts in exposure, then the dose–effect or exposure–response relationship is said to exist. As another example, a small dose of carbon monoxide is likely to cause drowsiness, whereas a very large dose can be fatal. Another often-cited example of dose–effect is related to lead exposures; the dose-effect increases from possible low-level exposures and small effects on IQ to large exposures resulting in neurotoxicity and death.[5] Population rates can also demonstrate relationships between specific levels of exposures and outcomes. One could look at grouped data to determine the overall level of cigarette usage in a population and correlate that with the resulting rates of lung cancer—although not as dramatic a statistic as the relative risk, such regression coefficients generated from using population cause-specific rates can indicate possible adverse impacts before more complicated studies are completed and also provide information for hypothesis generation.

Toxicological Concepts

There are many toxicological concepts relevant to environmental epidemiology. They include forms of hazard, types of exposures, and toxicity or potency of hazard. Typical hazard forms include chemical agents (e.g., dust, fume, gas, vapor, and mist), physical agents (radiation, heat, cold, and noise), biological agents (virus, bacteria, and allergens), and psychosocial hazards. Exposures generally include intensity, duration, and frequency. Another key toxicological concept relevant to environmental epidemiology is route of entry, including inhalation, ingestion, and dermal absorption. The field of toxicokinetics includes absorption, distribution, metabolism, and excretion. *Dose* is a critical concept, and issues of whether a substance is absorbed, bio-availability, and target organ dose are commonplace in the field of toxicology. In addition, the following two concepts are important: onset of effect, whether due to acute or chronic exposure, and the induction period, which is similar in some sense to the incubation period for infectious diseases. It is worth noting that the induction period starts after a disease process begins and lasts until the illness or condition is manifested. Variations in susceptibility and the concept of reversibility are also important issues.

Monitoring Chemicals in the Environment

Key to the study of environmental epidemiology is the classification and monitoring of chemicals in our environment. Although more than 80,000 chemicals are registered for use in the United States, with over 50,000 used in consumer products, the 1976 Toxic Substances Control Act does not require chemicals to be proven safe or to be registered before use. The Toxic Chemicals Safety Act of 2010 and the Safe Chemicals Act of 2010 will change that with increased safety criteria. At this time, however, relatively few chemicals are thought to pose significant health risks to humans.[6] The *Fourth National Report on Human Exposure to Environmental Chemicals* (2009) provides an ongoing assessment of the exposure of the U.S. population to environmental chemicals by the use of biomonitoring. The report's Web site (http://www.cdc.gov/exposurereport) is a good source for recent updates.[7]

Historical Events of Note

Sir Percival Pott, a London surgeon, was one of the first individuals to describe an environmental cause of cancer; he made the observation that chimney sweeps had a higher incidence of scrotal cancer compared with other young males in other occupations.[8] In the 1800s John Snow, an anesthesiologist, was able to find the source of a cholera outbreak in London by studying various water supplies used by healthy and sick households and thus to determine differences in risk between consumption patterns related to water sources.[9] Large-scale events such as the dropping of nuclear bombs at Hiroshima and Nagasaki have allowed the study of radiation effects among bomb survivors. Severe air pollution episodes beginning in London in the 1950s sparked interest in analyzing episodic-related deaths from heart and lung disease. Today, the WHO estimates that worldwide more than 600,000 premature deaths occur yearly related to air pollution.[10] A number of environmental studies based on gross contaminations have helped bring to light a variety of adverse health effects in the environment. Two such noteworthy events, both of which occurred in Japan, include "Minamata disease"—related to the consumption of fish contaminated with methyl mercury—and "Itai-Itai disease," related to the consumption of rice contaminated with cadmium. Finally, a third event was a mysterious disease first called Pontiac fever, which was identified years later as Legionnaires' disease. After careful environmental study this condition was determined to be an infectious bacterial disease spread through ventilation and water cooling systems; these reservoirs continue today to serve as a source of this disease.[11] Although it is not exhaustive, Table 8.1 lists a large number of studied diseases that may have causal association with environmental factors.[12]

TABLE 8.1
Diseases Linked with Environmental Factors

Disease Group	Subgroup	Environmental Factor
Gastrointestinal	Diarrheal disease	Bacteria and viruses in water
Cancer	Lung	Tobacco smoke, metals, radiation
	Blood cells	Hydrocarbons, radiation
	Soft tissue	Hydrocarbons, herbicides
	Liver	Hydrocarbons
	Urinary tract	Certain chemical compounds
Respiratory	Asthma	Climate, particles, ozone
	Bronchitis	Particulates, sulfates
	Emphysema	Smoke, sulfates, ozone
Reproductive	Malformations	Chemical mixtures, solvents
	Abortions	Metals, hydrocarbons
	Low birth weight	Tobacco smoke, chemical mixtures
Neurological	Development	Lead
	Transmission	Lead, organic solvents

Source: D. Baker, Professor of Clinical Medicine, University of California, Irvine, and Director, Center for Occupational and Environmental Health.

Study Endpoints Used in Environmental Epidemiological Research

It is necessary to understand the various clinical stages to be able to measure morbidity associated with the natural history of a disease. For some studies mortality is the outcome of interest, whereas for others a certain clinical or physiological measurement is used to evaluate adverse health effects. Finally, the weakest form of outcome measurements are considered to be those only obtained through self-reporting of conditions and symptoms by the affected individual.

Observational vs. Experimental Epidemiology

In examining the occurrence of health and disease in human populations, epidemiologists take two general approaches. The experimental design is probably most known to the reader and involves a controlled situation in which researchers control a variety of exposures and observe effects. However, due to dangers in testing humans, it is generally reserved for laboratory testing on animals or insects. Observational studies, which are less intrusive, include almost all other study designs available to the epidemiologist. Basically, these include a wide variety of methods wherein the investigator does not manipulate the exposure and can only observe or document exposures. These types of studies range from the simple, descriptive cross-sectional or prevalence surveys, to correlational or ecological studies, to the more sophisticated cohort and case-control studies. Another name for this group of studies is "analytic," although simple descriptive studies are sometimes excluded from this category. The level of analytic complexity also ranges widely depending on the study design employed, the number and types of variables involved, the limitations and potential biases in approach, and the analytic acumen of the investigators.

The Role of Statistics

Much of what we learn in environmental epidemiology comes from either observational studies or working with small samples (as in experimental trials); thus, we rely on the use of statistics and statistical modeling to help us provide some level of error in those estimates. In the literature there tends to be an overemphasis on the value of the p statistic as well as on confidence limits, both of which may give one a false sense of assuredness. The key issues to focus on are the validity of the study behind the numbers and a proper perspective related to the meaning of any statistically "significant" result. For example, the establishment of an arbitrary p value (e.g., 0.95) or a confidence limit (e.g., 95 percent) for expressing likelihood that a result is due to chance may give the mistaken "confidence" that a value is found within a range that is accurate 95 percent of the time. Such findings have little value if samples were nonrepresentative, there are biases in the sampling scheme, or problems in the study design are not taken into account. Another area that seems to have received much attention in environmental epidemiology is cluster investigations. Apparently the use of cluster analysis has led many to believe that they have found a causal relationship between an environmental factor and an observed outcome, when what they have done is only to isolate clusters of disease from their overall context. One is reminded of the Texas-sharpshooter fallacy, that a Texas sharpshooter shoots a large number of holes in the side of a barn and then draws a bull's-eye around the best clustered shots. Thus, what appears to be statistically significant (i.e., not due to chance events) is actually expected by the laws of chance. To date, of the large number of studies of cancer clusters investigated in the literature, very few have turned out to have a convincingly identified underlying environmental cause.[13] In summary, despite limitations, the full lineup of analytic epidemiologic methods continues to provide clarity and perspective on problems related to adverse health effects in the environment.

The Epidemiological Triangle and Concepts of Causation

One of the fundamental models of causation used in epidemiological studies has been the triangle of host, agent, and environment. The host in this context refers to the affected group or population and may also imply certain genetic characteristics. This could be either humans or animals, which in turn might also serve as a reservoir for an infectious agent. The agent refers to a factor in whose absence the disease may not occur, for example, a microorganism, chemical substance, or form of radiation. However, in terms of current causal models this framework might be looked upon as an oversimplification. Clearly, for many diseases such as tuberculosis or malaria, the bacteria or parasite in question is the *sine qua non* (the required cause); however, there are almost always other variables and co-factors. Although alone they are not sufficient to produce the disease, if removed they could leave the individual disease free. An example is using appropriate netting to prevent the mosquitoes that carry malaria from biting an individual. Thus, a more proper way to look at the causal pathways is that there are a large number of causally related events (sometimes referred to as a web), including genetic, infectious, and environmental; if any one event is removed it may prevent the disease or condition in question. This sort of thinking allows much more creative preventive approaches to be taken.

The Role of Critical Thinking

Logical fallacies are errors that occur in arguments. One of the most important fallacies in the area of environmental epidemiology is the post hoc fallacy. The *post hoc ergo propter hoc* (after this therefore because of this) fallacy is based on the mistaken notion that since one thing happens after another, or B follows A, then A caused B. To put it another way, the first event in any sequence is the cause of the subsequent event. This sort of post hoc reasoning is the basis for many erroneous associations. Such sequences do not establish the likelihood of causality any more than observed correlations do. Causal chains are sometimes much more complicated than that which is observed in linear time—for example, one needs to take into account incubation periods when trying to connect an intestinal infection to one's most recent meal consumed. Another important fallacy is known as the regressive fallacy, or regression to the mean. This is a special post hoc fallacy wherein there is a failure to take into account normal fluctuations of events when ascribing causes to them. Both human and animal variability has taught us that individuals take actions near the peaks of their problems; this is when changes may occur on their own and regress toward the mean.[14] One example is the installation of a camera in a high accident area; the following lower rate of accidents may not be due to the camera per se, but to the publicity of the high rates and/or the tendency of high accident rates in certain areas to decline over time independent of the camera placement. To establish the probability of a causal connection between two events, controls are generally required in order to rule out other factors such as chance or possible confounding variables (described below). In addition, there is always the need to replicate scientific studies.

Biases

All studies have the potential for various forms of bias, which is best defined as any deviation from the truth for any reason, whether recognized or not recognized by the investigator. This is a very broad area and can include problems in study design, sampling, data collection, measurement devices, analysis, interpretation, and even undue or undetected financial or investigator biases. Two types of biases commonly referred to are recall bias and selection bias. They are covered first, followed by less commonly referred to, but no less important, publication and funding biases.

Recall bias refers to the phenomenon in which those who have illnesses might have different recall levels of any exposures being assessed by a survey questionnaire than those who do not have those illnesses. This is sometimes referred to as differential misclassification, because accuracy of recall may be different between those who are ill and those who are healthy.

Selection bias occurs when those in a study group are so different from the general population out of which they arose that findings can no longer be generalized. This is especially true when such selection bias differs for controls in the same study and goes undetected or is not controlled for (another form of differential misclassification). One historical example of this is the "healthy worker effect," which held that employed workers were likely to be healthier than those who lost their jobs or could never get jobs.[15] For example, if one studies lifeguards, obviously a self-selected and screened group, and compares them with the same aged individuals in the general population, and observes differences in health issues, such differences might be due to the lifeguards' greater level of fitness at time of employment. It is often stated that great conductors live very long lives, since so many great conductors are very old. However, this phenomenon might be due solely to selection bias and to the healthy worker effect; in other words, it takes many years to become great, and thus only those who survived to old age had the chance of being recognized as great. Possibly the best statement of this type of survivorship bias was made by Ambrose Bierce, the American satirist and critic, who in his *Devil's Dictionary* defined the word "dawn" as:

The time when men of reason go to bed. Certain old men prefer to rise at about that time, taking a cold bath and a long walk with an empty stomach, and otherwise mortifying the flesh. They then point with pride to these practices as the cause of their sturdy health and ripe years; the truth being that they are hearty and old, not because of their habits, but in spite of them. The reason we find only robust persons doing this thing is that it has killed all the others who have tried it.[16]

Publication bias is the tendency for most researchers and journals to submit/publish research that has positive outcomes instead of research with negative results. Lack of submission and publication of such research might be due to either lack of statistically significant findings or a poorly explained causal sequence. In some instances, unpublished studies might even have had positive findings, but these results were not felt to be important or reflected low-level associations. Even researchers who conduct their research post hoc (also known as data dredging) may discard negative results as they appear in their analyses and never report them; however, they might be more likely to report any positive results observed, whether or not they had been guided by an a priori hypothesis.[17]

Funding biases can exert overt or subtle and even subconscious pressures to publish only positive results and to suppress negative results.[18] This might be especially true when one's livelihood or source of funding is at stake. To reduce the undue influence of funding sources on research findings, many journals and professional associations require that investigators disclose any potential conflicts of interest.

Confounding

Confounding is probably the most important consideration when evaluating an epidemiological study. It occurs when an association is "confused" or "distorted," that is, when an association between an exposure and outcome is obscured by some third factor. For example, if we say that the city of Phoenix is a very unhealthy place to live because crude mortality rates are higher than those of Los Angeles, we might be very mistaken to attribute unhealthy living conditions to Phoenix on the basis of such crude death rate comparisons alone. The reason is that crude rates do not take into account the possibility of confounding by age or sex or differences in the age and sex structure between the two populations. In the case of Phoenix or similar cities in the desert, it is possible that older people might migrate to a drier and hotter climate as they become sick; thus one would expect to see an overall higher death rate in such cities. This higher death rate might have been caused by this migration pattern alone. In this case both age and health status at time of immigration, if not accounted for, would confound the situation or association between the cities as to which is "healthier." When one conducts occupational studies with regard to various exposures and outcomes such as lung cancer, one needs to take into account other likely or known causes of lung cancer, the most obvious being smoking history. Once the variable of smoking history is taken into account, that variable will no longer operate as a confounder; however, it may still act as a co-factor or contributor to the outcome being studied. In asbestosis among asbestos workers, it is well demonstrated that smoking among exposed workers produces a combined risk greater than exposure to either asbestos or cigarettes alone; this relationship is generally referred to as a *synergistic effect*.

Measure of Disease Frequency and Risk

As quantification is a very important aspect of almost all epidemiological studies, there are a number of commonly used terms to best express distributions of both disease and risk. Probably the most common are *prevalence* and *incidence*. Sometimes these two measures are used as interchangeable terms (although not always appropriately so) and generally give an idea of the size of a specific problem and sometimes even the risk of an adverse outcome in a population. Prevalence is generally defined as a fraction in which the number of cases/conditions at or during a specific period of time (numerator) is divided by the number (denominator) in the population during that same time period. If only a single point in time is involved, such as a survey carried out on a specific day, then prevalence is generally referred to as *point prevalence*; if a period of time is involved, prevalence is referred to as *period prevalence*.

The term *incidence* adds new dimensions to both the numerator and the denominator. The denominator includes the number in the population among which the cases or conditions arose, that is, those persons capable of developing the specific disease or condition in question. In the numerator an incidence rate deals with new cases that arise in those populations capable of developing the disease or condition. Incidence rates are central to the study of causal mechanisms and the concept of risk. They provide some measure of risk to the individual or society from which the rates arise. There are a number of ways one can construct an incidence rate, and generally the differences in the methods are related more to the denominator (population) than to the numerator (cases). For example, one can take the average number of persons at risk for the denominator (incidence rates)—or if the data are available, one can include the specific periods of time persons are available for being included in the denominator. Thus if a person becomes ill six months into a year, he or she could count as one-half person-year time in the denominator (this measure is sometimes called *incidence density*).

There are several other measures of risk that can be obtained from epidemiological studies. One is the *absolute risk*, which is the amount of disease or condition usually expressed per 10,000 population (this can also be per 1,000, or 100,000—just move the decimal over) and not related to a specific exposure. For example, there might be a 5 per 10,000 population lifetime risk of having lung cancer regardless of whether one smokes (these numbers are estimates and are used for illustration only). Furthermore, they do not take into account exposure to secondhand smoke—for more precise estimates see the American Lung Association report referenced here.[19]

There is also a measure known as *attributable risk*—the risk that can be "blamed on" (related to) a certain exposure. For example, if without smoking an individual has a 0.5 per 10,000 chance of getting lung cancer, then the amount of lung cancer attributable to smoking would be 9.5 per 10,000 (or 10.0–0.5). Such risks are generally referred to as population risks. However, in epidemiology, in order to assess the relative strength of a risk factor we use relative risk or estimates of relative risk. Simply expressed, this is the measure of the excess or deficit of risk in one group compared with another. One way to express this risk is that those who smoked two packs of cigarettes a day are twenty times more likely to develop lung cancer than those who did not smoke—or a 20:1 relative risk. Preferably this risk rate is derived from the incidence in the exposed group divided by the incidence in the unexposed group. For example, if lung cancer among smokers is 200 per 100,000 population and among nonsmokers is 10 per

100,000, the relative risk would be 20:1, or twenty times greater risk. If this estimate was derived from a study that sampled the population, then this number would only be a point estimate, for which the often over-used p value is usually assigned and is formulated something like p<.05 or other p value. However, one has to take into account statistical error, so that something called a confidence interval is developed to express a range of possibilities—for example, we are 95 percent confident that the likely risk of lung cancer is between fifteen and twenty-five times greater for smokers than for nonsmokers. Such elevated risk ratios and even narrow confidence limits may have little meaning if the study did not take into account any major biases or confounders that might have occurred in either the design or the analysis of the study.

Elsewhere in this chapter the author covers how the measures of risk pertain to various types of study design, such as cohort and case-control studies. For example, a cohort study may result in a direct measure of relative risk (this is covered in more detail below). Another study design frequently employed in epidemiology is the case-control study, which does not usually provide a direct measure of relative risk, but rather uses an indirect measure known as an odds ratio, which in some cases can approximate the relative risk. *Odds ratios* are defined as the odds of being exposed among ill individuals divided by the odds of being exposed among well individuals (this concept is not intuitive and might require the reader to work out a number of exercises before fully understanding the differences between relative risk and odds ratios). However, the term is used for the same purpose: to quantify risk to individuals in a population.

Methods of Epidemiological Studies

Major steps in conducting epidemiological studies include correctly formulating the study questions along with establishing hypotheses and objectives; selecting an appropriate study design; determining the sampling frame; selecting the study population; and measuring relevant variables, including exposures, health outcomes, and covariates. The underlying objective of study design methods is to make appropriate and fair comparisons among populations studied. Considerations of study design are particularly crucial to environmental epidemiology. Two major methodological problems in environmental epidemiology are selection of study subjects (sampling frame) and carrying out exposure assessments. In addition, the researcher may have to contend with multiple hazards as well as multiple levels of risk. Finally, small effects in large populations can have significance but be difficult to detect. These concerns are common throughout all of epidemiological research but present major challenges to environmental epidemiology.

Sampling Frames

Typical sampling frames for environmental studies are commonly found in community studies. Sampling frames sometimes make use of census data or various multilevel cluster sample techniques (e.g., all households on selected blocks). Other sampling frameworks include schools, workplaces, primary care clinics, and community groups. For case-control studies, cases may be obtained from registries or hospitals; controls may be selected from the same source as the cases or by other techniques (such as matching). In the past random digit dialing was popular, although with the reduction in the use of landline telephones this procedure is quickly losing favor as a valid approach. Use of cell phone numbers or Web site surveys may introduce other forms of response bias. For larger cohorts, various methods are used to study groups over long periods of time, for example, using persons in a selected occupation (e.g., firemen or policemen) or selecting populations in small cities (e.g., Framingham, Massachusetts).[20]

Determining Exposure Status

There are several ways to accomplish exposure assessments. One can use indirect methods, which include obtaining existing environmental data sets or even diaries when available. Questionnaires and surveys can be developed to collect information regarding recalled exposures and possibly time-related activity patterns. Various environmental monitoring tools, which may permit area sampling, are also available to the investigator. For example, there are many potential lead exposure pathways that could ultimately result in an increase in human lead burden. The lead sources might be airborne, from combustion or evaporation (inhalation uptake), in contaminated soil or dust (direct contact), or in water, for example, in particulate runoff or leaching (ingestion, excretion).

Finally, direct methods for measuring sources of exposure include personal monitoring and use of various laboratory tests to examine individuals for biological markers based on pharmacokinetic and pharmacodynamic models. Overall, there is a hierarchy of the quality of exposure measurements, ranging from quantifiable personal measurements, of the highest quality to geographical location of the residence or employment, the weakest. For example, an approximation of intermediate quality of actual exposure to trichloroethylene (TCE) via consumption of locally contaminated drinking water might employ available estimates of average water consumption and use.

Exposure Assessment Concepts

With regard to pathways of exposure, the most common are respiratory, skin, and alimentary. In general, there is a hierarchy of exposure measurement techniques that pertain to hazards, ranging from those involved with human activities to those occurring as a result of natural phenomena. Regardless of the source of exposure, the dispersion or environmental concentration of the hazardous substances may occur in air, water, food, or soil. In many cases there have to be integrated exposure measurements, taking into account in addition to dose and target organ, exposures via multiple pathways, exposures to multiple hazards, and exposures over time in different microenvironments. All of this information leads to an Integrated Personal Exposure (IPE) and to possible identification of potential adverse health effects. Valid (nonbiased) exposure assessments are critical in conducting environmental and occupational epidemiological studies. Personal measurements and biological monitoring are preferable to use of questionnaires, whereas existing environmental monitoring data as well as surrogate measures, are often necessary to support integrated exposure models. Model building based on such information may play an important role in the assessment of exposures to multiple hazards that occur via multiple pathways over time. However, error rates increase as the number of assumptions increase when such models are utilized.

Epidemiological Study Designs

Most epidemiological studies use the individual as the unit of study, although there are a few exploratory designs that attempt to study populations as a group or as combined entities. Some epidemiology texts artificially divide these studies into analytic versus nonanalytic. Such overlapping divisions may be confusing, because even in descriptive studies there is a need to conduct appropriate analyses. Descriptive approaches tend to be used for those studies that are more likely to answer the *Who, When,* and *Where,* rather than the *How.* Such studies are likely to be used to describe patterns and frequencies of disease or to make comparisons of group summary data. These types of studies are useful for purposes of hypothesis generation. Simply put, these studies range from the very rudimentary case reports and case series, to registry summaries, surveys, and even cross-sectional studies (another name for a prevalence study).

Case Series, Case Reports, and Case Studies

Case series, case reports, and case studies may be anecdotal or formally reported to health authorities via a reporting system. Clinical information is described in detail and in general may be related to diseases of unknown etiologies. Although the individuals who gather the information may think that they are coming up with causal clues based on histories obtained, this stage in disease investigation is very exploratory, and researchers should not jump to premature conclusions. For example, a doctor who has seen in his or her practice several cancer patients enter remission while taking a new unproven therapy (e.g., Laetrile)[21] may be led to believe that this unproven drug is now causally related to the patients' improvements. In reality, the doctor might be seeing a chance event of an excess number of remissions or some other random occurrence that has nothing to do with the unproven therapy.

Cross-Sectional Studies (Prevalence Surveys)

For cross-sectional studies, the sampling frame for selecting study participants must be carefully considered. Methods for measurement of exposures and health outcomes must be clearly defined. Health outcome information collected based on prevalence and not incidence of conditions is subject to many biases, as previously mentioned, especially recall bias. Thus there are a number of limitations inherent in such cross-sectional studies, including, for example, the necessity of distinguishing incident versus prevalent cases, correctly interpreting temporal sequences, and taking into account the selective survival of study subjects. In these types of studies, descriptions of the health status of a population (e.g., a community or nation), including variables of person, place, and time, are most reasonable to study. From such studies there should be no formal attempts to test hypotheses or link exposures with health outcomes. Studies can also be based on existing data, such as birth or death statistics, hospital discharge data, and prior health surveys. Surveys in particular are often used to describe frequency of behaviors as well as exposures or other health conditions reported by individuals who participate in the survey. If appropriately sampled, and the disease or condition is nonfatal, this approach may yield good descriptive data and support a certain level of hypothesis generation. However, it should be noted that both the exposures and outcomes are being measured at the same time, with a number of potential limitations and biases at play, not to mention the fallible memories of the study participants. Such recall bias can be in any direction and may be difficult to take into account in trying to analyze the results of such surveys meaningfully.

An example of a use of cross-sectional studies is to provide information for planning health services; for example, health planners might conduct a survey to decide whether or not to build a new wing in a hospital in a community with a high rate of arthritis. Other major advantages of such studies are the lower costs that are sometimes involved and the relatively quick analyses that can be carried out. In addition to the limitations described previously, there is the question of extrapolation or producing generalizations, which are dependent on the sampling methods employed, as well as the issue of response rates (response rates under 80 percent are generally not looked upon favorably).

Ecological Studies
Ecological studies fall under the category of descriptive studies. They are correlational studies and basically provide a way to analyze a number of prevalence or cross-sectional studies at one time—while looking for gross associations. Commonly this type of study uses summary data based on both group exposures and group disease rates. This procedure might be something like comparing average alcohol consumption by country with rates of heart disease by country. Associations are then looked for and examined for trends. Positive correlations between exposures and

outcomes when observed (sometimes simply by plotting disease and exposure rates on a graph) must be interpreted cautiously. Scatterplots and correlation coefficients can be used to bring such trends to one's attention. This information can be useful for preliminary testing of hypotheses, as a wide range of exposures and outcomes can be plotted. Of course, the more plots one makes a posteriori (after the fact), the more likely it is that any observed association may be due to chance. Jumping to an early causal conclusion can lead to the ecological fallacy (i.e., making inappropriate generalizations to individuals from information obtained at the group level). This might occur if one stipulates that among individuals there is a causal link between development of a particular condition and exposures measured. In its most absurd form, one could blame the observed reduction in birth rates in cities in Belgium on the decline of the stork populations observed in those same cities during the same time period. The confounder here is that a change in the roof styles made the city inhospitable for the storks during the same time period, and they nested elsewhere. Nevertheless, the study of spatial and temporal patterns has recently been aided by the increasing use of spatial mapping (also known as geographic information systems, or GIS), time series studies, ecological time series, and panel studies. Correlations observed between changes in exposures and changes in outcomes over time can serve as a first-level warning system or a hypothesis generator. To some degree, communities observed over time serve as their own controls, thereby reducing confounding. Some other examples of ecological studies are those that examine variations in mortality, hospitalizations, and emergency room visits along with changes in air pollution by using aggregate data in major cities. Panel studies that examine variations in asthma symptoms, medication use, and respiratory peak flow along with changes in exposure to air pollutants and fungal spores (measured via air and personal monitoring) are variations of this type of study design.

Case-Control Studies
Case-control studies have also been referred to as retrospective, case-comparison, or case-referent studies. Basically such studies consist of some form of studying cases (those who already have a disease or condition under question), examining a number of potential risk factors, and

then comparing the results with those from a group of controls. The latter are those who do not have the disease but are in general similar to the cases with respect to basic characteristics unrelated to the disease or outcome. There are many studies of this type reported in the literature; recent examples include the study of possible adverse health effects among people exposed to magnetic fields and the study of the relationship between cell phone use and cancer risks.[22] The first step in a case-control study is establishing a case definition. Second, it is necessary to identify a sampling frame for selecting cases and controls. Controls may be selected in a number of ways, including a method known as "matching"; this is the process of selecting controls so that they are similar to the cases with respect to potentially confounding characteristics. As noted, these characteristics may include variables such as age, race, gender, and SES (socioeconomic status). If they are not controlled for when the controls are selected, such potentially confounding variables may be dealt with in the analysis. Third, one compares prior exposure experiences of cases and controls in order to estimate the association between exposures and health outcomes. Examples of environmental health topics investigated by this type of study design are lung cancer and residential exposure to radon, lung cancer and indoor air pollution in China, and leukemia among individuals exposed to radioactive fallout from nuclear testing facilities in the United States. There are a number of advantages to case-control studies. They are relatively efficient and can be done with small numbers of study subjects. Although in the past they have been thought to be inexpensive, this perception has changed as the complexities in carrying out such studies have increased. Case-control studies are thought to be particularly efficient for investigating rare diseases as well as diseases with long latencies. Such studies can assess the potential effects of several exposures on a disease. There are a number of disadvantages to case-control studies, including temporal ambiguity (i.e., when current exposure information is collected, it is not always clear whether the exposure preceded the onset of disease), difficulties in selecting appropriate comparison groups; potential bias in measuring the exposure (e.g., the outcome may affect the subject's recollection of the exposure and the measurement or recording of the exposure), and the investigator's inability to directly estimate attributable risk. However, the exposure odds ratios do provide a good estimate of the relative risk when certain conditions are fulfilled, for example, that the ill and non-ill persons sampled are reasonably representative of the underlying population being studied.

Cohort Studies

Other names for cohort studies are prospective, follow-up, incidence, and longitudinal. A cohort is a clearly defined group of individuals who can be followed together over time, in either real time or time reconstructed in the past (a nonconcurrent cohort study). Most important, cohort studies enable the investigators to categorize groups (cohorts) who are without the outcome of interest (the disease or condition) at the starting period, and also to classify the study population according to various exposure status factors and other risk factors. Following cohort members over time takes place to determine the occurrence of health outcomes based on the exposure status of subgroups. Issues related to changes in exposure status over time must also be measured carefully.

Examples of subjects examined in cohort studies include populations exposed to radiation, such as Japanese people who experienced World War II nuclear bomb explosions, children who underwent radiation treatment for an enlarged thymus, women treated with radiation for post-partum mastitis, and communities that were affected by the Chernobyl nuclear accident. A number of environmental poisoning episodes have been studied using the cohort approach. They concerned Yusho disease from ingestion of rice oil contaminated with polychlorinated biphenyls;[23] populations exposed to acute releases of toxic chemicals from industrial facilities, for example, children living near Minamata, Japan, who were exposed to methyl mercury and subsequently developed Minamata disease;[24] and a pesticide factory in Bhopal, India, from which methyl-isocyanate leakage killed at least 3,800 people and left over 100,000 with permanent disabilities in 1984.[25] Advantages of cohort studies are that they can determine incidence rates and attributable risks directly. Furthermore, exposures can be determined with less potential for bias than if outcomes were already known. In some instances this study design can also be efficient for studying rare exposures. Disadvantages include the need to examine large numbers of individuals, because usually such studies must follow many more subjects than will experience the event of interest. Such cohorts are expensive to keep track of, and the results may not be available for a long period of time. Major biases may result from attrition or loss to follow-up.

Experimental Studies

Experimental studies are also cohort studies, but they differ in that there is some control, either naturally or by design, over who is being exposed. There are a number of names for such types of studies, ranging from the most common, clinical trials, to randomized controlled trials (RCTs), as well as interventional and community trials (often the experimental studies are referred to as the "gold standard" of study designs). "Natural experiments" monitor changes in health outcomes associated with changes in environmental conditions or regulations. For example, if one water system uses fluoride additives and another does not, we can look for outcome differences between the two populations (for dental caries and fluorosis, for example).

A variation of the experimental design is the community trial. The advantages of community trials are that the "experiment" takes place within the field or community and not in the laboratory or a controlled clinical setting. Thus it is possible to evaluate effects of dose and temporal relationships as well as measures associated with exposures occurring in the real world. Both experimental studies and community trials, in particular, share the disadvantage of generally being limited to the study of short-term, reversible effects that usually are related to treatment or preventive interventions. Furthermore, there may be problems in generalizing results, as such studies look at volunteers or atypical communities willing to participate. In addition to the potential for high costs, these types of studies may be impacted by loss to follow-up of study subjects. However, the advantage of an experimental design is the investigator's ability to control a large number of confounding variables and to better test a causal hypothesis. Significance is generally measured statistically, and differences are evaluated for whether they are likely to be due to chance. A double-blind condition exists when neither the test groups nor the evaluator know which group receives the real intervention. Preferably groups are randomly assigned, thus reducing the possibility of bias in allocating individuals to the test or control groups. The overall purpose of controls, double-blinding, and randomization is to reduce error, self-deception, and bias.

Conclusion

Environmental epidemiology provides methods for researching many of the fundamental questions regarding the role of environmental exposures and human health. This discipline has a long history, ranging from the writings of Hippocrates in the fourth century BC to the nineteenth, twentieth, and twenty-first centuries, in which a variety of important breakthroughs occurred, first in infectious disease control and then in occupational health and safety issues. A number of challenges remain for the environmental epidemiologist: identifying appropriate sampling frames; studying populations in affected communities, particularly those that require monitoring; and further refining exposure assessment tools and making them available to field researchers. There is a need to better separate multiple hazards and determine multiple levels of risks. Clearly small effects in large populations may be significant, but they are also difficult to detect. Finally, there are some hazards that may be difficult to identify and thus are not amenable to control. New study designs, such as two-stage phasing, in which exposure and disease information are collected in the first stage, and covariate information is collected on a subset of subjects in the second stage, might provide answers to a number of exposure-related questions. Future improvements in exposure assessment, such as the use of molecular biology, especially biomarkers, might be helpful in identifying biological footprints left behind by environmental exposures.

Notes

1. Hippocrates, *On airs, waters, and places*, Part 7, translated by Francis Adams (The Internet Classics Archive by Daniel C. Stevenson, Web Atomics), http://classics.mit.edu//Hippocrates/airwatpl.html (accessed July 21, 2011).
2. R. Kipling, The elephant's child, in *Just so stories*, http://www.onlineliterature.com/poe/165/ (accessed July 21, 2011).
3. A. Prüss-Üstün and C. Corvalán, *Preventing disease through healthy environments: Towards an estimate of the environmental burden of disease* (Geneva: World Health Organization, 2006).
4. J. M. Samet and F. E. Speizer, Sir Richard Doll, 1912–2005, *American Journal of Epidemiology* 164 (1) (2006): 95–100.
5. Agency for Toxic Substances and Disease Registry, *ToxGuide™ for lead Pb*, CAS no. 7439-92-1 (October 2007), http://www.atsdr.cdc.gov/toxguides/toxguide-13.pdf (accessed July 21, 2011).

6. National Toxicology Program (NTP), *Current directions and evolving strategies: Good science for good decisions*, http://ntp.niehs.nih.gov/files/CurrentDirections2005.pdf (accessed July 21, 2011).
7. Centers for Disease Control and Prevention, *Fourth national report on human exposure to environmental chemicals* (2009), http://www.cdc.gov/exposurereport/pdf/FourthReport.pdf (accessed July 21, 2011).
8. P. Pott, *Chirurgical observations relative to the cataract, the polypus of the nose, the cancer of the scrotum, the different kinds of ruptures, and the mortification of the toes and feet* (London, England: printed by T. J. Carnegy, for L. Hawes, W. Clarke, and R. Collins, 1775).
9. D. P. Sandler, John Snow and modern-day environmental epidemiology, *American Journal of Epidemiology* 152 (2000): 1–3.
10. A. Gulland, Air pollution responsible for 600,000 premature deaths worldwide, *BMJ* 325 (2002): 1380.
11. M. Keramarou and M. R. Evans for the South Wales Legionnaires' Disease Outbreak Control Team, A community outbreak of Legionnaires' disease in South Wales, August–September 2010, *Eurosurveillance* 15 (42) (2010): pii=19691, http://www.eurosurveillance.org/ViewArticle.aspx?ArticleId=19691 (accessed July 27, 2011).
12. D. Baker, Environmental and occupational epidemiology (PowerPoint slide), University of California, Irvine, Center for Occupational and Environmental Medicine, http://www.epi.uci.edu/media/docs/2005%20Environ%20epi%20-%20handouts.pdf (accessed January 7, 2011).
13. S. Benowitz, Busting cancer clusters: Realities often differ from perceptions, *JNCI* 100 (9) (2008): 614–621.
14. T. Gilovich, *How we know what isn't so: The fallibility of human reason in everyday life* (New York: The Free Press, 1993).
15. D. Shah, Healthy worker effect phenomenon, *Indian Journal of Occupational and Environmental Medicine* 13 (2009): 77–79.
16. A. Bierce, *The devil's dictionary* (New York: Neale, 1911).
17. J. D. Scargle, The "file-drawer" problem in scientific inference, *Journal of Scientific Exploration* 14 (1) (2000): 91–106.
18. S. Greenland, Accounting for uncertainty about investigator bias: Disclosure is informative, *Journal of Epidemiology and Community Health* 63 (2009): 593–598.

19. American Lung Association, *State of lung disease in diverse communities* (2010), http://www.lungusa.org/assets/documents/publications/solddc-chapters/lc.pdf (accessed July 27, 2011).
20. G. W. Comstock, Commentary: The first Framingham study—a pioneer in community-based participatory research, *International Journal of Epidemiology* 34 (6) (2005): 1188–1190.
21. B. Wilson, *The rise and fall of laetrile* (2004), http://www.quackwatch.com/01QuackeryRelatedTopics/Cancer/laetrile.html (accessed July 27, 2011).
22. R. B. Dubey, M. Hanmandlu, and S. K. Gupta, Risk of brain tumors from wireless phone use, *Journal of Computer Assisted Tomography* 34 (6) (2010): 799–807.
23. D. Onozuka, T. Yoshimura, S. Kaneko, et al., Mortality after exposure to polychlorinated biphenyls and polychlorinated dibenzofurans: A 40-year follow-up study of Yusho patients, *American Journal of Epidemiology* 169 (1) (2009): 86–95.
24. M. Harada, Minamata disease: Methyl mercury poisoning in Japan caused by environmental pollution, *Critical Reviews in Toxicology* 25 (1) (1995): 1–24.
25. E. Broughton, The Bhopal disaster and its aftermath: A review, *Environmental Health* 4 (2005): 6.

9

Using Epidemiology as a Tool to Study Environmental Health

Keith B. G. Dear and Anthony J. McMichael

Introduction

The phrase "environmental health" includes many fields of study, all concerned with the influence of the environment on human health. There is no clear definition of *environment* itself. It is defined by some as all entities and associated "exposures" that originate outside the physical human body. For purposes of environmental health research, it is usually defined to include the physical, chemical, microbiological, and ecological conditions with which humans come in contact. The word *ecological* also challenges easy definition. It refers here to natural ecosystems and associated relationships between nonliving and living components of the environment, including the human species—and also, therefore, to the built environment as part of "human ecology." While social structures and relationships are critical determinants of human health in their own right, and of human-influenced environmental conditions and responses to them, the "social environment" per se is not included in "the environment" in this chapter.

This chapter is about the science of environmental health, broadly conceived, and how it is pursued through the application of epidemiology. As in any field of science, there are three aspects: (a) science as a body of knowledge about the natural world; (b) science as method, the body of research tools used to achieve and expand that knowledge, tailored to each field; and (c) science as applied in practical life, which in some fields is called technology.[1] For the science of environmental health, these three aspects become

- our growing knowledge and understanding of how the environment, in its many aspects, influences the health of human individuals and populations;
- the research methods we use to identify, test, and measure those influences; and
- the use of that knowledge and understanding to design and evaluate preventative interventions that will modify the environment in ways beneficial to health.[2]

Other aspects of environmental health, addressed elsewhere in these volumes, include issues of policy; advocacy; education both of the public and of the health workforce; and the interface with other areas, chief among which are energy, transport, urban planning, agriculture, and water.

The chapter begins with an overview of what we mean by "environmental epidemiology." We then lay out the principal challenges that the study of environmental health presents to the epidemiologist and show through examples how these challenges have been met in particular studies. We conclude by discussing the novel and special challenges of climate change epidemiology, then suggest one way in which epidemiology may adapt to meet them through the use of complexity science.

Environmental Epidemiology

Epidemiology is the principal method for investigating environmental impacts on health. Its main purpose is to provide conceptual frameworks, study designs, and analytical tools that accept and encompass the multilevel nature of most environmental exposures. We must, however, ask: How successful can epidemiology be in investigating the natural science of environmental health? What are its principal strengths and weaknesses, as a tool applied in this field?

Further, social needs extend beyond risk identification and assessment into risk management. Modern epidemiology has made great strides over the past three or four decades in relation to the first two of these needs. The formulation and then evaluation of risk management may seem less

Using Epidemiology as a Tool to Study Environmental Health 183

intellectually exciting to researchers, but if that part of the work is not done, then the antecedent research may not be used to good account for prevention.

The World Health Organization (WHO) divides environmental impacts on health as direct and indirect. Direct impacts are considered to be those in which a state of the environment affects the health of exposed individuals through recognized pathways; examples are the presence of harmful chemicals, ionizing radiation, and some biological agents. Indirect impacts arise from aspects of the physical, social, psychological, and aesthetic environment, such as are determined by urban design, housing, land use, and transport. To date, environmental epidemiology has largely been concerned with direct impacts, though a growing trend is for multidisciplinary teams to also address the more subtle social and psychological influences, as recognition grows of their importance for the mental health of populations.

Today, a further and different type of environmental hazard to human health is emerging. The scale and intensity of the "human enterprise" and its demands upon the wider environment are now so great that many of the world's regional and global environmental systems are being disrupted. Note that the central process at work in this case is imposed change or disruption, not contamination. The environmental systems at risk include Earth's climate system, the UV-filtering stratospheric ozone layer, the hydrological cycle and associated freshwater flows and supplies, the circulation and biological activity of key elements (especially nitrogen and phosphorus), the dynamics and chemistry (and hence vitality) of the oceans, the productivity of forests and their ecosystems, and the fertility and regeneration of soils.

These natural biogeophysical systems underpin life support everywhere; the sustained health of human populations depends on them absolutely.[3] What should and can be the role of epidemiology, working in an interdisciplinary setting, in contributing to the elucidation, description, quantification, and prospective modeling of the current and future health impacts of these more systemic environmental changes?

The Limitations of Classic Reductionist Thinking

Historically, in high-income countries epidemiology has successfully identified health hazards arising from high levels of exposure, such as to severe air pollution (the "London Fog" of 1952), to heavy metals (lead, mercury) in water and food, or to asbestos in the workplace. Although such classical studies of physicochemical environmental exposures are still of importance in low- and middle-income countries, many of the environmental health questions now being addressed in high-income countries refer to more subtle exposures. Examples are electromagnetic fields as a putative cancer risk and persistent organic pollutants (POPs), which act cumulatively over decades on fertility and reproduction and on the functioning of the central nervous system and the immune system.[4, 4A] These exposures may encompass several fields: toxicology, infectious diseases, climate, and so on. For many diseases, these influences interact, whether synergistically or antagonistically,[5] and may also be impacted concurrently by an environmental change; for example, the health consequences of climate change include effects mediated by toxic agents (e.g., some air pollutants) and by communicable disease vectors.

Feingold et al. argue for "the need to expand the scope of environmental health research, which now focuses largely on the study of toxicants, to incorporate infectious agents."[6] They conclude that "[i]f basic research is to increase our ability to predict the consequences of exposure to environmental chemicals, we must embrace non-reductionist thinking and design experimental models that emulate human experience." We must not be beholden to the textbook litany relegating population-level studies to "curtain-raiser," hypothesis-generating status, nor be drawn toward the rocks of ultra-reductionism by the sirens of molecular genetics. As the scale of influences on population health increases in today's interconnected, intensifying world, environmental epidemiologists must achieve a balanced research agenda employing *all* the tools of modern epidemiology.

Levels of Observation and Effect

Epidemiology is "the study of the distribution and determinants of disease in human populations, and the application of findings for the prevention of disease." The phrase "in human populations" is ambiguous: Is our task to understand what changes disease rates in whole populations/communities, or to account for the distribution of cases of disease *within* a population? Are we interested in explaining the bobbing of corks among the surface ripples or identifying the currents that shift the corks along the shoreline of disease risk?[7] Geoffrey Rose argued that our prime task is the latter: the policy need is to understand the causes of changes in *rates* of disease in populations, not merely the causes of *cases* within a population.[8] There is strong historical precedent for this, especially from the early stirrings of epidemiology in nineteenth-century Europe. Farr compared rates of death in England's geographic regions, in cities versus towns.[9] Engels compared death rates in rich and poor urban groups.[10] Virchow attributed the typhus epidemic in Silesia to the poverty and oppression of the farmworkers.[11]

Many environmental exposures, by dint of their pervasive nature, impinge approximately equally on all members of a community, even a whole urban population. This raises a tantalizing question for epidemiologists: What is the appropriate unit of observation? A related, practical issue is the feasibility of assessing exposure at smaller scales, at the subgroup level or even at the individual level.

Where there are likely to be nontrivial differences in individual exposure, the situation holds promise of estimating the dose–response (exposure–effect) relationship through comparisons between individuals or by comparison across a set of well-defined subgroups. Examples of such studies include the Port Pirie cohort study of blood lead and child neurocognitive development[12] and more recently, the "Ausimmune" case-control study of neural demyelination (a precursor to multiple sclerosis) in relation to lifetime sun exposure measured by individual skin casts.[13]

Meanwhile, for many other research and public health policy purposes, the most important analysis is at the aggregated or whole population level. A prime example comes from the domain of air pollution epidemiology, wherein the central question is how the population rate of hospitalization or death varies, over time, in relation to changes in exposure levels. This category of research has spawned a number of conceptual and methodological developments over the past two to three decades. Similar questions are now being framed in relation to the impacts on population health outcome rates of different levels of thermal stress exposure from heat waves.

Both exposures and outcomes exist at different levels, essentially either in individuals or in groups (the "population level"). This is easier to appreciate in the case of exposures: individuals are exposed to varying levels of pollution (air, water, noise), whereas legislation, from local road rules to federal tax, applies to populations. Health outcomes are most easily thought of as individual (deaths, hospital admissions), but some are intrinsically statistical and so must be measured in populations, such as life expectancy[14] and average quality of life.[15] The emphasis has changed over the nearly 200 years of epidemiological activity. The trend for most of this time has been downward in scale, from populations, to subgroups, to individuals, to molecules and genes. More recently, with the surge of interest in global environmental issues, the trend has reversed, or rather the front has widened.

With modern epidemiological methods and sophisticated biomedical theory and techniques, we can better determine the causes of cases within populations. That is always important and may elucidate causal mechanisms, but it may also distract or mislead. Within a population of smokers, smoking would be an undetectable disease risk factor, apparent only through external comparison with other (healthier) populations. Uniquely among the health sciences, epidemiology now operates at all levels, from genes to populations. It is possible, indeed necessary and productive, to study environmental health effects using all levels.

All health-related exposures ultimately operate at the individual level. However, they may arise, and be most readily measured, at other levels. For example, laws that limit exposure to environmental tobacco smoke (ETS) in public places vary at the state or national level. The efficacy of those laws, in terms of adherence and enforcement, may be measurable at the community level. But their impact on health operates through reducing individual exposure and thereby risk. The key principle is that environmental effects are best studied at the level—individual, local, or state—at which they principally vary.

With many environmental exposures, this consideration supports the use of ecological analyses comparing aggregate outcomes to group-level exposures. There is then an unavoidable urge to infer individual-level causation from group-level associations: the ecologic fallacy. The fallacy is best avoided by measuring relevant individual-level variables and conducting multilevel analyses.[16]

Meeting the Challenges of Environmental Epidemiology

Conventional, single risk-factor epidemiology investigates the health effects of individual behaviors such as smoking and of individual-level exposures such as to food-borne infections. In comparison, the assessment of environmental health risks and impacts often—whether by preference or necessity—investigates relationships at the group or population level. Further, environmental health research faces several particular challenges, including these (explored further in following text): (a) difficulties in obtaining accurate, relevant data; (b) subtlety of the causal processes; and (c) the difficulties of accounting for exposures that have arisen gradually, or even that have as yet barely begun to be felt.

Difficulties in Obtaining Accurate, Relevant Data

All epidemiologic methods relate exposures to outcomes. Exposure assessment in particular usually involves applying methods from other disciplines (see Figure 9.1).

Environmental exposures may be difficult to quantify, especially at the individual level. Most studies of the health effects of air pollution have used ambient concentrations—those measured at the population level—as proxies for individual exposure, though some researchers make strenuous efforts to use individual measurements. For example, Gehring et al. used land-use regression models to assign individual levels of exposure to atmospheric pollutants.[17] A different approach to the problem is to use individually measured biomarkers as indicators of probable exposure; for example, Moller and Loft surveyed the association of particulate matter with molecular markers of oxidative damage and proposed that the latter be used as a biomarker for air pollution.[18] A danger here is that because the biomarker will be subject to influences other than the imputed environmental exposure (diet, for example), any associations with disease may be misattributed.

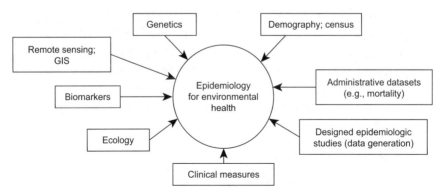

Figure 9.1 Sources of data for epidemiologic studies.

A general problem for environmental epidemiologists is the lack of good-quality, fine-scaled, georeferenced data on climatic and environmental conditions; indices of human well-being and health (e.g., specific infectious diseases, measures of undernutrition, mental health indices); and the many social, cultural, political, and material factors that affect the vulnerability of exposed populations. (Populations with a high capacity to make adaptive responses will be able to offset their intrinsic, or constitutional, vulnerability.)

Subtlety of the Causal Processes

The importance of studying population-level exposures may result, unavoidably, in comparing exposed and unexposed populations that differ in other important characteristics. It is therefore common for many exposures to be present at once (or to be highly correlated), leading to mutual confounding and the danger of misattributing risk. Moreover, exposures may change slowly over time and vary gradually in space, making it difficult to identify the spatiotemporal borders of the relevant populations.

Further, the complexity of many of the causal processes that impinge on human health as a result of global environmental and climatic changes often precludes conventional, straightforward analysis. That complexity makes difficult both the attribution of cause to particular components of the system or stages of the process and the quantitative estimation of those attributable health impacts. Careful consideration of effect modification is required; for example, Roberts analyzed the joint actions of air pollution and temperature on mortality in Pennsylvania and found that results were highly sensitive to details of the models.[19]

The difficulties of teasing out complicated, interacting effects are aggravated by the importance of investigating even small relative risks, because of widespread exposure to entire local populations. Further, where effective adaptive, risk-reducing responses occur (both spontaneous and deliberate), part of the "true" unavoided risk is then masked. Again, this may result in risks being underestimated or missed altogether.

The Importance of Time

The fact that the "change" processes are mostly ongoing (e.g., climate change) means that the resultant exposures are continually changing quantitatively and also, often, qualitatively. The risk estimation task then has to deal with a moving target. In societies in which the rate of technological and consumer-behavior change is ever faster, many of the recent past exposures that might be studied (e.g., cell phone use over the past decade) are being superseded by different levels and forms of exposure. Can epidemiology keep up with modernity, or do we provide a historical record of recent past environmental risks?

This is one main reason that extended and improved surveillance systems will be needed, to keep track of future changes in population- and community-level exposures and health consequences. That, in turn, will be much more informative if, as mentioned above, information across a range of factors and indices is collectable, and is collected, at a finer scale.

For these things to happen there is a new and urgent need for epidemiologists, other health researchers, and health policy people to engage more widely and purposefully with other research disciplines and policy sectors. The larger-scale environmental and social influences on the health of population in today's increasingly interconnected, populous, and economically intensive world cannot be tackled on a more traditionally narrow, even mono-disciplinary, basis.

The Future Profile of Environmental Epidemiology

The methods and capacities of current mainstream epidemiology are being challenged by the advent of even larger-scale and more complex forms of environmental exposure. Although strenuous efforts are being made to address these exposures through conventional means, entirely new methods may need to be added to the arsenal of environmental epidemiology.

The Epidemiology of Climate Change

The estimation of the health risks due to climate change raises other interesting issues. The necessary use of scenario-based modeling of future risks to health is one such key example. That type of modeling is not a routine part of epidemiology.

Climate change may be the only health-endangering "exposure" variable we (the world community) have so far had to deal with that (a) will inevitably continue to change, both quantitatively and qualitatively, over many future decades; (b) will change in some ways, and with some environmental consequences, that are not foreseeable, as thresholds are passed or feedback processes emerge; and (c) for which the changes in climate are the end product of a long chain of other incompletely understood geophysical processes (e.g., saturation of carbon "sinks," modulation by altered cloud formation) and unknowable future human trajectories (e.g., population growth, new technologies, intergovernmental decision making). Scenarios therefore deal in plausible ranges, not future likelihoods. The currency of analysis duly shifts toward handling uncertainty and providing indicative plausible ranges of outcomes, not computing formal probabilities and risks. This is unsettling for many scientists, baffling for the public (who expect scientists to "know"), and awkward for policy makers who want "facts" and/or "costings" and who find it difficult to sell seemingly expensive insurance policies to the electorate.

Several other particular methodological problems arise from the need to assess the health impacts of climate change—clearly a research task for environmental epidemiology. First, it is not clear what is meant by "exposure" to climate change. It is clear that *weather* has impacts on health, for example, via extreme weather events, and climate is merely the pattern of weather in a certain place. Therefore climate change implies a change in weather, either in the frequencies with which particular conditions occur or in the range of weather that ever occurs—and these are in fact not dissimilar, since a wider range of possible weather merely means a change from negligible to measurable frequency. But there is more to it than this. Climate itself, as distinct from weather, can be viewed as directly affecting health, and it is in this sense that novel and difficult epidemiological methods are called for. It is climate, not weather, that determines the viability of crops and hence rural livelihoods. It is climate, not weather, that limits the range of disease vectors such as bats, snails, and mosquitoes.

Further difficulties emerge when we distinguish, as we must, between the effects of climate per se and of climate change. Climate change may impact health by bringing adverse climates (and hence higher levels of heat stress and disease, for example), but also through the process of change itself. Climate is central to the sense of personal environment, and it is climate *change* that leads to the mental distress identified as solastalgia by Albrecht.[20] Climate may determine what crops are commercially viable in a given location, but it is climate *change*, combined with lack of economic agility, that leads to farm failures, community breakdown, and the associated mental health disorders.

In studying the association of climate with health, the risks from confounding are often severe, essentially because people already exposed to extremes of climate, and therefore already living in marginal conditions, may be subject to other risk factors such as poor diet and endemic disease. A modest degree of climate variation may then have amplified apparent impact: studies linking geographical variation in climate extremes to health outcomes will tend to reveal spurious associations. Although these can be controlled through conventional epidemiologic techniques, essentially based on adjusting for individual-level covariates, the danger remains that associations of *present* climate with health may not be very informative about the health risks due to actual future climate *change*.

Even in the limited context of measuring exposure to extreme weather such as heat waves there are difficulties, as evidenced by the variety of definitions used in heat wave warning systems. It is still not clear, for example, whether hot days are more dangerous as a sequence than they are individually; that is, whether the concept of a "heat wave" has any epidemiological validity at all. Recent work in the United States suggests that "most of the excess risk with heat waves . . . can be simply summarized as the independent effects of individual days' temperatures."[21] Due to the spatial scale on which climate varies, it may be difficult to identify unexposed subpopulations to compare with exposed individuals in the equivalent locations.

Adaptation to the threat of climate change is as much a social as an individual matter. Individual adaptation may occur through physiological acclimatization to heat stress or through more mundane mechanisms such as installing air-conditioning. But community response is also important, potentially leading to a kind of climate-related "herd immunity."[22] Thus multilevel modeling is likely to be particularly important in teasing out the causes of susceptibility to extremes of weather.

Particularly in the case of communicable disease, climate–health relationships will involve complicated webs of causal pathways. Ross River virus disease rates in Australia depend on local population densities of mosquitoes and macropods (the principal vertebrate hosts), as well as on human behavior and susceptibility. Xun et al. discuss the web of factors underlying the health impacts of sea-level rise in Bangladesh, in which feedback loops involving shrimp and prawn farming operate as well as the expected effects of salinity on rice farming.[23] Complexity is discussed further in the next section.

Epidemiology and Complexity Science

The property that distinguishes a "complex system" from a merely complicated one is that its large-scale properties "emerge" from the local interactions of semiautonomous components.[24] The emergent, self-organizing, intrinsically unpredictable behavior of such systems tends to feature nonlinearity, oscillation arising from negative feedback loops, and tipping-points arising from positive feedback. It is reasonable to expect all natural systems, including those underlying human health, to exhibit these characteristics to some extent, and under that view the only reasonable way to understand and predict them is through the methods of complexity science: primarily, computer-based simulations of mathematically defined networks of components. Such a model is first tuned so as to approximate the observed behavior of chosen aspects of a real system and then used to make causal inferences and to predict the effects of interventions by modifying its input parameters.

These methods have been applied in fields such as ecology, economics, and neuroscience, where the importance of interacting autonomous entities is obvious, but they are now also being extended to epidemiology,[25] not only in communicable disease (where again, a system of interacting units is a natural model), but also, for example, in chronic disease.[26] This approach may be of particular value in environmental epidemiology because of the complexity and large scale of the causal systems typically involved.[27]

Conclusion

Epidemiology is, at least for the present, the principal research tool of environmental health, but it is neither a perfect nor a sufficient tool. We have surveyed the primary methodological challenges that an epidemiologist involved in environmental health research faces: difficulties of data collection, appropriate multivariable modeling, and even conceptualizing the research question, which often demands to be framed in terms of the population as well as the individual. Environmental epidemiologists have risen to this challenge by developing correspondingly complex measurement and analysis methods and by accepting that the appropriate inferential conclusions in many cases will not be as simple as A (individual exposure) causes B (clinically defined disease). Instead, the premise of modern environmental epidemiology is that interacting aspects of the environment in which a population lives may have multiple, interwoven impacts on the health of that population.

The timescales of environmental health impacts are in one sense a familiar problem: exposures may change slowly, at rates comparable to, and therefore confounded with, other societal changes that also affect the health of populations. Changes in the *quality* of the typical Western diet have been paralleled by the change in its *quantity*, so that the recognized effects of obesity on health are in danger of concealing more subtle influences such as mild vitamin deficiencies. But we are now faced with a further time-related challenge: assessing the likely future impact of exposures, ultimately related to climate or other aspects of a changing environment, aspects that are foreseeable but not yet measurable. This is giving rise to a rapid growth in the application of methods based on simulation and computer modeling rather than experiment, survey, and data analysis. This trend seems likely to take environmental health research still further from its single risk-factor roots.

Notes

1. Richard P. Feynman, *The pleasure of finding things out: The best short works of Richard P. Feynman* (New York: Perseus Books, 1999).
2. H. K. Armenian, Epidemiology: A problem solving journey, *American Journal of Epidemiology* 169 (2) (2008): 127–131.
3. S. S. Myers and J. A. Patz, Emerging threats to human health from global environmental change, *Annual Review of Environment and Resources* 34 (2009): 223–252.
4. E. Diamanti-Kandarakis, J.-P. Bourguignon, L. C. Giudice, et al., Endocrine-disrupting chemicals: An Endocrine Society scientific statement, *Endocrine Reviews* 30 (4) (2009): 293–342.
4A. United States Environmental Protection Agency, *Persistent organic pollutants: A global issue, a global response* (Washington, DC: Office of International Affairs, 2002 [updated 2009]), http://www.epa.gov/international/toxics/pop.html.

5. L. S. Birnbaum and P. Jung, Evolution in environmental health: Incorporating the infectious disease paradigm, *Environmental Health Perspectives* 118 (2010): a327–a328, doi:10.1289/ehp.1002661.
6. B. J. Feingold, L. Vegosen, M. Davis, et al., A niche for infectious disease in environmental health: Rethinking the toxicological paradigm, *Environmental Health Perspectives* 118 (8) (2010): doi:10.1289/ehp.0901866.
7. A. J. McMichael, The health of persons, populations and planets: Epidemiology comes full circle, *Epidemiology* 6 (6) (1995): 633–636.
8. G. Rose, Sick individuals or sick populations? *International Journal of Epidemiology* 14 (1985): 32–38.
9. J. Eyler, *Victorian social medicine: The ideas and methods of William Farr* (Baltimore, MD: Johns Hopkins University Press, 1979).
10. F. Engels, *The conditions of the working class in England* [1845] (London: Penguin, 1987).
11. R. Virchow, Report on the typhus epidemic in Upper Silesia, *Social Medicine* 1 (2006): 11–98.
12. P. A. Baghurst, A. J. McMichael, N. R. Wigg, et al., Environmental exposure to lead and children's intelligence at the age of seven years—The Port Pirie cohort study, *New England Journal of Medicine* 327 (1992): 1279–1284.
13. R. M. Lucas, A.-L. Ponsonby, K. Dear, et al., Sun exposure and vitamin D act independently in risk of central nervous system demyelination, *Neurology* 76 (2011): 540–548.
14. C. A. Pope, M. Ezzati, and D. W. Dockery, Fine-particulate air pollution and life expectancy in the United States, *New England Journal of Medicine* 360 (2009): 376–386.
15. A. E. de Hollander, J. M. Melse, E. Lebret, and P. G. Kramers, An aggregate public health indicator to represent the impact of multiple environmental exposures, *Epidemiology* 10 (5) (1999): 606–617.
16. J. Pekkanen and N. Pearce, Environmental epidemiology: Challenges and opportunities, *Environmental Health Perspectives* 109 (1) (2001): 1–5.
17. U. Gehring, A. H Wijga, M. Brauer, et al., Traffic-related air pollution and the development of asthma and allergies during the first 8 years of life, *American Journal of Respiratory and Critical Care Medicine* 181 (6) (2010): 596–603.

18. P. Moller and S. Loft, Oxidative damage to DNA and lipids as biomarkers of exposure to air pollution, *Environmental Health Perspectives* 118 (8) (2010): 1126–1136.
19. S. Roberts, Interaction between particulate air pollution and temperature in air pollution mortality time series studies, *Environmental Research* 96 (2004): 328–337.
20. G. Albrecht, Solastalgia, a new concept in human health and identity, *Philosophy Activism Nature* 3 (2005): 41–44.
21. A. Gasparrini and B. Armstrong, The impact of heat waves on mortality, *Epidemiology* 22 (1) (2011): 68–73.
22. A. J. McMichael, Prisoners of the proximate: Loosening the constraints on epidemiology in an age of change, *American Journal of Epidemiology* 149 (1999): 887–897; M. Susser, *Thinking in the health sciences: Concepts and strategies of epidemiology* (New York: Oxford University Press, 1973).
23. W. W. Xun, A. E. Khan, E. Michael, and P. Vineis, Climate change epidemiology: Methodological challenges, *International Journal of Public Health* 55 (2009): 85–96.
24. J. Finnigan, The science of complex systems, *Australasian Science* (June 2005): 32–34.
25. N. Pearce and F. Merletti, Complexity, simplicity and epidemiology, *International Journal of Epidemiology* 35 (2006): 515–519.
26. R. B. Ness, J. S. Koopman, and M. S. Roberts, Causal system modeling in chronic disease epidemiology: A proposal, *Annals of Epidemiology* 17 (7) (2007): 564–568.
27. D. Loomis, Taking environmental epidemiology to the next level: Understanding complex systems, *Occupational and Environmental Medicine* 67 (2010): 577.

10

Solutions to the Growing Solid Waste Problem

Robert H. Friis

Introduction

The problem of how to dispose of municipal solid waste (MSW)—the term used to refer to trash or garbage—has confronted humanity since ancient times. MSW consists of materials that people discard after use. Wastes from construction sites, sludge from wastewater treatment, and industrial wastes that are nonhazardous usually are not termed MSW, although such wastes may be placed in landfills.[1] Waste has always been a by-product of human activities and can have a major adverse impact on public health.[2] Despite some recent improvements, the worldwide volume of MSW has increased with the growth of the population and a rising standard of living. During the early history of the United States, MSW often was simply deposited on the land, which was formerly abundant. During this time little thought was given to the consequences of uncontrolled dumping. Presently, with disposal sites filling to capacity and concern growing about the contribution of MSW to global warming, alternatives have been developed to cope with the threat of humanity being inundated by its own garbage. These other methods of coping with MSW include source reduction, recycling, and composting.

The history of solid waste disposal in Europe and North America spans the period from the early origins of civilization to recent developments during the present era.[3,4] Consult Figure 10.1 for an overview of the history of solid waste disposal.

198 The Praeger Handbook of Environmental Health

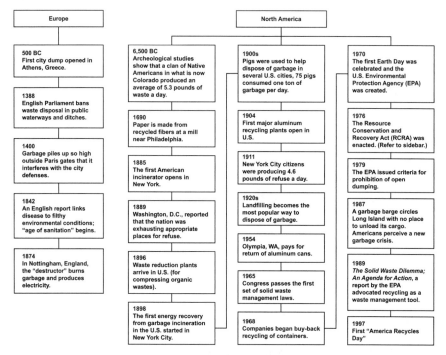

Figure 10.1 History of solid waste disposal in Europe and North America. (Data from [3], [4].)

The following section highlights some of the crucial developments in the disposal of MSW in Europe and North America:

- Europe:
  ◊ 500 BC—The first city dump opens in Athens, Greece.
  ◊ 1842—A report in England links disease to filthy environmental conditions; about this time the age of sanitation is said to have begun.
  ◊ 1874—A new technology called "the Destructor" provides the first systematic incineration of refuse in Nottingham, England. Until this time, much of the burning was accidental and caused by methane gases released by putrification of garbage.
- North America:
  ◊ 6500 BC—Native Americans in the present state of Colorado produce an average of 5.3 pounds per person of waste per day.

Solutions to the Growing Solid Waste Problem

◊ Early 1900s—Pigs are used to dispose of garbage in several cities; about seventy-five pigs can consume one ton of garbage a day. Later, public health officials discover that when pigs consume garbage, they can acquire diseases such as trichinosis. Therefore, garbage must be cooked before it is fed to swine.

◊ 1920s—Landfilling becomes the most popular way to dispose of garbage.

◊ 1965—The Solid Waste Disposal Act (SWDA) is created to improve solid waste disposal methods in the United States.

◊ 1970—An important year for waste disposal:
- The first Earth Day is celebrated.
- The U.S. Environmental Protection Agency (EPA) is created.

◊ 1976—The Resource Conservation and Recovery Act (RCRA), which amends the SWDA, is enacted; RCRA provides the legal basis for the disposal of solid and hazardous waste in the United States in a safe manner (see the sidebar for a history of RCRA).[5]

History of RCRA

The Resource Conservation and Recovery Act—commonly referred to as RCRA—is our nation's primary law governing the disposal of solid and hazardous waste. Congress passed RCRA on October 21, 1976, to address the increasing problems the nation faced from our growing volume of municipal and industrial waste. RCRA, which amended the Solid Waste Disposal Act of 1965, set national goals for

- protecting human health and the environment from the potential hazards of waste disposal,
- conserving energy and natural resources,
- reducing the amount of waste generated, and
- ensuring that wastes are managed in an environmentally sound manner.

To achieve these goals, RCRA established three distinct, yet interrelated, programs:

- The solid waste program, under RCRA Subtitle D, encourages states to develop comprehensive plans to manage nonhazardous industrial solid waste and municipal solid waste, sets criteria for municipal solid waste landfills and other solid waste disposal facilities, and prohibits the open dumping of solid waste.
- The hazardous waste program, under RCRA Subtitle C, establishes a system for controlling hazardous waste from the time it is generated until its ultimate disposal—in effect, from "cradle to grave."
- The underground storage tank (UST) program, under RCRA Subtitle I, regulates underground storage tanks containing hazardous substances and petroleum products.

RCRA banned all open dumping of waste, encouraged source reduction and recycling, and promoted the safe disposal of municipal waste. RCRA also mandated strict controls over the treatment, storage, and disposal of hazardous waste. The first RCRA regulations, "Hazardous Waste and Consolidated Permit Regulations," published in the *Federal Register* on May 19, 1980 (45 FR 33066), established the basic "cradle to grave" approach to hazardous waste management that exists today.

RCRA was amended and strengthened by Congress in November 1984 with the passing of the Federal Hazardous and Solid Waste Amendments (HSWA). These amendments to RCRA required the phasing out of land disposal of hazardous waste. Some of the other mandates of this strict law include increased enforcement authority for EPA, more stringent hazardous waste management standards, and a comprehensive underground storage tank program.

Source: Adapted and reprinted from United States Environmental Protection Agency, *History of RCRA* (2010), http://www.epa.gov/wastes/laws-regs/rcrahistory.htm (accessed April 14, 2011).

Current Status of MSW in the United States

According to the Environmental Protection Agency (EPA), the United States—residents, businesses, and institutions—produced approximately 243 million tons of MSW (before recycling) in 2009; this amount equates to 4.34 pounds of waste per person per day.[6] The per capita waste produced

Solutions to the Growing Solid Waste Problem 201

in 2009 was almost twice the amount produced in 1960 (2.68 pounds).[6] The methods for disposal of MSW include combustion, temporary storage in transfer stations, deposition in landfills, and environmentally friendly methods such as recycling.[7]

Along with the increasing amounts of solid waste that are being produced by U.S. citizens, the expense of disposing of MSW is increasing over time. For some time the eastern United States has had inadequate space for landfills. In areas of the Midwest that at one time had abundant space for waste disposal, landfills are filling up and closing.[8] Shutting down landfills has caused disposal fees, known as tipping fees, to increase substantially. Average tipping fees per ton in the Midwestern United States increased from about $25 in the late 1990s to approximately $30 in 2003.

The Global Production of MSW

The problem of waste disposal is not confined to developed countries. The less-developed areas of the world also have shown an increase in the amount of solid waste production. Although there are abundant examples, take the case of the world's two most populous countries, China and India. With respect to China, urbanization, population expansion, and increasing industrialization have caused a rapid spike in the amount of garbage produced annually. In 1980 the level of waste generation was approximately 0.50 kg/capita/day. By comparison, in 2006 the average had doubled to 0.98 kg/capita/day.[9] At the national level, the entire country of China is estimated to have produced about 200 million tons of MSW in 2002 and is projected to generate about twice as much MSW as the United States by 2030.[2] In India, which has been experiencing an improving standard of living, the amount of MSW produced is much smaller at present and is estimated to total about 42 tons of MSW each year.[10] In addition, the amount of per capita waste produced varies from one Indian city to another. Refer to Table 10.1 for a comparison of MSW generation in selected developed and developing countries. Among developed countries the volume of MSW produced (kg/capita/day) ranged from a low of 0.85 kg/capita/day (Greece) to a high of about 2.00 kg/capita/day (U.S.). The production of MSW in Greece was similar to that of Mexico. In the Southeast Asian regions, MSW produced ranged from 0.40 (Bangladesh) to 3.90 in the commercial cities of the special administrative region of Hong Kong.

TABLE 10.1
International Comparisons in MSW Generation

Name of Country	MSW Generation (kg/per capita/per day)
Developed Countries (North America, Europe, and Australia)	
USA	2.00
Australia	1.89
Canada	1.80
Ireland	1.83
Belgium	1.10
Switzerland	1.10
Spain	0.99
Italy	0.96
Greece	0.85
Developing Countries and Cities in Southeast Asian Regions	
Changging, China	1.20
Shanghai, China	0.60
Beijing, China	0.88
Hong Kong, China (residential cities)	1.17
Hong Kong, China (commercial cities)	3.90
Tokyo, Japan	1.50
Osaka, Japan	2.70
Jakarta, Indonesia	0.66
Kuala Lumpur, Malaysia	1.29
Metro Manila, Philippines	0.53
Khulana, Dhaka, Chittagong–Bangladesh	0.50
Sylhat, Bangladesh	0.40

Source: Data from R. Rajput, G. Prasad, and A. K. Chopra, Scenario of solid waste management in present Indian context, *Caspian Journal of Environmental Science* 7 (2009): 45–53.

Components of the Municipal Solid Waste Stream

Solid waste encompasses the subset of materials referred to as MSW plus other types of solid waste. Among the producers of these wastes are private residences, educational facilities, industries, and healthcare settings. The components of the MSW stream include packaging, furniture, appliances, clothing, bottles, food waste, papers, batteries, and organic materials such

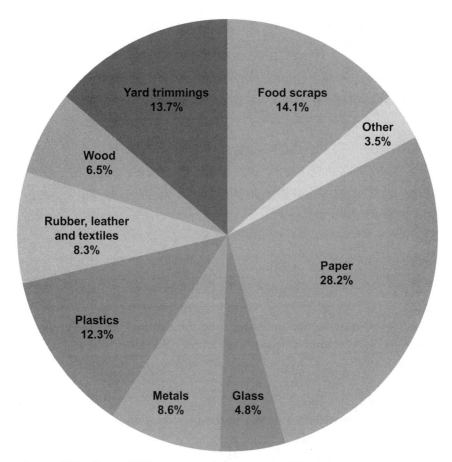

Figure 10.2 Total MSW generation (by material), 2009; 243 million tons (before recycling). (Adapted and reprinted from [6].)

as grass clippings from landscaping.[1] Figure 10.2 shows the relative proportion of materials discarded in the municipal solid waste stream. The figure shows that paper and paperboard make up the largest percentage of solid waste, about 28 percent. Three other major components of the solid waste stream are food scraps, yard trimmings, and plastics.

Solid Waste Management

Municipal solid waste disposal is an industrial enterprise that has four main dimensions: recycling, landfilling, composting, and combustion.[11] Note that when MSW decomposes in landfills or is combusted, a by-product can be the production of useful energy. This energy may be produced from methane gas that seeps from disposal sites or by combustion.

The EPA has developed a hierarchy for the management of MSW.[12] Procedures at the top of the hierarchy are favored over those at the bottom. The components of the hierarchy are as follows:

- Source reduction (or waste prevention), including reuse of products and onsite, or backyard, composting of yard trimmings.
- Recycling, including offsite, or community, composting.
- Disposal, including waste combustion (preferably with energy recovery) and landfilling (p. 11).[12]

Source Reduction

The term *source reduction* refers to "*reducing* the amount of waste created, reusing whenever possible, and then recycling whatever is left" (p. 12).[1] Through source reduction, the volume of solid waste that must be deposited in landfills is limited. Two important components of source reduction are waste reduction and waste recycling. *Waste reduction* aims to reduce the amount of waste produced at the source. *Waste recycling* refers to reuse of materials in the waste.[1]

As noted previously, as a result of source reduction, some wastes never enter disposal channels. Following are examples of source reduction methods:

- Financial incentives to homeowners, for example, PAYT, "pay-as-you-throw."[13] Municipalities that use the PAYT system require that the amount that residents pay for trash pickup be proportional to the amount of trash[14] that they generate.
  ◊ To illustrate, in a hypothetical PAYT system, residents purchase stickers for a set price such as $1 each and affix them to each trash bag being disposed of. Residents could save money by disposing of fewer bags of trash.[2] (Refer to the sidebar for more information.)
- Improved packaging designs that reduce the amount of materials that must be discarded (e.g., the use of smaller packages for products).
- Design of products, such as refillable bottles, that can be reused.
- Design of products that have longer service lives so that they will not need to be disposed of so often (e.g., longer-lasting tires for vehicles).

Pay-As-You-Throw

In communities with pay-as-you-throw programs (also known as unit pricing or variable-rate pricing), residents are charged for the collection of municipal solid waste—ordinary household trash—based on the amount they throw away. This creates a direct economic incentive to recycle more and to generate less waste.

Traditionally, residents pay for waste collection through property taxes or a fixed fee, regardless of how much—or how little—trash they generate. Pay-as-you-throw (PAYT) breaks with tradition by treating trash services just like electricity, gas, and other utilities. Households pay a variable rate depending on the amount of service they use.

Most communities with PAYT charge residents a fee for each bag or can of waste they generate. In a small number of communities, residents are billed based on the weight of their trash. Either way, these programs are simple and fair. The less individuals throw away, the less they pay. EPA supports this new approach to solid waste management because it encompasses three interrelated components that are key to successful community programs:

1. **Environmental Sustainability:** Communities with programs in place have reported significant increases in recycling and reductions in waste, due primarily to the waste reduction incentive created by PAYT. Less waste and more recycling mean that fewer natural resources need to be extracted. In addition, greenhouse gas emissions associated with the manufacture, distribution, use, and subsequent disposal of products are reduced as a result of the increased recycling and waste reduction PAYT encourages. In this way, PAYT helps slow the buildup of greenhouse gases in the earth's atmosphere that leads to global climate change. For more information on the link between solid waste and global climate change, go to EPA's Climate Change Web site.

2. **Economic Sustainability:** PAYT is an effective tool for communities struggling to cope with soaring municipal solid waste management expenses. Well-designed programs generate the revenues communities need to cover their solid waste costs, including the costs of such complementary programs as recycling and composting.

Residents benefit, too, because they have the opportunity to take control of their trash bills.

3. **Equity:** One of the most important advantages of a variable-rate program may be its inherent fairness. When the cost of managing trash is hidden in taxes or charged at a flat rate, residents who recycle and prevent waste subsidize their neighbors' wastefulness. Under PAYT, residents pay only for what they throw away.

4. EPA believes that the most successful programs bring these components together through a process of careful consideration and planning. This Web site was developed as part of EPA's ongoing efforts to provide information and tools to local officials, residents, and others interested in PAYT.

Source: Adapted and reprinted from United States Environmental Protection Agency, *Wastes—resource conservation—conservation tools—pay-as-you-throw,* http://www.epa.gov/epawaste/conserve/tools/payt/index.htm (accessed April 14, 2011).

Recycling

The EPA defines recycling (reuse) as the process of "minimizing waste generation by recovering and reprocessing usable products that might otherwise become waste (i.e., recycling of aluminum cans, paper, and bottles, etc.)."[15] In 2009, of the total of 243 million tons of MSW produced in the United States, 82 million tons were recycled, which is about a 34 percent recycling rate. This recycling rate translates to approximately 1.46 pounds of waste per person per day. The rate of recycling of MSW in the United States has increased from 6.4 percent (5.6 million tons) in 1960.[6] Table 10.2 shows the amounts and percentages of recovery though recycling of various materials. The highest recovery rates in 2009 were for the category of metals, including nonferrous metals such as lead (primarily lead from lead-acid batteries), paper and paperboard, and yard trimmings. Table 10.3 shows the recovery of materials by product category and demonstrates that the highest recovery rate among the four categories in the table was for containers and packaging. From the information shown in the tables, it may be concluded that the materials

TABLE 10.2

Generation and Recovery of Materials in MSW, 2009*

Material	Weight Generated	Weight Recovered	Recovery as Percent of Generation
Paper and paperboard	68.43	42.50	62.1%
Glass	11.78	3.00	25.5%
Metals			
Steel	15.62	5.23	33.5%
Aluminum	3.40	0.69	20.3%
Other nonferrous metals†	1.89	1.30	68.8%
Total metals	20.91	7.22	34.5%
Plastics	29.83	2.12	7.1%
Rubber and leather	7.49	1.07	14.3%
Textiles	12.73	1.90	14.9%
Wood	15.84	2.23	14.1%
Other materials	4.64	1.23	26.5%
Total materials in products	171.65	61.27	35.7%
Other wastes			
Food, other‡	34.29	0.85	2.5%
Yard trimmings	33.20	19.90	59.9%
Miscellaneous inorganic wastes	3.82	Negligible	Negligible
Total other wastes	71.31	20.75	29.1%
Total municipal solid waste	242.96	82.02	33.8%

*In millions of tons and percent of generation of each material. Includes waste from residential, commercial, and institutional sources.

† Includes lead from lead-acid batteries.

‡ Includes recovery of other MSW organics for composting.

Details might not add to totals due to rounding.

Negligible = Less than 5,000 tons or 0.05 percent.

Source: Reprinted from United States Environmental Protection Agency, *Municipal solid waste generation, recycling, and disposal in the United States: Facts and figures for 2009*, EPA-530-F-010-012 (Washington, DC: U.S. EPA, 2010).

TABLE 10.3
Generation and Recovery of Products in MSW, 2009*

Products	Weight Generated	Weight Recovered	Recovery as Percent of Generation
Durable goods			
Steel	13.34	3.72	27.9%
Aluminum	1.35	Negligible	Negligible
Other nonferrous metals†	1.89	1.30	68.8%
Glass	2.12	Negligible	Negligible
Plastics	10.65	0.40	3.8%
Rubber and leather	6.43	1.07	16.6%
Wood	5.76	Negligible	Negligible
Textiles	3.49	.044	12.6%
Other materials	1.61	1.23	76.4%
Total durable goods	46.64	8.16	17.5%
Nondurable goods			
Paper and paperboard	33.48	17.43	52.1%
Plastics	6.65	Negligible	Negligible
Rubber and leather	1.06	Negligible	Negligible
Textiles	9.00	1.46	16.2%
Other materials	3.25	Negligible	Negligible
Total nondurable goods	53.44	18.89	35.3%
Containers and packaging			
Steel	2.28	1.51	66.2%
Aluminum	1.84	0.69	37.5%
Glass	9.66	3.00	31.1%
Paper and paperboard	34.94	25.07	71.8%
Plastics	12.53	1.72	13.7%
Wood	10.08	2.23	22.1%
Other materials	0.24	Negligible	Negligible
Total containers and packaging	71.57	34.22	47.8%
Other wastes			
Food, other‡	34.29	0.85	2.5%
Yard trimmings	33.20	19.90	59.9%
Miscellaneous inorganic wastes	3.82	Negligible	Negligible
Total other wastes	71.31	20.75	29.1%
Total municipal solid waste	242.96	82.02	33.8%

*In millions of tons and percent of generation of each product. Includes waste from residential, commercial, and institutional sources.
† Includes lead from lead-acid batteries.
‡ Includes recovery of other MSW organics for composting.
Details might not add to totals due to rounding.
Negligible = Less than 5,000 tons or 0.05 percent.
Source: Reprinted from United States Environmental Protection Agency, *Municipal solid waste generation, recycling, and disposal in the United States: Facts and figures for 2009*, EPA-530-F-010-012 (Washington, DC: U.S. EPA, 2010).

Solutions to the Growing Solid Waste Problem 209

that are most amenable to recycling (and consequently are recycled most frequently) are automobile batteries, paper, and landscaping materials; automobile batteries are the products that are recycled most frequently (99+ percent of auto batteries).

Among the advantages of recycling are that it

- reduces emissions of greenhouse gases;
- prevents pollution generated by the use of new materials;
- decreases the amount of material shipped to landfills, thereby reducing the need for new landfills;
- preserves natural resources;
- opens up new manufacturing employment opportunities and increases U.S. competitiveness; and
- saves energy[16]

Examples of accomplishments achieved through recycling follow:

- Missouri (U.S.) has developed a recycling program that has resulted in diverting 41 percent of its solid waste stream.[8]
- Japan: The problem of solid waste disposal is very difficult, given the country's dense population, productive economy, and limited availability of landfill space. In April 2000 Japanese municipal governments began a recycling program for plastic and paper packaging waste.[17]
- Germany: A very effective program for waste management and recycling of waste is in place. The German city of Freiburg has developed a noteworthy system for waste management.[17]

Composting

Composting is defined as "the aerobic biological decomposition of organic materials [e.g., leaves, grass, and food scraps] to produce a stable humus-like product. . . . Biodegradation is a natural, ongoing biological process that is a common occurrence in both human-made and natural environments" (pp. 7–8).[18] Composting produces a useful material that resembles soil and can be used in gardening. According to the EPA, composting has the potential to greatly reduce the amount of materials that must be disposed of in landfills, as about one-quarter of household wastes consists of clippings from gardens and food waste. Some types of materials are not appropriate for composting. According to the EPA, some food products such as meats may yield poor-quality compost or be a magnet for vermin.

Composting breaks down organic materials through the action of physical and chemical processes. If you dig through your own backyard compost pile, you may notice that the temperature of the materials in the compost pile can become quite high. In compost piles that have been maintained properly, optimal temperatures can reach 32°–60°C (90°–140°F) as a result of microbial action.[18] These high temperatures are believed to kill off many harmful components such as pathogenic bacteria, weed seeds, and insect larvae. Nevertheless, a criticism of composting is that the heat may be insufficient to eliminate pathogenic agents present in the materials that are being composted.[19]

Various types of microorganisms are very important in composting. Other organisms such as invertebrates (e.g., beetles, sowbugs, and earthworms) aid in decomposing the materials when they have adequate oxygen and water (provided by well-designed compost bins); however, they play a lesser role in composting than do microorganisms.[18] A specific type of composting, known as vermicomposting (composting by a special kind of earthworm), puts red wigglers to work consuming food scraps and uses their castings as a rich garden fertilizer.

The state of Massachusetts has one of the most successful composting programs in the United States.[20] Yard and food waste are composted, preventing 37,500 tons of waste from entering the disposal process and saving approximately $2 million each year. The state's Department of Environmental Protection Recycling Equipment Grants Program supports the purchase of compost bins.

The island of Crete, located in the Mediterranean, has developed a noteworthy procedure for composting organic solid wastes.[21] The island has been faced with increasing desertification (gradual conversion of land into desert), exhaustion of landfill capacity for waste disposal, and a chronic lack of organic material for agricultural soils. The island operates programs for composting agricultural organic wastes (e.g., olive leaves, branches, and pressed grape skins), manure, and sewage sludge. The resulting compost has potential applications for agriculture and landscaping.

Conclusion

As a result of human activities, the worldwide volume of MSW is projected to increase dramatically in the future. Since the 1960s the per capita amount of MSW generated in the United States has increased gradually each year and has declined slightly only in the last few years. The impetus for increases in the generation of solid waste include excessive amounts of packaging to make consumer products more attractive and convenient, using a new plastic bag each time for grocery purchases instead of recycling used plastic bags, and the careless discarding of unwanted products that may have a remaining service life. Landfilling is one of the most common methods for disposal of MSW. From the worldwide perspective, overpopulation, urbanization, and conversion of land for human uses limit the availability of new disposal sites at the same time that existing sites have been filled to capacity. The volume of materials that are destined for landfill can be reduced significantly by policy initiatives that encourage reuse of materials; recycling as many materials as possible; and the use of compositing, which can produce valuable materials for agriculture. These initiatives can provide solutions to the growing solid waste problem in the United States and worldwide.

Notes

1. United States Environmental Protection Agency, Office of Solid Waste (5306P), *Municipal solid waste in the United States: 2007 facts and figures* (November 2008), EPA530-R-08-010.
2. L. Giusti, A review of waste management practices and their impact on human health, *Waste Management* 29 (2009): 2227–2239.
3. R. C. Barbalace, The history of waste: Do you want to be a garbologist? *EnvironmentalChemistry.com* (2003), http://EnvironmentalChemistry.com/yogi/environmental/wastehistory.html (accessed April 4, 2011).
4. Wisteme, *What is municipal solid waste (MSW)?* (2010), http://www.wisteme.com/question.view?targetAction=viewQuestionTab&id=3739 (accessed April 4, 2011).
5. United States Environmental Protection Agency, *History of RCRA* (2010), http://www.epa.gov/wastes/laws-regs/rcrahistory.htm (accessed April 14, 2011).

6. United States Environmental Protection Agency, *Municipal solid waste generation, recycling, and disposal in the United States: Facts and figures for 2009*, EPA-530-F-010-012 (Washington, DC: U.S. EPA, 2010).
7. United States Environmental Protection Agency, *Wastes—non-Hazardous waste—municipal solid waste*, http://www.epa.gov/epawaste/nonhaz/municipal/index.htm (accessed April 3, 2011).
8. M. Fickes, Gateway to recycling, *WasteAge* (2003), http://wasteage.com/mag/waste_gateway_recycling/index.html (accessed April 6, 2011).
9. D. Q. Zhang, S. K. Tan, and R. M. Gersberg, Municipal solid waste management in China: Status, problems and challenges, *Journal of Environmental Management* 91 (8) (2010): 1623–1633.
10. R. Rajput, G. Prasad, and A. K. Chopra, Scenario of solid waste management in present Indian context, *Caspian Journal of Environmental Science* 7 (2009): 45–53.
11. J. Carlin, Municipal solid waste profile, in *Renewable energy annual 1996*, U.S. Department of Energy, Energy Information Administration, Office of Coal, Nuclear, Electric and Alternate Fuels (Washington, DC: U.S. Department of Energy, 1996).
12. United States Environmental Protection Agency, Office of Solid Waste and Emergency Response (5305W), *Municipal solid waste in the United States: 2001 facts and figures executive summary*, EPA 530-S-03-011 (Washington, DC: U.S. EPA, 2003).
13. R. Denison, *Pay-as-you-throw: A cooling effect on climate change*, United States Environmental Protection Agency, Office of Solid Waste and Emergency Response (5306W), EPA 530-F-03-008 (Washington, DC: U.S. EPA, 2003).
14. United States Environmental Protection Agency, *Wastes—resource conservation—conservation tools—pay-as-you-throw*, http://www.epa.gov/epawaste/conserve/tools/payt/index.htm (accessed April 14, 2011).
15. United States Environmental Protection Agency, *Terms of environment: Terms beginning with "r,"* http://www.epa.gov/OCEPAterms/rterms.html (accessed April 17, 2011).
16. United States Environmental Protection Agency, *Wastes—resource conservation—reduce, reuse, recycle*, http://www.epa.gov/osw/conserve/rrr/recycle.htm (accessed April 7, 2011).

17. K. Uela and H. Koizumi, Reducing household waste: Japan learns from Germany, *Environment* 43 (2001): 20–32.
18. United States Environmental Protection Agency, Chapter 7: Composting, in *Decision maker's guide to solid waste management, volume II*, EPA-530-R-95-023 (Washington, DC: U.S. EPA, 1995).
19. G. Hamer, Solid waste treatment and disposal: Effects on public health and environmental safety, *Biotechnology Advances* 22 (2003): 71–79.
20. United States Environmental Protection Agency, *Wastes—information resources: Massachusetts makes composting a way of life*, http://www.epa.gov/osw/inforesources/news/2003news/11-comp.htm (accessed April 17, 2011).
21. T. Manios, The composting potential of different organic solid wastes: Experience from the island of Crete, *Environment International* 29 (2004): 1079–1089.

11

Population Trends and the Environment

Colin D. Butler

Introduction

It is obvious that humans depend on the environment that exists here on Earth, in the habitable "Goldilocks" zone around our sun. Indeed, the region of the galaxy thought suitable to form terrestrial planets and old enough to allow biological evolution of complex multicellular life is comparatively limited.[1] This nurturing milieu of Earth, atmosphere, and solar energy provides space, breathable air, liquid water, food, energy, and other resources necessary, useful, and desirable for humanity. Human life also depends on nonhuman life-forms. Not only did humans evolve from earlier forms of life, but all of the food we eat was once alive. Many other species contribute to "ecosystem services" that benefit humans.[2] Less obviously, many microscopic forms of life coexist with, on, and within humans (e.g., in the human gut); though some of these are harmful, many are benign, but the majority may be symbiotic and thus beneficial.

As humans have increased in number (see Figure 11.1), so too has their impact on the environment.[3] This is true even if additional people live in poverty. Poor people still require food, clothing, housing, and fuel. However, the environmental impact of wealthy people is much greater, whether one considers diet (higher on the food chain, greater wastage of calories), housing (quality and size), or energy usage (electricity, travel). Technological improvements (e.g., electricity powered by renewable energy, the driving of hybrid electric cars) only partially alleviate this impact. Increased efficiency is lowered by Jevons paradox, the tendency

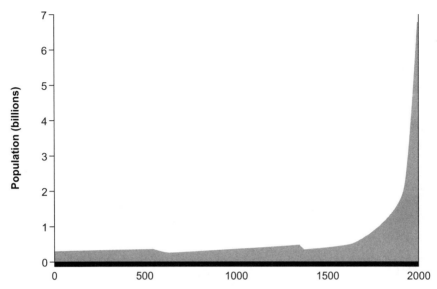

Figure 11.1 Global population: 0-2000 CE. (Author, based on datasets obtained at http://esa.un.org/UNPP/ and http://www.digitalsurvivors.com/archives/World%20Population%20Estimates%20Interpolated%20and%20Averaged.pdf.)

to increase the use of products if they become more efficient (and hence often less costly).

Global human population is now approaching seven billion and continues to climb by a million people every few days, as it has done now for several decades. Between 2,000 years ago and AD 1650, global population increased by only 200 million, during which time two serious setbacks occurred, the Justinian plague in the sixth century and the Black Death in the fourteenth century. In 1900 global population was only 1.6 billion; it is now almost 7 billion. The rate of global population growth peaked in the late 1960s, but in absolute terms it is still very high, increasing by at least 70 million a year. Almost all of this growth is among poor people in low-income countries.

In comparison to the size of a person, the surface of the earth is enormous. The fact that surface area is fixed suggests that the expansion in human numbers will one day be forced to slow. However, well before space limitations force any such tapering, many resources upon which humans depend will become scarce, depleted, or degraded. This will necessitate radically new ways of life or a slower rate of population increase. It may even cause a steep decline in global population size.

A conscious human understanding of the relationship between population and environment is likely to be many thousands of years old and to far predate written records. This is difficult to prove, but considerable evidence can be detected. Writing consistent with this understanding survives from ancient Babylon, Rome, Greece, and China.[4] Ecclesiastes (5:11) records that "when goods increase they are increased that eat them."[5] In China, Hung Liang-Chi (1744–1809), independently of Thomas Malthus (1766–1834), wrote about population outstripping food supply, survival of the fittest, and reliance on the natural "checks" of flood, drought, plague, pestilence, and warfare to limit population growth.[6]

An unconscious understanding of this relationship can be deduced by consideration of the phenomena of migration and ancient forms of fertility restriction. Hundreds of thousands of years ago human ancestors and species related to humans, such as Neanderthals, migrated from Africa. Today, large-scale human migrations continue, including from Africa. The reasons for ancient and modern migration are easily understandable: people were and are looking for opportunity and reward, and they thought, hoped, and believed that this might exist over the horizon. Often they were correct.

Written records survive from comparatively recent times of abundant food species available to humans, especially in marine and coastal areas, lightly populated by modern standards. For example, stocks of cod were so rich and dense in some parts of the North Atlantic that they seemed as though they could be walked on.[7] The paleontologist Tim Flannery describes, but also laments, how human populations often mismanaged and even squandered abundant resources upon arriving in an uninhabited ecosystem. A good example is the moa, a large flightless bird that existed by the thousands when Polynesians first migrated to Aoetoroa (New Zealand) around 1200.[8] Within a hundred years humans were well-established in that country, but the moa was extinct. Indeed, throughout the Pacific, many species of birds perished, fairly soon after human colonization.[9] Flannery characterizes behavior such as overhunting of the moa as "future eating," because such profligacy reduced the resources available to future human generations.[8]

In some cases local elimination of species may have been intended, such as the retreat of large carnivores that threatened humans and their domesticated animals. In other cases, such as the "megafauna," extinction may have been unintended but still inevitable, given the growing sophistication of hunting and the fragility and scale of the environmental resources required to support them.

Nonetheless, future historians are likely to lament the wastefulness of our own time, especially in the way that we are consuming fossil fuels and several other nonrenewable resources.[10] We are also critically reducing biodiversity (including highly prized species used for food, such as fish)[2] and polluting, to the cost of our descendants, the atmosphere and oceans, with greenhouse gases, particulate matter, and other toxic substances such as mercury, plastics, and persistent organic pollutants.[11]

The Debate: Are Humans the Exceptional Animal?

Readers with backgrounds in ecology, geography, or anthropology may find the introduction above entirely unexceptionable. In fact, debate about the significance of the human impact on the environment has become in recent decades extremely contentious, especially in the social sciences. For example, in 1982 two demographers summarized the issue of population in the then recent demography literature by stating: "From a high point some 10–15 years ago, intellectual concern about population has steadily waned to a position where it falls now somewhere between ocean mining and acid rain."[12]

For much of the following two decades a widespread view prevailed, not only in demography but also in economics, that the issues raised by those concerned about limits to population and economic growth were invalid. A representative view was expressed in *The Economist*:

> It is argued that predictions of ecological doom, including recent ones, have such a terrible track record that people should take them with pinches of salt instead of lapping them up with relish. For reasons of their own, pressure groups, journalists and fame-seekers will no doubt continue to peddle ecological catastrophes at an undiminishing speed. These people, oddly, appear to think that having been invariably wrong in the past makes them more likely to be right in the future. The rest of us might do better to recall, when warned of the next doomsday, what ever became of the last one.[13]

This lessening of anxiety permeated public opinion. A global backlash flared against family planning, also fueled by the clumsy, corrupt, and coercive practices of many "family planners."[14] Yet in the scientific literature, apprehension about the danger of too many people abated only slightly. Some writers continued to warn that a high rate of population growth, especially among poor and badly governed nations, was harmful to economic development,[15] although this debate was far less evident after 1980 compared to its heyday in the 1950s.[16, 17]

Perhaps the most significant evidence for this ongoing fear was that in 1992 more than 1,700 leading scientists, including over 100 Nobel Laureates, signed the *World Scientists' Warning to Humanity*, which argued that human population growth threatened the viability of the global environmental support mechanisms for humanity and called for urgent action to slow human population growth.[18] In contrast to the urgency of this warning and to the stature of the signatories, the paucity of press attention paid to it was shameful, revealing, and unsettling.

In 1993 fifty-eight science academies issued a similar statement, nominating human population growth as a critical issue if civilization is to flourish. The African Academy of Science dissented:

> Whether or not the Earth is finite will depend on the extent to which science and technology is able to transform the resources available for humanity. There is only one earth—yes, but the potential for transforming it is *not necessarily finite* (emphasis added).[19]

Yet in 1994 lethal genocidal conflict erupted in Rwanda, then (as now) the most densely populated nation in Africa. Within a few months about 800,000 people were killed, representing almost 10 percent of the Rwandan population.[20] Though many analysts have focused on the rivalry between the two main ethnic groups (Hutu and Tutsis) as the "cause" of this conflict, that is a simplistic explanation. A perfect storm occurred to ignite this genocide. Land scarcity in Rwanda[21] combined with a scarcity of urban employment opportunities and declining prices for its major export, coffee.[22] Unlike many developing countries, Rwanda received very few remittances (i.e., foreign exchange transferred home by Rwandans working internationally). The high fertility rates in this largely Catholic nation led to a "bulge" of youth,[23] many of whom were unemployed, unmarried, and squashed together in Rwanda's main city, Kigali.

These young men became the main implementers of the genocide. It is widely accepted that male youth are the most violent of any human age group, but of course youth bulges elsewhere have not led to conflict on any scale similar to that of Rwanda at that time. Without excusing the "genociders" for their collective responsibility, the Rwandan butchery can also be viewed as a systemic phenomenon, an example of an "irruptive" population trajectory. This phenomenon is well known in ecology, a pattern of a comparatively steep rise in population followed by a sharp decline. This was first described for some herbivores in ecosystems that lack predators, such as wolves.[24] It is also common for some invertebrate populations, especially for introduced species.[25]

However, the possibility and application of irruption to human societies and populations is far less accepted, for several reasons. One such reason is that most humans do not like to consider themselves as animals. We prefer to think of ourselves as a special case, liberated from the biological constraints that may have burdened our ancestors. Our collective capacity to reason, to invent, and to migrate helps to disguise and hide human dependence upon environmental (including ecological) resources. But this dependency remains absolutely fundamental and unchangeable. Indeed, the more we collectively assert our independence from nature, the more vulnerable we are.

Many authors point to a deep human capacity for denial, perhaps one that is "hard-wired."[26] Such denial is also evident in the human capacity to ignore the suffering of people who are far away in distance or time or unlike us in other ways. However, humans also have a disconcerting capacity to become conditioned to ignore the suffering of people close to them;[27] this is unsettling in the context of the future.

Many other examples exist of conflict among humans that have a basis partially derived from resource scarcity; in fact, some might argue that it is hard to find situations of human conflict that do *not* have an environmental basis, whether the Israeli–Palestinian struggle,[28] Darfur (Sudan),[29] the 2003 invasion of Iraq, or the German invasion of Russia in World War II, when the Nazis desperately strove to secure the oilfields around Stalingrad.

On the other hand, a vigorous literature denies any substantial relationship between resources and conflict, instead focusing on political, religious, and ethnic elements that are claimed to be unrelated to control over resources. A prominent recent example of this disconnect is the 2003 invasion of Iraq, when the U.S. government, and some of its allies, strenuously and repeatedly denied any oil-related motivation for the invasion.

The History of Contraception

In recent history, and perhaps even now, there has been a widespread perception that contraception is a relatively recent addition to human knowledge.[30] In fact, there is substantial evidence of ancient forms of restricting fertility, that is, well before the era of oral contraception, and even before widespread knowledge of the "fatal secrets" (including withdrawal and condoms) that underpinned the first marked decline in fertility in modern Europe, observed in France in the eighteenth century.[31]

Coitus interruptus was not invented by the French. There are numerous references to this practice in Jewish, Christian, and Islamic texts, but this method appears to have lost favor in the early Christian period. Mohammed approved the use of *al-azl*, mentioning that the man's wife should also give her permission.[32] Its practice was widespread in Europe, though officially frowned upon. To this day it retains its Latin name, at least in English. Herbs have also been used for millennia, both as contraceptives and to induce abortion, though with varying success.[32, 33]

Many customs have been followed that do not rely on pills, potions, or douches. In the seventeenth century in England (where excellent records survive), the average age of marriage reached thirty, and it was higher among poorer than wealthy families.[34] Western societies had no monopoly on spacing out and sparing children. The Koran recognizes that prolonged lactation delays fertility.[32] More than 1,000 years after that book was written, Charles Darwin observed that prolonged breastfeeding likely contributed to small families.[32] But where birth could not be prevented, infanticide and neglect were common.[32]

The anthropologist Virginia Abernethy describes a variety of customs and practices that helped to regulate human populations, including in isolated valleys of New Guinea, keeping the populations within the limits of "human carrying capacity." In some valleys with a comparative scarcity of people, widows could easily marry; in others, with a comparative oversupply of people, various taboos limited opportunities for sexual intercourse.[35] Abernethy also describes how the Roman Empire, under Constantine (the first Christian emperor), turned away from pronatalist legislation, at a time when unemployment was rising, agricultural production was falling, and there was a growing sense that "the world was full."[35] In many parts of Africa, prolonged breastfeeding (up to four years among the Bushmen), postpartum abstinence, and pathological sterility modified fertility.[36] In parts of the Himalayas and Tibet, polyandry has long been practiced, another custom that reduces fertility.

Perhaps the most extraordinary method of reducing human fertility is sub-incision, a traditional ritual practice long observed among adult indigenous men in the most arid and harsh environment of Australia. Early European commentators developed rather fanciful psychoanalytic interpretations for this custom,[37] and there is no clear evidence that indigenous men understood its purpose, beyond that of bonding and initiation. However, it is almost certain that sub-incision had a powerful contraceptive effect, by causing an artificial hypospadius (a condition in which the urethra opens along the shaft of the penis).[38] The coincidence of this custom with the driest part of Australia provides very strong circumstantial evidence that sub-incision co-evolved in the human population as a mechanism to limit human environmental impacts and to maintain prolonged and tolerable human existence.[39]

Malthus, the Green Revolution, and the Cornucopians

In the West the debate that concerns Limits to Growth[40] and the tension between population and food supply is often summarized as Malthusianism, an eponymous term derived from Thomas Robert Malthus. However, Malthus was but one of many intellectuals who recognized the general principle. Western forebears of Malthus included Benjamin Franklin, Adam Smith, and the pioneering political scientist Giovanni Botero (1544–1617). Similar principles were recognized by such non-Westerners as Japanese, Chinese, and Arabic scholars.[6]

Simply expressed, the principle that often bears Malthus's name states that population growth is constrained by the availability of food and that, in the absence of deliberate checks on population growth (such as delayed marriage or the use of contraception), other "checks" will occur, like famine, epidemic. or war. Though Malthus appears to have referred only to food, his thinking can be applied to resources in general, such as land, water, energy, and other raw materials required to grow and distribute food. Caldwell notes that Jevons, in the mid-nineteenth century, indeed did extend the ideas of Malthus beyond food.[41]

Such thinking has long been controversial in both West and East. Many nations saw power and prestige flowing from large populations,

Population Trends and the Environment 223

irrespective of the health, wealth, and vitality of those populations. Mao Tse Tung, the first Communist ruler of China, held such views,[42] at least prior to the terrible famine from 1959 to 1962, in which about 30 million people perished.[43] Remnants of this "mercantile" thinking persist today. For example, Ugandan President Museveni is on record as recently calling for a larger Ugandan population, which he equates with success.[44] Yet such high population growth in Uganda and some other countries of East Africa is likely to delay achievement of the Millennium Development Goals[45, 46] and may, if it continues long enough, lead to other examples of irruptive population trajectories.

Even in his day, Malthus was controversial. Malthusian thinking influenced British economic and population policy, including its response to the Irish famine of the hungry 1840s. Marx was very critical of Malthus, starting a long tradition of suspicion among the Left of any advocate or policy that stresses population rather than distribution or production. Malthusian thinking began to be challenged in Europe as early as the "optimistic 1860s" but persisted far longer in less-industrialized, low-income, and famine-prone countries such as India.[47]

However, support for Malthus periodically strengthened, including from Lord Maynard Keynes, the most influential economist of his period, from the 1930s until his death soon after he had cofounded the World Bank. Bowen, defending Malthus in 1930, wrote: "It did not occur to those who contributed to this torrent of abuse that a fire which could not be put out must be the very fire of truth itself."[48] A chief motivation used by Hitler for the expansion of the German Reich was his internally popular call for more *Lebensraum*, or living space. Hitler also craved oil security. However, Hitler's policies had two fatal flaws. The first is that *Lebensraum* could only be acquired at the expense of the living space of other people, such as the Poles and the Russians. The second flaw is that, in fact, the resources available in Germany in the late 1930s were adequate for a relatively high standard of living.

Although the expansion of the European empires preceding the Third Reich was not justified in terms of *Lebensraum* or carrying capacity, a concept that had not been applied at that time to human populations,[24] there should be little doubt that territorial expansion and conquest both provided a relief valve for millions of Europeans and enabled the appropriation of vast extra-European resources from the colonies. This contributed to a rapid increase in European living standards.

In the first decades after World War II, concern about the Malthusian check of famine rose. In 1959 the then comparatively new Food and Agricultural Organization of the UN (FAO) warned that population was outstripping food. Such concerns were widely held, not only by Paul Ehrlich (whose book title, *The Population Bomb*,[49] was taken by permission from a pamphlet earlier published by the heir of the Dixie Cup fortune, William Draper[14]), but also by U.S. President Lyndon B. Johnson and many other writers.

Yet as global population increase crested (as a proportion) at just over 2 percent in 1969, relief was in sight, in the form of the Green Revolution. Though much attacked by its critics, the Green Revolution led to a remarkable increase in the productivity of much cultivated land area, for example by the use of new strains such as dwarf wheat. It is true that these new crops were comparatively homogenous, fertilizer-hungry, and pesticide-reliant. There is also no doubt that some human actors used the Green Revolution to increase inequality, for example, by encouraging indebtedness and perhaps by overcharging for fertilizer. Further, the enormous and rapid increase in per capita food supply did not eliminate global undernutrition, though it did lay the foundation for a reduction in both proportion and absolute terms, which continued until about 2000. But the failure of global society to achieve a steeper and more substantial decline in hunger is certainly not the fault of the technologies used in the Green Revolution, but of the failure of numerous human actors to collectively remedy the numerous other factors that are associated with and help to perpetuate inequality. A co-factor for the failure to better capitalize on the Green Revolution was the continued high rate of population increase, especially in the 1980s, the lost "decade of development." The lagging demographic transition not only reduced the global per person abundance of food but also hindered economic development in low-income countries,[46] of which Rwanda is the most tragic example.

In 1970 the agricultural scientist Norman Borlaug was awarded the Nobel Prize for Peace for his seminal role in fostering the Green Revolution. Borlaug (later a signatory of the *World Scientists' Warning to Humanity*) was well aware of the consequences of rapid population growth. In his acceptance speech, he warned:

> The Green Revolution has won a *temporary* success in man's war against hunger and deprivation; it has given man a breathing space. If

fully implemented, the revolution can provide sufficient food for sustenance *during the next three decades*. But the frightening power of human reproduction must also be curbed; otherwise the successes of The Green Revolution will be ephemeral only. (emphasis added)[50]

It is sobering to realize that in late 1984, almost midway through the period of respite forecast by Borlaug, President Ronald Reagan became the first U.S. president in a series to specifically deny the importance of population.[51] The Catholic U.S. President John F. Kennedy has been credited with advancing the global agenda to slow population growth in the early 1960s, particularly through U.S. support for key UN resolutions on population.[52] Between Kennedy and Reagan, U.S. Presidents Johnson, Nixon, and Carter were explicitly Malthusian in their outlook. In 1968 Johnson authorized the emergency shipment of grain to India, then facing a famine, on condition that India increase its family planning program.[53]

Toward a Theory for Human Carrying Capacity

The science of ecology largely concerns the interaction of many species, a mixture of predators and prey, all sharing different degrees of competition and cooperation in the context of always limited resources. All ecologists recognize that there are limits to the growth of any one species and also of the total biomass supported by any area and climate. Yet in ecology, the theory of carrying capacity has fallen from favor, perhaps because more precision was anticipated from the concept than could be delivered.[24] Ideas about human carrying capacity (the maximum population supportable by a region for an indefinite period) are even more contested, not least because the concept can and has been used to justify coercion,[24] invasion, and "ethnic cleansing." Nevertheless, problems with the precise definition of human carrying capacity do not justify abandonment of the entire concept. This is because its denial would obviate limits to population growth, which is untenable, despite legitimate debate concerning the proximity of those limits.

One way to conceptualize global human carrying capacity is as an emergent phenomenon arising from the interaction of different forms of wealth. Humans will remain perpetually in need of resources from the physical environment, such as food, water, and energy. But people are not ciphers, programmable to live in perfect harmony, irrespective of their social milieu. Though some groups (such as the citizens of Hong Kong or Tokyo) are able to live largely peaceful, cooperative lives in intense proximity, these forms of group behavior cannot easily or quickly be transplanted to the entire globe.

Ample evidence for this includes the high fraction of wealth used for military purposes, largely insulated from spending cuts even during recession. History teaches that war and conflict are part of the human condition. Natural capital (both nonrenewable resources such as fossil fuels or uranium and renewable resources such as forests, water, and fisheries), and the degree of social cohesion are important determinants for the human carrying capacity of any region. However, the resources that generate human carrying capacity can be used not only to support a given number of humans, but also to provide wealth for any given population size. For example, the carrying capacity of Rwanda at the time of the genocide in 1994 would have been higher if there had been only one main ethnic group. But total homogeneity would not have indefinitely postponed the Rwandan calamity. The ethnically homogenous (and also largely Catholic) country of Ireland suffered an even greater proportional population crash between 1848 and 1852, following the cool, wet summer of 1848, when the potato blight *Phytopthora infestans* flourished. A more plausible mechanism by which genocide may have been deferred in Rwanda is via an earlier introduction of education, health care, and family planning supported by the Rwandan Catholic Church and promoted by good government. This in turn is likely to have facilitated economic takeoff, generating a virtuous circle leading to the establishment of export industries, more tourism, and increased skills enabling significant remittances. Instead, the people of Rwanda were effectively trapped, with diminishing marginal productivity that heralded the barbarity of its genocide.

In addition to the foundations of natural and social capital, human carrying capacity can be expanded by ingenuity and by inherited wealth: the quality and quantity of built and financial resources. Another example concerns the Caribbean island of Hispaniola, of which about two-thirds is allocated to the Dominican Republic, with the balance allocated to the more densely populated Haiti. The average income in the larger eastern nation is about seven times that of Haiti. There are, of course, multiple interlocking reasons for this difference, but an important component is the rate of population growth.

With a sufficient expansion in resources (such as an infusion of capital), both nations could support more people at a higher living standard. The tiny, natural capital-poor island of Hong Kong shows that ingenuity, cooperation, and high human, built, and financial resources can support and sustain a densely settled population at a high living standard. However, Hong Kong cannot be extrapolated to the entire world. Many of its natural resources are supplied from beyond its boundaries, purchased with capital generated by the sale of manufactured goods and of ideas and services that are exported. Similarly, the wealth and total population size of Europe expanded in the nineteenth century, due in part to the European import of natural capital from its numerous colonies; again a strategy that could not be applied globally.

The concept of human carrying capacity is necessarily imprecise; it does not mean that any simple number exists for how many people the earth could support. Such an answer depends not only on the human, financial, natural, social, and built resources available to any society, but also on any particular society's preference for wealth, numbers, and population growth rates.

A clearer understanding of these issues will reduce the chance that nations fruitlessly pursue wealth via a high birth rate. It also reduces the chance of population crashes, not only on a local and regional scale, but perhaps even on a global or subglobal scale.

Conclusion

Doctors, like ecologists, also recognize that endless growth is impossible. Indeed, trends in that direction, beyond an optimum, are harmful to a patient, whether as obesity or cancer. Some of our most eminent economists, including John Stuart Mill (1806–1873), have recognized that in the future limits to growth will impose a steady state economy.[54] Yet the prospect of Limits to Growth continues to be vigorously resisted. Its denial and that of the fundamental dependency of human well-being on limited environmental and ecological resources is every day leading the world closer to an eco-social precipice. Climate change, costlier energy, and the consequences of food insecurity and inequality could well push us over this precipice.

If we continue on this pathway, we risk a massive eruptive population trajectory, manifesting via a "blowback" world,[55] in which nuclear-armed terrorist groups and rogue states strike out, leading to confusion in the "fog of war." A more plausible future, however, is the gradual intensification of our existing "enclave world," in which "seas" of order increasingly strengthen the barriers—financial, legal, military, and psychological—against the "islands" of disorder. The size and influence of these islands are continuing to grow, whether manifest as pirates from Somalia, favelas in Rio de Janeiro, terrorists from Afghanistan, or train-derailers in the growing chaos and insurgency of northeast India.

To prevent these futures, we need to work toward a genuinely fairer world, in which we "muddle through" to a viable future. This is not a utopian vision of "Health for All," and the term "muddle through" should not be taken to mean a relaxed, laissez faire pathway. But it recognizes that, even if we need a massive technological campaign to accelerate the technological and sustainability transition, we will still need considerable agility and some luck. Irruptions are inevitable but need not become overwhelming. Complacency will, however, be a fatal error.

Notes

1. C. H. Lineweaver, Y. Fenner, and B. K. Gibson, The galactic habitable zone and the age distribution of complex life in the Milky Way, *Science* 303 (2004): 59–62.
2. Millennium Ecosystem Assessment, Living beyond our means, *Natural assets and human well-being* (Washington, DC: Island Press, 2005).
3. P. R. Ehrlich and J. P. Holdren, Impact of population growth, *Science* 171 (1971): 1212–1217.
4. J. E. Cohen, *How many people can the earth support?* (New York: WW Norton and Co., 1995).
5. P. Demeny, Demography and the limits to growth, *Population and Development Review* 14 (1988): 213–244.
6. L. Silberman, Hung Liang-Chi: A Chinese Malthus, *Population Studies* 13 (1960): 257–265.
7. M. Harris, *Lament for an ocean: The collapse of the Atlantic cod fishery; a true crime story* (Toronto: M&S, 1999).
8. T. F. Flannery, *The future eaters* (Melbourne: Reed Books, 1994).

9. J. Diamond, Easter Island revisited, *Science* 317 (2007): 1692–1694.
10. C. A. S. Hall, W. John, and J. Day, Revisiting the limits to growth after peak oil, *American Scientist* 97 (2009): 230–237.
11. J. Rockström, W. Steffen, K. Noone, et al., A safe operating space for humanity, *Nature* 461 (2009): 472–475.
12. G. McNicoll and M. Nag, Population growth: Current issues and strategies, *Population and Development Review* 8 (1982): 121–139.
13. Anonymous, Plenty of gloom, *The Economist* 345 (1997): 19–21.
14. J. Kasun, *The war against population: The economics and ideology of population control* (San Francisco: Ignatius Press, 1988).
15. M. King, Health is a sustainable state, *The Lancet* 336 (1990): 664–667.
16. A. C. Kelley, The population debate in historical perspective: Revisionism revised, in *Population matters: Demographic change, economic growth, and poverty in the developing world*, edited by N. Birdsall, A. C. Kelley, and S. W. Sinding (Oxford; New York: Oxford University Press, 2001), 24–54.
17. R. R. Nelson, A theory of the low-level equilibrium trap in underdeveloped economies, *American Economic Review* 46 (1956): 894–908.
18. Union of Concerned Scientists, *World scientists' warning to humanity* (Cambridge, MA: Union of Concerned Scientists, 1992).
19. African Academy of Sciences, The African Academy of Sciences on population, *Population and Development Review* 20 (1994): 238–239.
20. P. Uvin, Tragedy in Rwanda: The political ecology of conflict, *Environment* 38 (1996): 7–15.
21. C. André and J.-P. Platteau, Land relations under unbearable stress: Rwanda caught in the Malthusian trap, *Journal of Economic Behavior and Organization* 34 (1998): 1–47.
22. M. Chossudovsky, *Economic genocide in Rwanda: The globalisation of poverty* (Penang: Third World Network, 1997), 111–122.
23. C. G. Mesquida and N. I. Wiener, Human collective aggression: A behavioral ecology perspective, *Ethology and Sociobiology* 17 (1996): 247–262.
24. N. F. Sayre, The genesis, history, and limits of carrying capacity, *Annals of the Association of American Geographers* 98 (2008): 120–134.

25. D. Simberloff and L. Gibbons, Now you see them, now you don't!—population crashes of established introduced species, *Biological Invasions* 6 (2004): 161–172.
26. R. Ornstein and P. Ehrlich, *New world, new mind* (London: Methuen, 1989).
27. S. Cohen, *States of denial: Knowing about atrocities and suffering* (Cambridge, UK: Polity, 2001).
28. D. E. Orenstein, Population growth and environmental impact: Ideology and academic discourse in Israel, *Population and Environment* 26 (2004): 40–60.
29. United Nations Environment Programme, *Sudan: Post-conflict environmental assessment* (Nairobi: UNEP, 2007), http://postconflict.unep.ch/publications/UNEP_Sudan.pdf (accessed October 13, 2011).
30. M. Raymond, The birth of contraception, *Nature* 444 (2006): 685.
31. E. van de Walle and H. V. Muhsam, Fatal secrets and the French fertility transition, *Population and Development Review* 21 (1995): 261–279.
32. M. Potts and M. Campbell, History of contraception, in *Gynecology and obstetrics,* vol. 6, edited by J. J. Sciarra (Philadelphia: Lippincott Williams and Wilkins, 2002), 1–27.
33. J. M. Riddle, *Contraception and abortion from the ancient world to the Renaissance* (Cambridge, MA: Harvard University Press, 1992).
34. E. A. Wrigley and R. S. Schofield, T*he population history of England 1541–1871* (Cambridge MA: Harvard University Press, 1981).
35. V. D. Abernethy, *Population politics: The choices that shape our future* (New York: Plenum Press, 1993).
36. J. Bongaarts, O. Frank, and R. Lesthaeghe, The proximate determinants of fertility in sub-Saharan Africa, *Population and Development Review* 10 (1984): 511–537.
37. P. Singer and D. E. Desole, The Australian subincision ceremony reconsidered: Vaginal envy or kangaroo bifid penis envy, *American Anthropologist* 69 (1967): 355–358.
38. D. A. M. Gebbie, *Reproductive anthropology—Descent through woman* (Chichester, UK; New York: John Wiley & Sons, 1981).
39. J. B. Birdsell, Some environmental and cultural factors influencing the structuring of Australian aboriginal populations, *American Naturalist* 87 (1953): 171–207.

40. D. Meadows, D. Meadows, J. Randers, et al., *The limits to growth* (New York: Universe Books, 1972).
41. J. C. Caldwell, The global fertility transition: The need for a unifying theory, *Population and Development Review* 23 (1997): 803–812.
42. W. D. Borrie, China's population struggle: Demographic decisions of the People's Republic 1949–1969 (review), *Demography* 11 (1974): 702–705.
43. J. Becker, *Hungry ghosts: China's secret famine* (New York: Henry Holt, 1996).
44. W. Wakabi, Population growth continues to drive up poverty in Uganda, *The Lancet* 367 (2006): 558.
45. A. C. Ezeh, B. U. Mberu, and J. O. Emina, Stall in fertility decline in Eastern African countries: Regional analysis of patterns, determinants and implications, *Philosophical Transactions of the Royal Society B* 364 (2009): 2991–3007.
46. M. Campbell, J. Cleland, A. Ezeh, et al., Return of the population growth factor, *Science* 315 (2007): 1501–1502.
47. J. C. Caldwell, Malthus and the less developed world: The pivotal role of India, *Population and Development Review* 24 (1998): 675–696.
48. E. Bowen, Malthus, a re-evaluation, *The Scientific Monthly* 30 (1930): 465–471.
49. P. R. Ehrlich, *The population bomb* (London: Ballantyne, 1968).
50. D. Tribe, Feeding and greening the world (Wallingford, UK: CAB International in association with the Crawford Fund for International Agricultural Research, 1994).
51. J. L. Finkle and B. Crane, Ideology and politics at Mexico City: The United States at the 1984 international conference on population, *Population and Development Review* 11 (1985): 1–28.
52. J. L. Finkle and C. A. McIntosh, The new politics of population: Conflict and consensus in family planning, *Population and Development Review* 20 (1994): 3–34.
53. J. Califano, *Governing America* (New York: Simon & Schuster, 1981).
54. H. Daly, *Beyond growth* (Boston: Beacon Press, 1996).
55. C. Johnson, *Blowback: The costs and consequences of American empire* (New York: Metropolitan Books, 2000).

12

International Environmental Law

Margaret Alkon

Introduction

As international law regimes become more far-reaching and intertwined, environmental health professionals need to understand how international law is created to advocate for environmental health priorities. This chapter discusses basic features of international agreements affecting environmental health, including new directions taken in response to tensions between trade agreements and international agreements aimed at protecting human health and the environment. It also discusses several important characteristics of the U.S. approach to international agreements. Health issues have a prominent place in foreign policy agendas,[1] and global health law can include a wide range of issues.[2] The goal of this chapter is to provide a basic orientation to international environmental agreements, since environmental health professionals increasingly need to influence and monitor the multitude of developments in international law.[3]

Types of Treaties

The most basic building block of international law is an agreement between two nations. Bilateral environmental treaties dominated until the twentieth century and remain important.[4] For example, the United States and Canada have over a century of cooperation on water management built on the Boundary Waters Treaty (BWT) of 1909. The need to coordinate issues of navigation fueled the BWT, but the agreement also contains an early articulation of the important principle that shared waters should not be polluted on one side to the injury of health or property on the other.[5, 6] When pollution and ecosystem protection emerged as critical areas for cooperation to restore and maintain the chemical, physical, and biological integrity of the Great Lakes Basin ecosystem, Canada and the United States negotiated and then periodically amended the Great Lakes Water Quality Agreement (GLWQA) of 1978. The GLWQA includes annexes addressing specific issues such as atmospheric deposition of toxic pollutants, development and implementation of remedial action plans, and management plans to control critical pollutants.[5] The GLWQA facilitates stakeholder input and coordination among states, provinces, and the line agencies of Canada and the United States.

Many of the bilateral environmental agreements that the United States negotiates facilitate coordination between government agencies. The 2007 agreement on food safety between China's General Administration of Quality Supervision, Inspection and Quarantine and the U.S. Department of Health and Human Services concerning inspections of food imports from China is an example of such coordination, although these efforts to harmonize government agency activities can be subject to criticism.[7]

Also important are regional treaties involving nations that share a geographically distinct environmental resource, such as the Barcelona Convention for the Protection of the Marine Environment and the Coastal Region of the Mediterranean.[8] The North American Agreement on Environmental Cooperation, signed by the United States, Mexico, and Canada as a side agreement to the North American Free Trade Agreement (NAFTA), is another example of a regional environmental treaty dealing with environmental health issues in a specific geographic region.[9]

Modern bilateral and regional agreements are negotiated against a backdrop of multinational environmental treaties that have variable levels of participation and influence. These multilateral environmental treaties address a wide range of issues, including global climate change;[10] fisheries;[11] stratospheric ozone protection;[12] transboundary air pollution;[13] loss of biodiversity;[14] transboundary transportation and disposal of hazardous wastes;[15] prior informed consent;[16] environmental assessment;[17] desertification;[18] protection and use of transboundary watercourses and international lakes;[19] oceans;[20] and even access to information, public participation, and environmental justice.[21] Some of these treaties have near-universal participation, while others articulate important values to the environmental health community but have as parties only a subset of nations in the world community. So what is a treaty or international agreement?

What Is a Treaty According to International Law?

Effective advocacy begins with understanding the difference between a treaty and the plethora of other documents that mark milestones—or impasses—on the international environmental law landscape. Of the major documents associated with the United Nations Conference on Environment and Development, held in Rio de Janeiro in 1992 (the so-called "Earth Summit"),[22] two are multilateral treaties, while three are "soft law."

International law considers a treaty to be any legally binding agreement between nations. An international agreement is generally considered to be a treaty and binding on the parties to that treaty if it meets four criteria:

1. The parties intend the agreement to be legally binding, and the agreement is subject to international law.
2. The agreement deals with significant matters.
3. The agreement clearly and specifically describes the legal obligations of the parties.
4. The form indicates an intention to conclude a treaty, although the substance of the agreement rather than the form is the governing factor. (pp. 3–4)[4]

Two Earth Summit documents are treaties. The first, the United Nations Framework Convention on Climate Change (UNFCCC), establishes a framework for considering what can be done to reduce anthropogenic climate change and to cope with whatever temperature increases are inevitable. Opened for signature at the Earth Summit, the UNFCCC entered into force in March 1994 and currently has 194 parties.[10] The second, the United Nations Convention on Biological Diversity, aims to promote sustainable development and the conservation of biological diversity, including the fair and equitable sharing of the benefits arising out of the utilization of genetic resources, although loss of biodiversity continues at an alarming rate despite the treaty.[8, 14] This treaty has only 193 parties; the United States has signed but not ratified it.[14]

Signing the Treaty Is Not Enough: Treaties Must Be Approved

Most states approve international agreements in the same way they approve domestic statutes, but the United States is different.[23] The United States differentiates between international agreements that are made with the advice and consent of the Senate, which are called treaties, and all other international agreements, which are called executive agreements under the law of the United States (but remain treaties under international law) (p. 1).[4]

The House of Representative has no role in the treaty approval process, and approval by the Senate requires a supermajority (p. 1021).[23] Although the U.S. Constitution allows for full and automatic incorporation of treaties into domestic law, treaties that involve environmental health typically require implementing legislation. Thus, the amended Clean Air Act authorizes and is the applicable law for implementation and enforcement in the United States of the Montreal Protocol. Even after the Senate consents to a treaty, the executive branch typically does not take the final steps necessary to make the treaty binding on the United States until the required implementing legislation has been enacted. For example, the Senate consented to the 1989 Basel Convention, which regulates the transboundary movement and disposal of hazardous waste, but since Congress has not enacted the required implementing legislation, the United States is still not a party to the Basel Convention.[4, 23] These structural reasons make the United States slow to ratify even treaties that it has actively negotiated. For example, the United States negotiated and signed but has not ratified the Kyoto Protocol to the UNFCCC. Negotiations under various treaty regimes can differentiate between parties and nonparties, so the United States is disadvantaged by its failure to become a party to so many treaties. Nevertheless, significant international agreements concerning the environment are still usually concluded as treaties.[4]

From 1939 to 2001, executive agreements comprised more than 90 percent of the international agreements concluded by the United States (p. 40).[4] Executive agreements can be concluded in several ways: without an underlying authorization by treaty or statute but using the president's independent constitutional authority; within the scope of an existing treaty or statutory authorization; or where Congress enacts legislation at the beginning of the negotiation process to specify the parameters for negotiation and approval.[4] These congressional-executive agreements, famously used for "fast track" trade agreements, use normal legislative processes, and the House is more likely to approve required implementing legislation after it has been consulted on the parameters of negotiation for the international agreement. Nigel Purvis has proposed using a congressional-executive agreement for climate change.[23] That proposal has not yet been embraced, but it illustrates that U.S. approval of international agreements can be pursued via treaty or executive agreement, and different approaches to this political choice can be advocated for practical or strategic reasons.

What Is Soft Law?

Many important articulations of environmental health priorities are "soft law": statements that are not intended to be legally binding, such as unilateral statements of intent, joint communiqués, and final acts of conferences (p. 4).[4] The Earth Summit adopted three soft law documents: The Rio Declaration on Environment and Development articulates principles intended to guide sustainable development;[24] Agenda 21 is an action plan for sustainable development;[25] and the Statement of Forest Principles articulates principles for the management, conservation, and sustainable development of forests.[26] Soft law documents are frequently cited and used for persuasive purposes, such as influencing the content and interpretation of binding agreements. Many of these documents articulate the right to health (pp. 464–465).[2] For example, Article 25 of the Universal Declaration of Human Rights states:

> Everyone has the right to a standard of living adequate for the health and well-being of himself and of his family, including food, clothing, housing and medical care and necessary social services, and the right to security in the event of unemployment, sickness, disability, widowhood, old age or other lack of livelihood in circumstances beyond his control.[27]

Soft law can be a prelude to treaties that make binding obligations. The United Nations Conference on the Human Environment Stockholm Declaration of 1972 articulated twenty-six principles for managing and protecting the human environment and set the stage for many multilateral environmental agreements.[28] Soft law can also be used to create significant moral or political pressures and to facilitate forward movement in areas where states may be reluctant to make binding commitments (p. 4).[4] For example, the United Nations Millennium Declaration of 2000 articulates goals for reducing extreme poverty and sets out a series of targets with a deadline of 2015, which have become known as the Millennium Development Goals. The eight targets to be achieved by 2015 are to eradicate extreme poverty and hunger; achieve universal primary education; promote gender equality and empowerment of women; reduce child mortality; improve maternal health; combat HIV/AIDS, malaria, and other diseases; ensure environmental sustainability; and develop global partnerships for development.[29] Although not binding law, the Millennium Goals have been very useful in facilitating coordinated action in these areas.

Voluntary codes of conduct and guidelines are another way to gap-fill or to create forward movement toward international agreements with binding obligations. For example, the 1985 Food and Agriculture Organization (FAO) International Code of Conduct on the Distribution and Use of Pesticides and the 1987 United Nations Environment Programme (UNEP) London Guidelines for the Exchange of Information on Chemicals in International Trade created voluntary procedures for prior informed consent for international shipment of pesticides or chemicals that are banned or severely restricted, in order to protect human health or the environment. These voluntary measures were superseded by the entry into force of the 1998 Rotterdam Convention on the Prior Informed Consent Procedure for Certain Hazardous Chemicals and Pesticides in International Trade (Rotterdam Convention), which creates binding obligations on its parties concerning procedures to ensure prior informed consent for international shipment of such pesticides or chemicals (p. 466).[30] Many nongovernmental organizations (NGOs) and professional groups propose model codes of conduct, guidelines, and model rules. Influential multilateral organizations can transform these proposals into soft law, and some ideas take the next step by being incorporated into binding treaty provisions. Since law is made by states, it is worth keeping in mind whether a particular code or guideline is the result of a law-making process or is at an earlier stage of development.

The United Nations System

The voluntary measures preceding the Rotterdam Convention were issued by multilateral organizations within the United Nations (UN) system. Because the UN system encourages international law regimes to become more far-reaching and intertwined, environmental health professionals need to understand its basic structure and limitations.

The UN Charter is the treaty that delineates the rights and obligations of member states. Member states get a vote in the General Assembly and agree to accept the obligations of the UN Charter. The Charter states that the UN works to maintain international peace and security, to develop friendly relations among nations, to cooperate in solving international problems and promoting respect for human rights, and to be a center for harmonizing the actions of nations.[31] The core entities of the UN are the General Assembly, Security Council, Economic and Social Council, International Court of Justice, and Secretariat.[32] The Secretariat is the administrative bureaucracy of the UN. In 2010 the UN Secretariat staff was 44,000, including almost 24,000 in field operations (such as peacekeeping missions) (p. 10).[33]

Linked to the UN, but with separate governing bodies, budgets, and secretariats, are over a dozen independent specialized agencies, including the World Health Organization (WHO), International Fund for Agricultural Development, International Labor Organization, International Maritime Organization, World Bank, and International Monetary Fund.[33] Together, these entities are the UN system.[34] The staff of the UN Secretariat and related UN entities totaled 74,816 in 2010 (p. 13).[32] These international agencies are created by treaties and also enter into treaties with national governments; in 2001, the United States had bilateral international agreements with approximately fifty international organizations (p. 17).[4]

It is important to understand the limits of the UN system and the multitude of documents that it generates. The UN Charter grants the Security Council some binding decisional powers in the realm of international peace and security, but sovereign nations are wary of ceding authority. The General Assembly has power to make decisions related to organizational matters internal to the UN legal order (including semi-external matters such as the budget, or admission, suspension, and expulsion of members), but its resolutions have no independent binding effect on external affairs.[35]

Although the UN system fosters coordination, international law can also be a universe of different systems wherein separate treaties can be a result of "stovepipe" negotiation processes; treaties addressing trade, investment, environment, health, and human rights law often not only function with separate international institutions, but can also have separate groups of negotiators, enforcement mechanisms, national ministries, NGOs, and academics (p. 21).[36]

Customary Law

Customary law is another factor pushing international treaty regimes to become more far-reaching and intertwined. Customary or general international law is used to resolve conflicts regarding treaty interpretation and application and conflicts between treaties.[4, 36] A state must explicitly consent for a treaty to be binding on that state. In contrast, customary international law is binding without explicit consent.[4, 36] Originally, customary law was determined by how states acted toward each other. In the wake of international treaties creating the UN system and various arbitration tribunals and international courts, these entities now also play a role in articulating customary law. In addition, the Vienna Convention on the Law of Treaties is an important source of customary law, acknowledged as influential even by nonparties such as the United States.[4]

A political decision of the UN Security Council with follow-through by member nations is the ultimate hard enforcement authority for a narrow category of superior or peremptory rules (*jus cogens*) that are asserted to preempt all other international law and international agreements that conflict, such as the condemnation of aggression in the UN Charter and of genocide in the Convention on the Prevention and Punishment of the Crime of Genocide (p. 195).[4] In the normal realm of treaty disputes that do not rise to the attention of the Security Council, environmental health advocates can point to various international agreements promulgated under the auspices of the UN system and elsewhere that articulate goals and rights relevant to the pursuit of environmental health. In some circumstances provisions contained in multilateral agreements may be binding on a state as customary law even if a state is not a party to the agreement,[4] but a treaty negotiated under UN auspices or in pursuit of an environmental health goal is not necessarily going to trump a conflicting provision in a different treaty or be read into a treaty that is silent on a particular point.

Dispute Resolution

Treaties can create dispute resolution mechanisms. The BWT created the first permanent bilateral entity, the International Joint Commission (IJC).[5,37] Originally the IJC was intended to resolve transboundary disputes between private litigants for which there was no good recourse in the national courts of Canada and the United States. An important early twentieth-century precedent concerning the responsibility of a state to control its transboundary pollution resulted. When citizens in Washington State complained of transborder damage caused by a Canadian smelter, the Trail Smelter II arbitration panel held that Canada was responsible for the transborder damage caused by the Canadian smelter, awarded compensation to the Americans, and ordered pollution reduction measures.[5] However, the Trail Smelter case has been the only use of binding arbitration under the BWT treaty regime. Binding arbitration results in a decision that the litigants must follow, whereas nonbinding arbitration is an advisory opinion. In part, because the U.S. Senate maintains control over whether disputes go to binding arbitration under the BWT, the IJC has instead engaged in nonbinding arbitrations with considerable success and has been of continued value by doing good science and promoting information exchange and the creation of consensus.[5] As a result, persons seeking to receive compensation for transboundary pollution or to compel behavior of transboundary polluters typically pursue their actions in the domestic courts of Canada or the United States.[37]

Because the United States is notoriously reluctant to cede control to supranational tribunals such as the International Court of Justice, its bilateral environmental agreements are unlikely to contain unilaterally invoked binding dispute resolution provisions. Instead, its bilateral environmental agreements typically coordinate government activities, with some agreements also creating processes for public input and participation.

Although environmental agreements do not typically create binding dispute resolution processes accessible to private citizens, some bilateral environmental treaties do contain binding dispute resolution mechanisms that can be unilaterally invoked by the national governments that are parties to the treaty regime. For example, Argentina and Uruguay jointly manage the River Uruguay, which forms the boundary between the two nations, via a bilateral treaty called the Statute of the River Uruguay,[38] and this bilateral agreement states that disputes not settled by direct negotiations may be submitted by either party to the International Court of Justice.[39] When Uruguay authorized pulp mills that Argentina feared would cause environmental damage to the river and to Argentine territory, and after political negotiations failed, Argentina referred the matter in 2006. Argentina pointed to the nonbinding UNEP Goals and Principles and the Espoo Convention on Environmental Impact Assessment in a Transboundary Context—a multilateral treaty to which neither Argentina nor Uruguay is a party—to argue that Uruguay has specific environmental assessment obligations under the Statute of the River Uruguay.

In 2010 the International Court of Justice decided that Uruguay had breached procedural obligations of the bilateral agreement, but that Argentina had failed to prove that significant environmental damage would occur, so the operation of the remaining mill could continue.[38] Although the International Court of Justice stated that general international law may now include the obligation to undertake an environmental impact assessment where there is a risk that a proposed industrial activity may have a significant adverse impact on a shared resource, the court declined to read into the nonspecific bilateral treaty language a requirement to undertake an assessment that met with the specific form and content obligations articulated by the international documents cited by Argentina.[38]

Stratospheric Ozone and Climate Change Treaty Regimes: Successful Models Help, But Each Fight Is Unique

Argentina's use of the Espoo Convention illustrates that a multilateral treaty with limited participation can be useful in advocating for recognition of a general principle in increasingly intertwined international law regimes, but it is more difficult to establish specific binding obligations. In determining customary law, one criterion examined is the extent to which a principle, norm, or treaty regime has been accepted by the world community. The treaty regime to protect stratospheric ozone is a model of successful global cooperation to identify and reduce a global risk to environmental health. In 2009 the Montreal Protocol to the Vienna Convention became the first treaty of any kind in the history of the UN system to achieve universal participation, by 196 parties (p. xiii).[40] This regime originated in the 1970s, when scientists became concerned that certain manufactured gases migrate to the stratosphere, remain there for very long periods, and break down the ozone layer.[41] The ozone layer protects humans and the environment from increased ultraviolet (UV) radiation at the earth's surface. Direct health risks caused by increased UV radiation include increased cataracts and skin cancers, while indirect risks include potential disruptions to agriculture and food security.[42]

Beginning in the late 1970s the UNEP coordinated international action with a World Plan of Action on the Ozone Layer that called for intensive international research and monitoring.[41] The next step was the 1985 Vienna Convention for the Protection of the Ozone Layer, a framework agreement that delineates stratospheric ozone protection as an issue of international concern and states that the parties agree to cooperate in research and scientific assessments of the problem, to exchange information, and to adopt appropriate measures to address the problem. Once a framework agreement is in place, the parties can negotiate protocols that specify more detailed commitments.

After new scientific evidence galvanized world attention on the need for concrete actions, the Montreal Protocol to the Vienna Convention went into effect on January 1, 1989. The Montreal Protocol controls the production and consumption of ozone-depleting substances (ODSs).[41] A state must be a party to the Vienna Convention in order to become a party to the Montreal Protocol. Parties must often join a framework agreement in order to join the protocols, but not every party to a framework agreement also joins all of the protocols to that agreement. This multistep approach allows the international community to start at the broad level of problem definition, where consensus is most easily achieved, before wading into the much trickier negotiations on binding obligations. However, a multistep approach also runs the risk of decreased participation as the obligations become more specific. The stratospheric ozone treaty regime includes the following key features:

1. Specific targets for reduction and timetables for action: The ODSs of highest concern were targeted for immediate phaseout, while the phaseout of the less active second tier (HCFCs) did not begin until 1996 and will not be completed until 2030.[42]

2. Common but differentiated responsibilities: Developed countries had a shorter timetable for action than developing countries. Also, developed countries contribute to a Multilateral Fund that helps developing countries comply with the chemical reduction deadlines of the Protocol.[42]

3. Use of the precautionary principle: Although discovery of a seasonal ozone hole over the Antarctic helped crystallize the need for a strong Protocol, the damage from current activities was not yet evident when the Convention and Protocol were being negotiated. Action was taken on the basis of probabilities and the best available science.

4. Mechanisms to regularly update the science: The Protocol includes Technology and Economic Assessment Panels to provide regular expert assessments.[41]

5. Reservations not allowed to the Protocol: A state joining the Protocol cannot pick and choose which parts are applicable to it.

6. Reporting and enforcement mechanisms: To make the Protocol effective, parties prohibited exports and imports of controlled substances with countries not parties to the Protocol. An Implementation Committee reviews annual reports from the parties and develops measures used in cases of noncompliance, including technical assistance to enable the country to comply.[41]

The decay in the ozone layer has stopped, but stratospheric ozone is not yet increasing, and recovery to 1980 levels will take decades outside the polar regions and even longer for recovery of the springtime ozone hole over the Antarctic.[43,44] Sustained success requires sustained efforts to deal with illegal trade, manage large stockpiles of controlled substances, ensure control of new chemicals that threaten the ozone layer, and monitor the ozone layer to ensure that recovery is taking place.[41,42] The latest scientific assessments of stratospheric ozone analyze the multiple and complex interactions caused by climate change.[43,45] Most ODSs are potent greenhouse gases.[43] The good news is that phaseout of CFCs under the Montreal Protocol reduced warming, far more than the measures taken under the Kyoto Protocol treaty.[45] However, substitutes for ODSs controlled by the Montreal Protocol are also greenhouse gases.[43] In part due to the Montreal Protocol's success in controlling ODSs, climate change effects are increasing in relative importance when evaluating the health of stratospheric ozone.[43, 45]

The world community has widely accepted the stratospheric ozone treaty regime, and the UNFCC and its Kyoto Protocol use many of its principles and strategies. But each treaty negotiation has its own context. Climate change is a significantly more complicated problem: regulated ODSs are all manufactured by a limited universe of companies and used in a comparably limited universe of applications. Global climate is affected by both anthropogenic and natural causes, and effective action to stabilize the global climate requires actions with far-reaching economic consequences. Many of the principles and approaches that the world used with success in the stratospheric ozone treaty regime are being critically reexamined, renegotiated, and protested in the context of climate change negotiations, including the specific targets for reduction and timetables for action; the meaning of common but differentiated responsibilities; and the role of reservations, reporting, and enforcement mechanisms.[4, 46, 47]

An Evolving Relationship: International Trade, Environment, and Health

The relationship between trade and environmental health treaty regimes best illustrates the far-reaching and intertwined aspects of international law. Trade liberalization can have positive effects by increasing wealth. But trade agreements that foster the free flow of goods and services encourage close scrutiny, harmonization, and even elimination of laws and regulations enacted in the name of protecting health and the environment, because such laws can also function as barriers to trade that protect local industries and economies.[3, 9, 48–52] The multilateral General Agreement on Tariff and Trade (GATT) and related trade agreements administered by the World Trade Organization (WTO), regional trade agreements such as NAFTA, and bilateral fair trade agreements have all been criticized for impinging upon environmental health efforts in a variety of ways.[3, 9, 48, 50, 52] Trade liberalization can increase pollution by increasing the scale of production or by consolidating economic activities, speed up changes that overwhelm the capacity of domestic regulators, and cause profound shifts in consumption patterns, some of which may have negative health consequences.[3, 50] Harmonization of government standards at the level least restrictive to trade can erode or hinder the development of rigorous standards. Also, the threat of challenges brought under trade regimes can deter the development of new environmental and health initiatives.[3, 52]

NAFTA and some bilateral agreements explicitly list certain environmental treaties that take precedence over the trade agreement.[51] Trade agreements also contain provisions similar to Article XX of GATT, which permits national and subnational "measures necessary to protect human, animal or plant life or health," but trade agreements add criteria to be considered in the application of these general health and safety provisions to a particular law or regulation such as a requirement that it be the least restrictive in regard to trade, that it be based on an international norm promulgated by a recognized standard-setting entity such as WHO or Codex, that it be based in science, or that it not be a disguised barrier to trade.[3, 51, 52] Trade challenges to health and safety measures create uncertainty and raise alarm, even if some challenges, such as Canada's WTO challenge to the EU ban on asbestos, are ultimately unsuccessful.[3]

A particularly worrisome aspect of the NAFTA regime is the ability of private investors to challenge a signatory state. Investors have disputed bans on gasoline additives, prohibition of activities such as PCB waste exports, and regulation of behavior that impacts health and the environment such as selling tobacco products and harvesting timber. Although more recent decisions have increasingly been decided in favor of the challenged state, over 70 percent of cases brought by investors under Chapter 11 of NAFTA can be characterized as challenges to measures aimed at protecting health or preserving the environment.[52]

After early adverse NAFTA and WTO decisions awakened the environmental community to the importance of these trade regimes, activists began pressing for trade negotiations and trade regimes to become more open. Although changes have occurred in response to these pressures, trade regimes are still not as transparent or inclusive of a wide variety of stakeholders as environmental treaty regimes.[51] Coordination with multilateral environmental treaty regimes is also now on the WTO's agenda (to little substantive effect in the Doha round of trade negotiations).[51] Challenges brought under trade regimes continue to be closely monitored by those concerned about the chilling effect on environmental and health laws.[3, 52] Since NAFTA, the United States has reformed some aspects of new bilateral trade agreements,[9, 51] but experts such as Ellen Shaffer recommend increased and sustained input by health professionals to advisory committees to the U.S. Trade Representative and to other forums so that national negotiation strategies are not so heavily weighted toward business concerns.[3]

Some treaties to protect health and the environment are being written with explicit consideration of how to withstand WTO scrutiny, but the threat of challenges under trade regimes remains a factor to be evaluated when considering even local environmental health initiatives.[3, 52] Trade regimes shift the burden of proof and the framework for analysis to one that places more of a burden on proponents of environmental or health measures. If challenged in a trade regime forum, environmental and health laws and regulations are more likely to survive scrutiny if such measures harmonize with standards set by international organizations such as Codex and WHO or are firmly based in science when more stringent than international standards.[3, 9, 48–52] Restrictions that tie tightly to the environmental or health goal being pursued are also more likely to withstand scrutiny, and some multilateral environmental treaty regimes take this into consideration when crafting enforcement mechanisms. For example, some fish treaties avoid discriminating on the basis of a ship's flag country and instead couple the question of whether a particular ship is allowed to land fish to whether that particular ship can show that it is adhering to the requirements of the fish conservation scheme.[49]

The World Health Organization: New Directions

The rise in importance of trade regimes has increased the importance of the WHO in articulating norms, setting standards, and coordinating global health initiatives. WHO faces these new challenges with an institutional culture shaped by the fact that it currently lacks the compulsory powers of the WTO.[53] WHO was proposed by the 1945 San Francisco conference that created the United Nations,[54] and most members of the United Nations are also members of the WHO. As the directing and coordinating authority for health within the UN system, the WHO provides leadership on global health matters, helping to shape the health research agenda, set norms and standards, articulate evidence-based policy options, provide technical support to countries, and monitor and assess health trends.[55] The governance structure of the WHO is similar to other organizations created to administer multiparty international agreements. Member states join the WHO by signing or accepting the WHO constitution and send delegations to the World Health Assembly, which appoints the Director-General of the WHO and meets annually in Geneva to govern and determine policy for the WHO.[56] Policy direction is facilitated by an Executive Board composed of thirty-four members technically qualified in the field of health.[57] The actual work of the WHO is done by a Secretariat working under the Director-General. The WHO Secretariat has a staff of around 8,000.[57] Since it came into existence in 1948, the WHO has facilitated international cooperation on a range of healthcare issues, from fighting to eliminate diseases such as smallpox and polio to addressing environmental causes such as maternal health care, sanitary engineering, and nutrition.[54] After the Earth Summit, the WHO launched initiatives to address the health hazards posed by environmental degradation.[54, 58]

International Health Regulations

For many years the WHO's major contribution to the legally binding framework of international law has been its International Health Regulations. Primarily intended to control communicable diseases, the reach of these regulations was expanded by revisions that came into force in 2007 (p. 461)[2] (p. 609).[53] The revised regulations' purpose is "to prevent, protect against, control and provide a public health response to the international spread of disease in ways that are commensurate with and restricted to public health risks and which avoid unnecessary interference with international traffic and trade" (p. 462)[2] (p. 609),[53] quoting from WHO IHRs (2005), at art. 1, WHA 58.3 (May 23, 2005). Member states must notify the WHO of events constituting a public health emergency of international concern within twenty-four hours of their receipt of evidence that identifies an extraordinary event (p. 610).[53] The WHO can issue nonbinding recommendations on the most appropriate ways to respond, but member states decide whether to follow this guidance (p. 611).[53] The WHO does not directly provide funds for member states to implement the regulations (p. 611),[53] and member states are explicitly allowed to pursue their own health policies under principles of national law, even if these national policies are inconsistent with the regulations' standards (p. 612).[53]

International Environmental Law 249

For example, the WHO declared a public health emergency of international concern and used its newly revised regulations during the H1N1 flu outbreak. Although the regulations forbid member states from unnecessarily restricting international traffic or trade, fear of disease caused some nations to disrupt the free flow of people and goods, even when those disruptions did not have a valid public health basis.[53] The regulations rely on member countries' various emergency and public health laws rather than imposing a detailed international code, but the WHO's reliance on national laws for responding to public health emergencies is not necessarily a weakness; since nations with different circumstances may require different responses, national laws can be significantly more flexible than international codes of behavior that require protracted multiparty negotiations to enact or amend. National laws can also encourage and authorize innovative, nontraditional responses (p. 623).[53]

At first blush the regulations' broad definitions of public health emergency and disease seem to allow their application to a broad array of environmental health threats, including health threats associated with climate change (p. 462).[2] However, as Lindsay Wiley notes, the entire structure of the regulations is built around the concept of event detection and response, but health threats associated with environmental degradation and climate change often cannot be properly characterized as events at all:

> Public health concerns like the intensification of vector-borne infectious disease risk or population-level increase in asthma severity or increasing vulnerability to a whole host of illnesses due to worsening malnutrition are gradually emerging processes that play out over the course of years or even decades. An internationally coordinated surveillance system geared toward rapid detection of newly emerging diseases is a tool well suited for covert, quickly developing threats like biological terrorism or pandemic influenza. It is far less useful in the face of a slow process of environmental degradation. (p. 474)[2]

The WHO's slow start in creating enforceable international agreements to respond to emerging environmental health threats has been criticized as timid regulatory behavior that does not match the broad conception of "health" embraced by its constitution and institutional culture (p. 461).[2]

However, the WHO has begun to expand upon its regulations in creating international agreements.

Beyond the IHRs: The Framework Convention on Tobacco Control

In 2005 the WHO embraced its key role in the trade/environmental health discussion with the entry into force of the WHO Framework Convention on Tobacco Control (FCTC), the first treaty negotiated under the auspices of the WHO.[59] The FCTC has 172 parties (the United States signed the FCTC but as of this writing had not ratified it), and its Secretariat operates within the WHO.[60]

Tobacco companies are willing to use the WTO and other international trade agreements to keep their products selling. When Uruguay put into place aggressive antitobacco policies (smoking is banned in public and private enclosed spaces, and 80 percent of every cigarette package must show graphic images of smoking consequences), Switzerland-based Philip Morris International Inc. responded by initiating a WTO arbitration against Uruguay that claims Uruguay's uneven enforcement of contraband cigarettes has damaged Philip Morris' investment.[61] But Uruguay and other countries seeking to regulate tobacco despite threats of investor actions under trade agreements may have a powerful new counterweight.

The Conference of the Parties to the FCTC showed its support for Uruguay in November 2010 by having its fourth meeting in that country.[61, 62] The substance of the FCTC, however, may provide the greatest counterweight to claims that tobacco control efforts are unfairly impinging on trade. The FCTC asserts the importance of demand reduction strategies as well as supply issues.[59] Demand reduction provisions discussed in the FCTC include both price and tax measures and nonprice measures to reduce the demand for tobacco, including regulation of exposure to tobacco smoke; the contents of tobacco products; tobacco product disclosures, packaging, and labeling; tobacco advertising, promotion, and sponsorship; as well as education, communication, training, and public awareness and measures to limit tobacco dependence and encourage cessation.[59] Supply reduction provisions of the FCTC address illicit trade in tobacco products, sales to and by minors, and provision of support for economically viable alternative activities.[59] A protocol to the FCTC addressing illicit trade in tobacco products is being negotiated.[63]

Commentators have pointed out that the final text of the FCTC was weakened with flexibilities and optional language,[1] but this new treaty provides an important forum for developing a global consensus and negotiating binding obligations for the control and regulation of tobacco to protect health. It also provides an example of the potential role the health community and the WHO can play in proactively responding to trade regimes and ensuring that international law reflects the needs and concerns of the environmental health community.

Conclusion

As international law regimes have become more far-reaching and intertwined, more environmental health priorities are being articulated in international agreements. Sustained involvement of environmental health professionals is needed in the development of international treaty regimes that might otherwise not fully consider such goals, such as trade agreements. Successful advocacy is aided by an understanding of the constraints of the international treaty process, the uses of both hard and soft law, and the difficulties in conforming domestic law with international agreements.

Notes

1. H. Feldbaum, K. Lee, and J. Michaud, Global health and foreign policy, *Epidemiology Review* 32 (1) (2010): 82–92, DOI 10.1093/epirev/mxq006.
2. L. Wiley, Moving global health law upstream: A critical appraisal of global health law as a tool for health adaptation to climate change, *Georgetown International Environmental Law Review* 22 (2010): 439–489.
3. E. Shaffer, H. Waitzkin, J. Brenner, et al., Global trade and public health, *American Journal of Public Health* 95 (1) (2005): 23–34.
4. Congressional Research Service, Library of Congress, *Treaties and other international agreements: The role of the United States Senate* (study prepared for the Committee on Foreign Relations, United States Senate, 2001).
5. N. Hall, Symposium—The Great Lakes: Reflecting the landscape of environmental law; Transboundary pollution: Harmonizing

international and domestic law, *University of Michigan Journal of Law Reform* 40 (Summer) (2007): 681–746.
6. International Joint Commission, *Treaties and agreements*, http://www.ijc.org/rel/agree/water.html (accessed December 2, 2010).
7. C. Liu, The obstacles of outsourcing imported food safety to China, *Cornell International Law Journal* 43 (spring 2010): 249–305.
8. R. Adam, Missing the 2010 biodiversity target: A wake-up call for the convention on biodiversity? *Colorado Journal of International Environmental Law and Policy* 21 (winter 2010): 123–166.
9. J. Knox, The neglected lessons of the NAFTA environmental regime, *Wake Forest Law Review* 45 (summer 2010): 391–424.
10. United Nations Framework Convention on Climate Change, http://unfccc.int/) (accessed November 28, 2010).
11. International Monitoring, Control, and Surveillance Network for Fisheries-Related Activities, *Fisheries treaties and governance bodies*, http://www.imcsnet.org/imcs/fisheries_treaties.shtml (accessed December 14, 2010).
12. United Nations Environment Programme Ozone Secretariat, http://ozone.unep.org/ (accessed November 28, 2010).
13. United Nations Economic Commission for Europe, Convention on Long-range Transboundary Air Pollution, http://www.unece.org/env/lrtap/ (accessed November 28, 2010).
14. Convention on Biological Diversity, http://www.cbd.int/ (accessed November 28, 2010).
15. Basel Convention on the Control of Transboundary Movements of Hazardous Wastes and Their Disposal, http://www.basel.int/ (accessed November 28, 2010).
16. Rotterdam Convention on the Prior Informed Consent Procedure for Certain Hazardous Chemicals and Pesticides in International Trade, www.pic.int/ (accessed December 1, 2010).
17. United Nations Economic Commission for Europe, Convention on Environmental Impact Assessment in a Transboundary Context (Espoo, 1991), http://www.unece.org/env/eia/welcome.html (accessed December 1, 2010).
18. United Nations Convention to Combat Desertification, http://www.unccd.int/ (accessed December 14, 2010).

19. United Nations Economic Commission for Europe, Convention on the Protection and Use of Transboundary Watercourses and International Lakes, http://www.unece.org/env/water/ (accessed December 14, 2010).
20. Division for Ocean Affairs and the Law of the Sea, United Nations, United Nations Convention on the Law of the Sea of 10 December 1982: Oceans and Law of the Sea, http://www.un.org/Depts/los/convention_agreements/convention_overview_convention.htm (accessed December 14, 2010).
21. United Nations Economic Commission for Europe, Aarhus Convention on Access to Information, Public Participation in Decision-making and Access to Justice in Environmental Matters, http://www.unece.org/env/pp/ (accessed December 14, 2010).
22. UN Briefing Papers, *The world conferences: Developing priorities for the 21st century*, UN Conference on Environment and Development (1992) summary on the earth summit, http://www.un.org/geninfo/bp/enviro.html (accessed November 22, 2010).
23. N. Purvis, Essay: The case for climate protection authority, *Virginia Journal of International Law* 49 (summer 2009): 1007–1062.
24. Rio Declaration, http://www.un.org/documents/ga/conf151/aconf15126-1annex1.htm (accessed November 22, 2010).
25. Agenda 21, http://www.un.org/esa/dsd/agenda21/ (accessed November 22, 2010).
26. Statement of Forest Principles, http://www.un.org/documents/ga/conf151/aconf15126-3annex3.htm (accessed November 22, 2010).
27. Universal Declaration of Human Rights, http://www.un.org/en/documents/udhr/ (accessed December 1, 2010).
28. Stockholm Declaration, http://www.unep.org/Documents.Multilingual/Default.asp?documentid=97&articleid=1503 (accessed November 22, 2010).
29. Millenium Goals, http://www.un.org/millenniumgoals/ (accessed November 22, 2010).
30. J. Vinuales, Legal techniques for dealing with scientific uncertainty in environmental law, *Vanderbilt Journal of Transnational Law* 43 (March 2010): 437–503.
31. United Nations, *The UN in brief: How the UN works*, http://www.un.org/Overview/uninbrief/about.shtml (accessed November 22, 2010).

32. United Nations, Item 136 of the provisional agenda, General Assembly Sixty-fifth session, Composition of the Secretariat: staff demographics, report of the secretary-general (September 8, 2010), http://daccess-dds-ny.un.org/doc/UNDOC/GEN/N10/510/57/PDF/N1051057.pdf?OpenElement (accessed November 22, 2010).
33. United Nations, *The UN in brief: The specialized agencies*, http://www.un.org/Overview/uninbrief/institutions.shtml (accessed November 22, 2010).
34. United Nations, *The UN in brief: How the UN works; the UN system*, http://www.un.org/Overview/uninbrief/unsystem.shtml (accessed November 22, 2010).
35. M. Öberg, The legal effects of resolutions of the UN Security Council and General Assembly in the jurisprudence of the ICJ, *European Journal of International Law* 16 (5) (2005): 879–906, http://www.ejil.org/pdfs/16/5/329.pdf.
36. R. Michaels and J. Pauwelyn, Conflict of norms or conflict of laws?: Different techniques in the fragmentation of international law, *Duke Law Working Papers*, paper 9 (2010), http://scholarship.law.duke.edu/working_papers/9.
37. S. L. Hsu and A. Parrish, Litigating Canada–U.S. transboundary harm: International environmental lawmaking and the threat of extraterritorial reciprocity, *Virginia Journal of International Law* 48 (fall 2007): 1–63.
38. International Law Association, International Committee on International Law on Sustainable Development, The Hague conference, international law on sustainable development (2010), http://www.ila-hq.org/en/committees/index.cfm/cid/1017 (accessed November 28, 2010).
39. UN Treaty Collection, Statute of the River of Uruguay, http://untreaty.un.org/unts/60001_120000/10/4/00018191.pdf (accessed November 28, 2010).
40. *Handbook for the Montreal protocol on substances that deplete the ozone layer*, 8th ed. (Secretariat for The Vienna Convention for the Protection of the Ozone Layer & The Montreal Protocol on Substances that Deplete the Ozone Layer, United Nations Environment Programme, 2009), http://www.unep.ch/ozone/Publications/MP_Handbook/MP-Handbook-2009.pdf (accessed November 24, 2010).

41. E. Brown Weiss, *Vienna convention for the protection of the ozone layer, Vienna, March 22, 1985; Montreal protocol on substances that deplete the ozone layer, Montreal, September 16, 1987*. United Nations Audiovisual Library of International Law, http://untreaty.un.org/cod/avl/ha/vcpol/vcpol.html (accessed November 24, 2010).
42. Secretariat for the Vienna Convention for the Protection of the Ozone Layer & The Montreal Protocol on Substances That Deplete the Ozone Layer, *Brief primer on the Montreal protocol*, http://www.unep.ch/ozone/Publications/MP_Brief_Primer_on_MP-E.pdf (accessed November 24, 2010).
43. Scientific Assessment Panel of the Montreal Protocol on Substances That Deplete the Ozone Layer, *Scientific assessment of ozone depletion: 2010*, WMO/UNEP, executive summary, http://www.unep.ch/ozone/Assessment_Panels/SAP/ExecutiveSummary_SAP_2010.pdf (accessed November 24, 2010).
44. New report highlights two-way link between ozone layer and climate change, press release WMO No. 898, http://www.unep.ch/ozone/Publications/ozone-day2010-press-release-WMO-UNEP.pdf (accessed November 24, 2010).
45. UNEP Environmental Effects Assessment Panel, *Environmental effects of ozone depletion: 2010 assessment; interactions of ozone depletion and climate change*, executive summary, http://www.unep.ch/ozone/Assessment_Panels/EEAP/EEAP_Executive_Summary_1_Nov_2010.pdf (accessed November 24, 2010).
46. International Law Association, Committee on the Legal Principles Relating to Climate Change, *The Hague conference (2010) legal principles relating to climate change*, http://www.ila-hq.org/en/committees/index.cfm/cid/1029 (accessed November 28, 2010).
47. R. V. Anuradha, *Demystifying the Cancun climate agreements* (Business Standard, December 19, 2010), http://www.business-standard.com/india/news/anuradha-r-v-demystifyingcancun-climate-agreements/418780/ (accessed December 19, 2010).
48. C. Wold, Taking stock: Trade's environmental scorecard after twenty years of "trade and environment," *Wake Forest Law Review* 45 (2010): 319–354.

49. E. DeSombre, *Flagging standards: Globalization and environmental, safety, and labor regulations at sea* (Cambridge, MA: The MIT Press, 2006).
50. D. Wallinga, J. Harkness, C. Hawkes, et al., Exporting obesity: U.S. farm and trade policy and its impact on diet-related health in developing countries (abstract) (presented at the American Public Health Association 138th Annual Meeting, Denver, Colorado, November 6–10, 2010).
51. K. Kennedy, The status of the trade-environment-sustainable development triad in the Doha round negotiations and in recent U.S. trade policy, *Indiana International & Comparative Law Review* 19 (2009): 529–552.
52. D. Farquhar, Trade, the environment, and state environmental health law (presented at the American Public Health Association 138th Annual Meeting, Denver, Colorado, November 6–10, 2010), available at www.NCSL.org. (accessed November 24, 2010).
53. J. G. Hodge, Global legal triage in response to the 2009 H1N1 outbreak, *Minnesota Journal of Law, Science & Technology* 11 (2010): 599–628.
54. M. McCarthy, A brief history of the World Health Organization (electronic article), *The Lancet* 360 (9340) (2002): 1111–1112.
55. World Health Association, About WHO, http://www.who.int/about/en/ (accessed November 22, 2010).
56. Constitution of the World Health Organization, http://apps.who.int/gb/bd/PDF/bd47/EN/constitution-en.pdf (accessed November 22, 2010).
57. World Health Association, Governance, http://www.who.int/governance/en/index.html (accessed November 22, 2010).
58. WHO, environmental health, http://www.who.int/topics/environmental_health/en/ (accessed November 22, 2010).
59. WHO Framework Convention on Tobacco Control, http://www.who.int/fctc/en/http://www.who.int/tobacco/framework/WHO_FCTC_english.pdf (accessed November 22, 2010).
60. Overview, WHO Framework Convention on Tobacco Control, http://www.who.int/fctc/text_download/en/index.html (accessed November 22, 2010).

61. Uruguay takes on tobacco giant Philip Morris, *AP: Fox News Latino*, November 23, 2010, http://latino.foxnews.com/latino/health/2010/11/22/uruguay-takes-tobacco-giant-philip-morris/ (accessed December 1, 2010).
62. T. Magarinos, Countries agree on tobacco controls at WHO health summit, *AFP*, http://news.yahoo.com/s/afp/20101120/hl_afp/health uruguaytobaccowho_20101120231630 (accessed December 15, 2010).
63. WHO Framework Convention on Tobacco Control, report of the chairperson of the intergovernmental negotiating body on a protocol on illicit trade in tobacco products to the fourth session of the conference of the parties, May 14, 2010, http://apps.who.int/gb/fctc/PDF/cop4/FCTC_COP4_4-en.pdf (accessed November 22, 2010).

13

Biomarkers Used in Environmental Health, with a Focus on Endocrine Disrupters

Tanja Krüger, Mandana Ghisari, Manhai Long, and Eva Cecilie Bonefeld-Jørgensen

Introduction

Endocrine-disrupting chemicals (EDCs) are compounds that either mimic or block endogenous hormones and can disrupt the normal functioning of the body. The European Commission has defined an EDC as "an exogenous substance or mixture, that alters the function(s) of the endocrine system, and consequently causes adverse health effects in an intact organism or its progeny or (sub)-population."[1] Growing evidence shows that they may also modulate the activity and/or expression of steroidogenic enzymes. These are expressed not only in the adrenal glands and gonads, but also in many tissues that have the ability to convert circulating precursors into active hormones.

Many EDCs are known to act as agonist/antagonists of estrogen (ER), androgen (AR), and aryl hydrocarbon receptors (AhR); to interfere with thyroid hormone (TH) function; and to inhibit the aromatase enzyme that converts testosterone to E2.[2] This is of particular concern for the developing organism, as it is highly sensitive to hormonal changes and EDC exposure, which can result in permanent changes.[3] The broad categories of human health effects that may be linked to exposure to environmental contaminants include cancer, birth defects, decreased fertility, altered sex hormone balance, immune system defects, neurological effects such as reduced IQ and behavioral abnormalities, altered metabolism, and specific organ dysfunctions.[4]

The EDCs may occur naturally, like phytoestrogens, or may be industrial chemicals such as Bisphenol A (BPA) and other phenols and plasticizers like phthalates and adipates, commonly used in the plastic industry. The EDCs also include the persistent organic pollutants (POPs) such as polychlorinated biphenyls (PCBs), polychlorinated dibenzo-p-dioxins/furans (PCDDs/PCDFs), organochlorine pesticides (OCPs), brominated flame retardants (BFRs), and perfluorinated compounds.

Biomonitoring is the assessment of internal doses of, for example, EDCs and has been used for decades to provide information about exposures to chemicals, giving information about a person's "body burden" of EDCs.

Human biomonitoring exploits the development and characterization of biomarkers, that is, "cellular, biochemical, or molecular alterations that are measurable in biological media such as human tissues, cells, or fluids," to measure direct and indirect human exposure to EDCs. In fact, biomonitoring can integrate environmental exposure, effects, individual sensitivity, and health risks, making it a useful instrument for risk assessment (see Figure 13.1). In particular, biomonitoring can identify trends and changes in current exposure patterns as well as establish distribution of exposure among the general population, identifying subgroups and/or populations with higher exposure levels.

According to the U.S. Environmental Protection Agency, biomarkers are observable properties of an organism, and three specific types of biomarkers are considered:

- Biomarkers of exposure: exogenous chemicals, metabolite(s), or the products of interactions between a pesticide and target molecules or cells that are measured in a compartment within an organism. This includes internal dosimeters of pesticide or metabolite concentrations and markers of biologically effective doses.
- Biomarkers of effects: measurable alterations of an organism that, depending on magnitude, can indicate a potential or established health impairment or disease.
- Biomarkers of susceptibility: indicators of inherent or acquired properties of an organism that may lead to an increase in the internal dose of a pesticide or an increased level of the response, resulting from exposure to a specific pesticide.

Biomarkers Used in Environmental Health   261

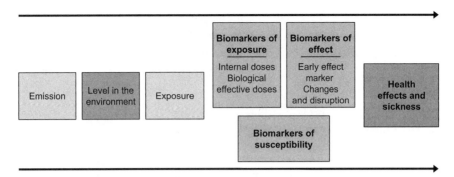

Figure 13.1 Biomonitoring and biomarkers. (From [4].)

We are concurrently exposed to a complex mixture of chemicals from various sources, and the toxicological assessment of EDCs in human is complicated. Toxicological studies have shown that individual POPs have very different biological effects and potentials, and many of the bioaccumulated POPs are estrogenic, while others are antiestrogenic and antiandrogenic, and some have dioxin-like potentials.[4, 6–11] As a result of these different and often opposed directed biological effects and potentials of the POPs, it is very difficult, if not impossible, to predict a given biological effect of the very complex mixtures of POPs that actually exists in the human body (blood). Furthermore, additive enhancement of hormone actions has been reported *in vitro* for xenoestrogen mixtures[12, 13] and recently *in vivo* for antiandrogens.[14] Therefore, the assessment of the integrated biological effect of the actual and complex chemical mixture in human blood is important, and *ex vivo* cell systems have recently been introduced to enable the assessment of the integrated level of xenobiotic receptor transactivity in human adipose tissue[15–17] or in human serum.[18–22] In this biomarker approach the lipophilic EDCs are separated from endogenous hormones, and the combined xenohormone receptor transactivities of the lipophilic extract are determined in an array of *in vitro* bioassays. Therefore, the final effect reflects the additive, synergistic, or antagonistic mechanisms that may account for the final effect observed.

This chapter provides an overview of our *in vitro* work, performed on single EDCs and their mixtures in five key bioassays for effects on the ER, AR, and AhR transactivation, TH action, and aromatase activity. In addition, the endocrine-disrupting bioactivity of wastewater extracts tested in the bioassays is addressed. Finally, *ex vivo* results for the combined potential of the complex POP mixtures in serum extracts to affect the function of ER and AR, as well as AhR mediated dioxin-like activity as biomarkers for effects, are reviewed.

Methods

In Vitro Cell Culture Models

Transactivation Assays

The transactivation assays are based on genetically modified cell lines in which specific receptor responsive DNA elements are linked to a firefly luciferase reporter gene that transcribes to the easily measurable luciferase protein. Basically, these transactivation assays measure receptor-induced gene expression, which gives information about the expected biological response to chemicals in humans. For example, elevated or reduced gene expression measured indicates whether specific chemicals would exert agonistic or antagonistic effects on the receptor function.

Several transactivation assays have been utilized to study the ER, AR, and AhR receptor-mediated transactivities of individual chemicals and mixtures[6, 7, 10, 23, 24] as well as for epidemiological purposes.[18–20, 25, 26]

ER, AR, and AhR Transactivation Assays: The ability of the xenobiotics to affect ER, AR, and AhR functions was assessed in the human mammary MVLN cells, Chinese Hamster Ovary (CHO-k1), and mouse hepatoma Hepa1.12cR cell lines, respectively, as described.[18–20, 23] The samples (single compounds, mixtures, or biological samples) were analyzed by cell exposure alone, reflecting the agonistic power. In addition, as a physiological mimic, the samples were analyzed by co-exposure with respective high potent receptor ligands (17β -estradiol [E2]) for ER, methyltrienolone [R1881] for AR, and 2,3,7,8-tetrachlorodibenzo-p-dioxin [TCDD] for AhR), reflecting the potential of the samples to compete/antagonize the potent receptor ligand induced activity as well as the ability to further increase the ligand induced receptor activity.

T-screen

The interference of xenobiotics with the TH action was studied using the T-screen assay, which is based on the TH dependent cell growth of a rat pituitary tumor cell line GH3. The GH3 cells express both endogenous TRs and ERs and can synthesize and secrete prolactin and growth hormone, in response to estrogen and TH, respectively. GH3 cells respond to physiological levels of triiodothyronine (T3) resulting in increased GH production and cell proliferation[27] and has been used for *in vitro* detection of agonistic and antagonistic properties of compounds at the level of the TR. The T-screen was carried out using rat pituitary GH3 cells essentially as described previously.[28] To assess the agonistic and antagonistic effects, the xenobiotics were tested in the absence or presence of the potent T3.

Aromatase Assay
Aromatase cytochrome P450 is the key enzyme in the biosynthesis of estrogens from cholesterol, as it catalyzes the final, rate-limiting step in which androgens are converted to estrogens. The conversion proceeds via three successive oxidation steps, and this is utilized in the aromatase assay measuring the release of tritium-labeled water.

The effects on aromatase enzyme activity were assessed using the human choriocarcinoma JEG-3 cell line, derived from malignant placental tissues. JEG-3 cells endogenously express aromatase activity at detectable levels. The catalytic aromatase activity was determined by the release of tritium as described previously,[29] with minor modifications.[24]

Principle and Practice in Mixture Analyses
Humans are exposed to a complex mixture of chemicals. Therefore, assessment of the combined effect of mixtures is very important. The method based on the principle of concentration addition (CA) has been shown to be a valid tool for assessing mixture effects of similarly and dissimilarly acting EDCs *in vitro*.[9, 30] The CA model assumes that one chemical can be replaced totally or partly by another in the mixture, without changing the overall combined effect, and that they act through a similar mechanism.[31] By applying the principle of concentration addition,[11] the concentration of the compounds in a mixture, at an observed mixture effect, can be predicted using the concentration-response data for each compound alone.[10, 32] Knowledge of the ratio of the compounds in the mixture is a prerequisite for using this method.

Extraction and Fractionation Methods

Wastewater

Water samples, collected from the influent and the effluent of two Danish sewage treatment plants (STPs), were extracted using solid phase extraction (SPE), as described.[33] The extracts were analyzed for the content of a range of industrial chemicals with endocrine disrupting (ED) properties and tested for effects on the ER, AR, and AhR transactivation as well as TH action. These effects were expressed as biological equivalents,[33] calculated using the fixed-effect-level quantification method.[34, 35] The fixed-effect-level toxicity equivalents are a suitable parameter for assessing the induction potency in complex environmental samples.[34]

Serum Samples

To obtain the serum fraction containing the actual mixture of bioaccumulated lipophilic POPs for determination of ER and AR transactivity measurements, a solid phase extraction (SPE) and high performance liquid chromatography (HPLC) fractionation was performed on 3.6 ml serum.[21] The first fraction (F1: 0.00 – 5.30 min.) was defined to include most lipophilic POPs and free of endogenous hormones.[21]

The extraction of the serum samples to obtain the fraction containing lipophilic POPs for AhR transactivity measurements was performed at a certified laboratory, Le Centre de Toxicologie, Sainte Foy, Quebec, Canada.[36]

The serum extracts were stored at $-80°C$, and on the day of analysis the extracts were thawed, processed, and analyzed as described in ER, AR, and AhR transactivation assays.[18–20]

Results

In Vitro Data on the Single EDCs and Their Mixtures

The following sections summarize the relative endocrine disrupting (ED) potencies of a range of EDCs that were analyzed in five key *in vitro* bioassays for ED effects, including ER, AR, and AhR transactivation, TH action, and aromatase activity.

PCBs

Research suggests that mechanisms of PCB toxicity are multifactorial, and appear to involve both AhR and other hormone receptor mechanisms.

The most frequently detected PCBs in humans are CB138, CB153, and CB180, which have been shown to contribute about 50–80 percent of total PCB body burden in, for example, the Swedish general population.[37] These three PCBs, individually and in a mixture, as well as other PCBs and PCB metabolites, were shown *in vitro* to interfere with cell proliferation as well as the function of the ER and AR.[7, 21, 22, 38]

Non- or mono-ortho PCBs are ligands to the AhR, whereas ortho-substituted PCBs are either weak ligands or do not bind to the AhR.[39] However, the higher chlorinated PCBs have the potential to inhibit the AhR function.[40]

The hydroxylated PCBs have a high degree of structural resemblance to THs, and the three OH-PCBs (OH-CB 121, OH-CB 69, OH-CB 106) were shown to be TH-like by stimulating the GH3 cell proliferation in the T-screen assay.[28] These OH-PCBs were previously reported to bind to TR and TH transport proteins TTR.[41]

Pesticides

Pesticides comprise a large number of different substances with dissimilar structures and diverse toxicity. Several pesticides exert ED activities through interaction with the hormone receptors. The organochlorine pesticide *o,p'*-DDT possesses estrogenic activities and its metabolite *p,p'*-DDE antiandrogenic activities,[42] whereas toxaphene is reported to be antiestrogenic.[7] Andersen et al.[8] have reported that several currently used pesticides, such as dieldrin, endosulfan, methiocarb, fenarimol, chlorpyrifos, deltamethrin, and tolclofos-methyl, possess estrogenic activity on the basis of cell proliferation assay and ER transactivation assay using MCF-7 human breast cancer cells. In the same study, dieldrin, endosulfan, methiocarb, prochloraz, and fenarimol possessed antiandrogenic activity on the basis of the AR transactivation assay using Chinese hamster ovary (CHO) cells. Furthermore, fenarimol and prochloraz were potent aromatase inhibitors, whereas endosulfan was a weak inhibitor.[8] In a subsequent study, prochloraz, iprodion, and chlorpyrifos were shown to exert dioxin-like AhR agonistic effects.[11] In the T-screen assay, chlorpyrifos weakly stimulated the GH3 cell proliferation, whereas prochloraz and iprodion inhibited the T3-induced proliferation.[28]

Phthalates and Phenols
As a part of the EU-sponsored research project ENDOMET, the potential of several widely used plasticizers in the industry and present in the environment were assessed *in vitro* to affect the following receptors: ER, AR, AhR, and T3 dependent growth of rat pituitary GH3 cells. The plasticizers investigated were BPA and BPA dimethacrylate (BPA-DM), alkyl phenols (n-octyl (OP), n-nonyl (nNP), tert-octyl (tOP)), bis-ethylhexyladipate (DEHA), phthalates (dibutyl, bis-ethylhexyl (DEHP), di-isononyl (DINP), di-isodecyl (DIDP), dioctyl (DOP), benzylbutyl (BBP), 2-phenylphenol (2-PP), 4-chloro-3-methyl phenol (CMP), resorcinol, and 2,4-dichlorophenol (2,4-DCP). The combined effect of an equipotent mixture of BPA, nNP, BBP, CMP, resorcinol, and tOP was also assessed in these assays.

BPA, BPA-DM, tOP, OP, nNP, BBP, and DBP elicited ER-mediated luciferase activity as determined by the ER transactivation assay.[24, 32] Using the concentration addition model, the observed effect concentrations of the mixture were equal to the predicted concentrations, suggesting additivity.

None of the tested plasticizers showed agonistic activities in the AR reporter gene assay, whereas, in presence of the synthetic androgen R1881, several compounds elicited AR antagonistic effects (BPA, BPA-DM, nNP, CMP, Res, tOP, 2-PP, DCP, and BBP).[10] The observed mixture effects antagonized the R1881-induced AR response. The predicted effect concentration was equal to the observed effect, suggesting additivity.

The plasticizers nNP, DBP, DEHP, and DIDP elicited weak agonistic AhR transactivity, whereas BPA and CMP inhibited the TCDD-induced AhR activity.[10] The mixture, composed of six compounds of which only nNP had a weak agonistic AhR potential, weakly induced the AhR transactivity compared to the individual compounds. This suggests that the non-AhR active compounds can act together with a weak AhR agonist to increase the AhR transactivity. In the presence of TCDD, the mixture dose-dependently inhibited the TCDD-induced AhR transactivity.

Generally, the tested plasticizers had the potential to induce the GH3 cell proliferation, except for nNP, DINP, and 2-PP. nNP inhibited the

T3-induced GH3 cell growth[32] and was also reported to antagonize the T3-induced luciferase activity.[43]

The TH-like effects of the mixture were less than additive in the absence of T3, indicating an antagonistic effect (interaction of compounds) in GH3 cells. This deviation from additivity might be due to antagonizing effects of NP in the mixture that potentially can obscure the overall additive effect of other mixture components.

Of the tested compounds, BPA, nNP, OP, BBP, DEHA, CMP, DBP, DOP, and 2,4-DCP were inhibitors of aromatase activity.[24, 44]

In summary, based on our studies,[10, 24, 32] the tested plasticizers have the ability to act via more than one mechanism, and this might enhance the biological effect in the intact organism, since the final response is likely to be determined by the interaction of all pathways implicated.

Wastewater Results

Industrial and municipal effluents are important sources of EDCs discharged into the aquatic environment. In addition to chemical analyses of the EDCs found in wastewater, our bioassays were used to elucidate the biological activity of EDCs released into environmental water streams.

The concentrations of all analyzed chemicals were reduced in effluents compared to influents, and for some to below the detection limit. Both influent and effluent samples from both STPs showed activation of the ER, AR, AhR, and TH, and the calculated biological equivalents were in the range of other studies.[33] The activities were higher for the influent than for the effluent in all the four bioassays, suggesting a reduction but not an elimination of EDCs from the two Danish STPs. In the applied bioassays a significant effect was detected in influent as well as effluent samples, indicating that even though the levels of the analyzed chemicals were reduced in the STP, the effluents still had endocrine disruption potentials. Thus the bioassays can be used for a more sensitive biomonitoring of effects of effluents from STPs as a supplement to chemical analyses.

Ex vivo Serum Xenohormone Activity Biomarkers

Laboratory studies on the effects of single chemicals or chemical mixtures in cell cultures and laboratory animals cannot fully elucidate the human health risks. Integration of epidemiological and biomarker studies on humans from exposed populations is needed in order to obtain information about the real health risks resulting from exposure to the accumulated mixtures of contaminants. The burden of POPs in Arctic people has been monitored since 1991, and a program for measuring the potential biological effects of these contaminants has been established (Arctic Monitoring and Assessment Programme [AMAP]). As part of the AMAP program, the levels of fourteen different PCBs, ten pesticides, selenium, lead, and mercury were determined for different Greenlandic districts. Furthermore, to determine the potential biological effects of these complex mixtures of contaminants, the integrated effects on the function of ER and AR, as well as AhR-mediated dioxin-like transactivity, were measured.

Biomarkers for POP and Metal Exposures

Figure 13.2 shows Greenland with geographical located columns showing the relative total sum of POP exposure biomarkers, and within the column the relative profile of each POP group, determined in human serum, typically including fifty men and fifty women from each district.[45]

Both the sum of PCBs and the sum of OCPs are highest in the East Greenlandic districts (EG) (Scoresbysund followed by Tasiilaq), the Thule district in Northern Greenland (NG), and Narsaq in South Greenland (SG). For Qeqertarsuaq (Godhavn, Disco Bay) and Nuuk (West Greenland, WG), the sums of PCBs and OCPs are very similar. However, it should be noted that the data for Nuuk men are atypical, as they were older and had a higher dietary intake of seabirds and, therefore, a higher level of serum POP concentrations. In general the POP levels in Nuuk and Sisimiut, the two most Westernized cities in terms of diet, are at the same level and lower than POP levels in the other Greenlandic districts. What also should be noted is that selenium and mercury serum concentrations were found at the highest level in the Thule area. With respect to serum lead, the highest levels were found in Scoresbysund.

Biomarkers for Receptor Effects

As a biomarker for effect, we determined the combined potential of the complex mixtures of POPs in serum extracts to affect the function of ER and AR, as well as AhR-mediated dioxin-like transactivity. Presently, the serum POP-related effects on receptor activities are determined for

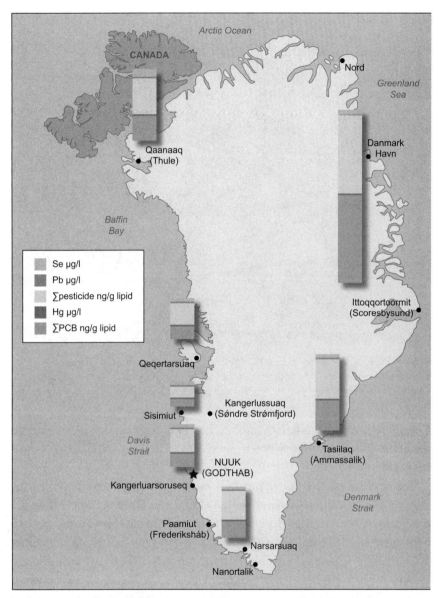

Figure 13.2 Human POP biomarkers for exposure in Greenland. The columns illustrate the relative $\sum$ POPs and the different POP profiles and metal concentrations within the columns in the different Greenlandic districts for the period 2000–2006. Note that the concentration units are different because $\sum$ PCBs and $\sum$ pesticides (details in Table 2) are given in ng/g lipid and metals in µg/l serum. (From [4].)

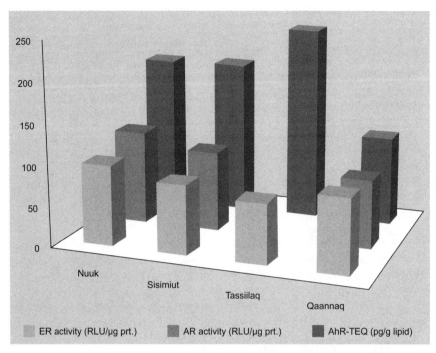

Figure 13.3 Diagram for serum POP related receptor activities in Greenlandic Inuit. ER, estrogen receptor; AR, androgen receptor; AhR, aryl hydrocarbon receptor; RLU/ug prt; relative light units per microgram cell protein. (From [4].)

Greenlandic Inuit in Nuuk, Sisimiut, Tasiilaq, and Qaanaaq,[25, 26] whereas the effect biomarkers for Scoresbysund (EG), Godhavn (WG), and Narsaq (SG) are in progress.

Figure 13.3 shows the serum POP-related receptor transactivities for Nuuk (WG), Sisimiut (WG), Tasiilaq (EG), and Qaanaaq (NG). For ER transactivity the serum POPs elicited in general an antagonistic effect (21–81 percent of the samples), with antagonistic activity in the order: Tasiilaq > Sisimiut > Qaanaaq > Nuuk. Few of the serum POP extracts elicited agonistic ER transactivity (< 6 percent, Qaanaaq > 14 percent). There were observed differences both among districts and between sexes, which in turn was district dependent.[25]

Preliminary data for serum POP-related ER transactivity for Scoresbysund, Godhavn, and Narsaq show as for Tasiilaq a trend of higher frequency of serum samples from East Greenland (Scorebysund) eliciting antiestrogenic activity compared with West Greenland (Godhavn and

Narsaq) (EC Bonefeld-Jørgensen; unpubl. data, 2011). The trend was similar for men and women. These data support observations on the antiestrogenic potentials of POPs as the highest ΣPOP concentrations were found in Scorebysund and Tasiilaq (Figure 13.1).

Also for the AR transactivity, a district-dependent difference in serum POP effect was observed with the following order of medians: Nuuk > Sisimiut > Qaanaaq (Figure 13.3). In general the AR transactivity was higher for men than women. For men from Nuuk and Sisimiut, AR agonistic activity was observed.[25] In contrast, a higher frequency of serum extracts elicited antiandrogenic activity for women in Nuuk and Sisimiut, and for both men and women in Qaanaaq.[25] Preliminary data for serum POP-related AR activity for Scoresbysund, Godhavn, and Narsaq show predominantly antiandrogenic activity (EC Bonefeld-Jørgensen; unpubl. data, 2011), supporting the antiandrogenic potentials of most POPs.

The POP-related dioxin-like serum activities are given by TCDD toxicological equivalence (AhR-TEQ). The following order of AhR-TEQ medians was observed: Tasiilaq ≥ Sisimiut ≥ Nuuk > Qaanaaq (Figure 13.3). The frequency of serum extracts with agonistic AhR activity was 71–100 percent, and similar levels were found in women and men.[26] Preliminary data for Scoresbysund, Godhavn, and Narsaq also show a high frequency of serum extracts with agonistic AhR activity (EC Bonefeld-Jørgensen; unpubl. data, 2011).

Comparing the effects of the serum POP mixtures on the three receptor transactivities (Figure 13.3), there is a trend toward a relationship between higher AhR-TEQ (dioxin activity) level and lower ER transactivity, which supports the reported antiestrogenic activity of dioxins.

The study was also intended to evaluate whether the transactivity was associated with serum POP concentrations (fourteen PCBs and ten OCPs) and/or lifestyle factors. The serum ER transactivities correlated negatively to the POPs for the combined female data, while the AR and AhR transactivity were inversely correlated to the levels of the POPs for the combined male data.

Associations between the ER, AR, and AhR transactivities and lifestyle factors such as age, intake of marine food, and smoker years were observed, indicating that comparison of different study populations requires the inclusion of age, diet, and lifestyle factors.

Also for Scoresbysund, Godhavn, and Narsaq, preliminary data show correlations between ER and AhR transactivities and POP levels as well as lifestyle factors (EC Bonefeld-Jørgensen; unpubl. data, 2011).

Using another cleanup method for serum samples, it was recently reported that high levels of PCBs in Slovakian male serum samples were associated with a decreased ER-mediated transactivity and increased AhR-mediated transactivity.[46]

Few studies using a different setup of measurement of xenoestrogenic activities in adipose tissue from Spanish women and serum samples from pregnant Danish and Faroese women have been reported, although in contrast to our studies, no correlations between POPs and MCF-7 proliferation were found.[15–17, 22] However, recent data showed, for nonpregnant female greenhouse workers in Denmark, that the E2-induced MCF-7 proliferation response was reduced, indicating an antiestrogenic effect of the serum extract containing major xenoestrogens, but without pharmaceutical and endogenously produced estrogens.[47]

Biomarkers for Receptor Effects: Comparison of Inuit and Europeans

The project Inuendo, supported by the European Union, aimed to examine whether there is a correlation between human fertility and exposure to POPs, using the POP exposure biomarkers PCB-153 and p,p'-DDE. The project involved population groups of fertile men from Europe (Kharkive, Ukraine; Warsaw, Poland; fishermen of the Swedish East Coast) and various Greenlandic districts.[48] A significantly higher serum concentration of the two POP exposure markers was found for Inuit. Concerning serum POP-related receptor effect biomarkers, we found a significantly lower ER and AhR activity level and a trend toward higher AR activity for Greenlandic serum samples than for the European samples.[18–20]

The POP-related serum xenoestrogenic transactivity of the three European study groups did not differ.[18] The xenoestrogenic transactivity correlated negatively to PCB-153 for the Inuit study group and positively to p,p'-DDE for the study group from Warsaw.[18] The highest frequency of antiandrogenic activity was observed for Ukraine, also having the highest level of p,p'-DDE, a well-known AR antagonist, and a significant negative association between the xenoandrogenic transactivity and p,p'-DDE across the combined European study groups was observed.[19] A higher median AhR activity level was observed for samples from Poland and Sweden compared to Ukraine.[20]

Unexpectedly, a significantly lower level of DNA damage was found in sperm DNA from Inuit compared with European samples.[49] Further studies are required to identify possible factors involved in the correlations found between POP-related receptor effects and sperm DNA damage.[50, 51]

Conclusion

The *in vitro* results showed that EDCs have the ability to act via more than one mechanism, and this might enhance the biological effects in the intact organism, since the final response is likely to be determined by the interaction of all pathways implicated. Furthermore, the mixture analyses suggested that the combined effect of all the compounds present in the human body must be taken into consideration for risk assessment.

The results from the wastewater analyses indicate that the bioassays can be used for a more sensitive biomonitoring of effects of effluents from STPs as a supplement to chemical analyses and as a tool to evaluate the effectiveness of treatment within the STP. A significant effect was detected in effluent samples, indicating that the wastewater treatment processes are not efficient enough to prevent contamination of environmental surface waters, and the release of the EDCs from STPs to the environment might have adverse effects on human health as well as the ecosystem.

Biomonitoring studies for exposure biomarkers have shown that there are geographical differences in the bioaccumulated POP levels in the blood of Europeans and Inuit from different districts. These differences are primarily attributable to diet and lifestyle factors such as smoking, as well as higher levels of traditional Greenlandic diet (e.g., seal, whale, polar bear, seabirds) reflected by a higher concentration of POPs in the blood of Inuit compared to Europeans and Inuit with a more Westernized diet. The highest POP values were found on the east coast of Greenland, where the inhabitants still primarily rely on traditional Greenlandic foodstuff.

Biomonitoring by studies on receptor effects showed a general correlation between high serum POP concentrations and inhibited ER and AhR transactivity; however, for men in the two West Greenlandic cities Nuuk and Sisimiut, a trend toward increased AR activity was observed. An inverse trend between dioxin-like AhR transactivity and ER transactivity supports the perception that dioxins exert an antiestrogenic effect. In conclusion, the actual cocktail of serum POPs has endocrine-disrupting potential.

In summary, the xenohormone transactivities can be used as an integrated biomarker of POP exposure and effects and lifestyle characteristics. The data suggest that geographical and sex differences might be caused by the variation in POP mixture profiles in concert with lifestyle and genetic factors. Biomarkers for POP exposure showed that men from Greenland (Inuit) had significantly higher serum POP levels compared with European men from Sweden, Ukraine, and Poland. Unexpectedly, in the same study groups it was observed that Inuit had a lower level of DNA damage in sperm than the Europeans. Further studies are needed to elucidate whether there is a relationship between the impact of serum POPs on the activity of hormone and/or dioxin receptors and the level of sperm DNA damage. However, it is known that selenium and n-3 unsaturated fatty acids are important factors in production of semen.

Notes

1. European Commission, *European workshop on the impact of endocrine disrupters on human health and wildlife*. Report of proceedings from a workshop held in Weybridge, UK, 2–4 December 1996. Report reference, EUR 17549, European Commission, DGXII, Brussels, Belgium.
2. M. Yang, M. S. Park, and H. S. Lee, Endocrine disrupting chemicals: Human exposure and health risks, *Journal of Environmental Science and Health Part C: Environmental Carcinogenesis & Ecotoxicology Reviews* 24 (2006): 183–224.
3. L. J. Guillette Jr., D. A. Crain, A. A. Rooney, et al., Organization versus activation: The role of endocrine-disrupting contaminants (EDCs) during embryonic development in wildlife, *Environmental Health Perspectives* 103 (Supp. 7) (1995): 157–164.
4. E. C. Bonefeld-Jørgensen and P. Ayotte, *Toxicological properties of POPs and related health effects of concern for the arctic populations*, chapter 6 (Oslo, Norway: Arctic Monitoring and Assessment Programme [AMAP], 2003), xiv + 137, www.amap.no.

5. E. C. Bonefeld-Jørgensen, Biomonitoring in Greenland: Human biomarkers of exposure and effects—a short review, *Rural Remote Health* 10 (2010): 1362.
6. E. C. Bonefeld-Jørgensen, H. R. Andersen, T. H. Rasmussen, et al., Effect of highly bioaccumulated polychlorinated biphenyl congeners on estrogen and androgen receptor activity, *Toxicology* 158 (2001): 141–153.
7. E. C. Bonefeld-Jørgensen, H. Autrup, and J. C. Hansen, Effect of toxaphene on estrogen receptor functions in human breast cancer cells, *Carcinogenesis* 18 (1997): 1651–1654.
8. H. R. Andersen, A. M. Vinggaard, T. H. Rasmussen, et al., Effects of currently used pesticides in assays for estrogenicity, androgenicity, and aromatase activity in vitro, *Toxicology and Applied Pharmacology* 179 (2002): 1–12.
9. M. Birkhoj, C. Nellemann, K. Jarfelt, et al., The combined antiandrogenic effects of five commonly used pesticides, *Toxicology and Applied Pharmacology* 201 (2004): 10–20.
10. T. Kruger, M. Long, and E. C. Bonefeld-Jørgensen, Plastic components affect the activation of the aryl hydrocarbon and the androgen receptor, *Toxicology* 246 (2008): 112–123.
11. M. Long, P. Laier, A. M. Vinggaard, et al., Effects of currently used pesticides in the AhR-CALUX assay: Comparison between the human TV101L and the rat H4IIE cell line, *Toxicology* 194 (2003): 77–93.
12. J. Payne, M. Scholze, and A. Kortenkamp, Mixtures of four organochlorines enhance human breast cancer cell proliferation, *Environmental Health Perspectives* 109 (2001): 391–397.
13. N. Rajapakse, E. Silva, and A. Kortenkamp, Combining xenoestrogens at levels below individual no-observed-effect concentrations dramatically enhances steroid hormone action, *Environmental Health Perspectives* 110 (2002): 917–921.
14. S. B. Metzdorff, M. Dalgaard, S. Christiansen, et al., Dysgenesis and histological changes of genitals and perturbations of gene expression in male rats after in utero exposure to antiandrogen mixtures, *Toxicological Sciences* 98 (2007): 87–98.
15. M. F. Fernandez, A. Rivas, F. Olea-Serrano, et al., Assessment of total effective xenoestrogen burden in adipose tissue and identification of

chemicals responsible for the combined estrogenic effect, *Analytical and Bioanalytical Chemistry* 379 (2004): 163–170.
16. Jm. J. Ibarluzea, M. F. Fernandez, L. Santa-Marina, et al., Breast cancer risk and the combined effect of environmental estrogens, *Cancer Causes and Control* 15 (2004): 591–600.
17. A. Rivas, M. F. Fernandez, I. Cerrillo, et al., Human exposure to endocrine disrupters: Standardisation of a marker of estrogenic exposure in adipose tissue, *Apmis* 109 (2001): 185–197.
18. E. C. Bonefeld-Jørgensen, P. S. Hjelmborg, T. S. Reinert, et al., Xenoestrogenic activity in blood of European and Inuit populations, *Environmental Health* 5 (2006): 12.
19. T. Krüger, P. S. Hjelmborg, B. A. G. Jönsson, et al., Xeno-androgenic activity in serum differs across European and Inuit populations, *Environmental Health Perspectives* 115 (2007): 21–27.
20. M. Long, B. S. Andersen, C. H. Lindh, et al., Dioxin-like activities in serum across European and Inuit populations, *Environmental Health* 5 (2006): 14.
21. P. S. Hjelmborg, M. Ghisari, and E. C. Bonefeld-Jørgensen, SPE-HPLC purification of endocrine-disrupting compounds from human serum for assessment of xenoestrogenic activity, *Analytical and Bioanalytical Chemistry* 385 (2006): 875–887.
22. T. H. Rasmussen, F. Nielsen, H. R. Andersen, et al., Assessment of xenoestrogenic exposure by a biomarker approach: Application of the e-screen bioassay to determine estrogenic response of serum extracts, *Environmental Health* 2 (2003): 12.
23. E. C. Bonefeld-Jørgensen, H. T. Grunfeld, and I. M. Gjermandsen, Effect of pesticides on estrogen receptor transactivation in vitro: A comparison of stable transfected MVLN and transient transfected MCF-7 cells, *Molecular and Cellular Endocrinology* 244 (2005): 20–30.
24. E. C. Bonefeld-Jørgensen, M. Long, M. V. Hofmeister, et al., Endocrine-disrupting potential of bisphenol A, bisphenol A dimethacrylate, 4-n-nonylphenol, and 4-n-octylphenol in vitro: New data and a brief review, *Environmental Health Perspectives* 115 (Supp. 1) (2007): 69–76.
25. T. Kruger, M. Ghisari, P. S. Hjelmborg, et al., Xenohormone transactivities are inversely associated to serum POPs in Inuit, *Environmental Health* 7 (2008): 38.

26. M. Long, B. Deutch, and E. C. Bonefeld-Jørgensen, AhR transcriptional activity in serum of Inuits across Greenlandic districts, *Environmental Health* 6 (2007): 1–17.
27. H. H. Samuels and L. E. Shapiro, Thyroid hormone stimulates de novo growth hormone synthesis in cultured GH1 cells: Evidence for the accumulation of a rate limiting RNA species in the induction process, *Proceedings of the National Academy of Sciences USA* 73 (1976): 3369–3373.
28. M. Ghisari and E. C. Bonefeld-Jørgensen, Impact of environmental chemicals on the thyroid hormone function in pituitary rat GH3 cells, *Molecular and Cellular Endocrinology* 244 (2005): 31–41.
29. H. J. Drenth, C. A. Bouwman, W. Seinen, et al., Effects of some persistent halogenated environmental contaminants on aromatase (CYP19) activity in the human choriocarcinoma cell line JEG-3, *Toxicology and Applied Pharmacology* 148 (1998): 50–55.
30. J. Payne, N. Rajapakse, M. Wilkins, et al., Prediction and assessment of the effects of mixtures of four xenoestrogens, *Environmental Health Perspectives* 108 (2000): 983–987.
31. A. Kortenkamp, Ten years of mixing cocktails: A review of combination effects of endocrine-disrupting chemicals, *Environmental Health Perspectives* 115 (Supp. 1) (2007): 98–105.
32. M. Ghisari and E. C. Bonefeld-Jørgensen, Effects of plasticizers and their mixtures on estrogen receptor and thyroid hormone functions, *Toxicology Letters* 189 (2009): 67–77.
33. K. O. Kusk, T. Kruger, M. Long, et al., Endocrine potency of wastewater: Contents of endocrine disrupting chemicals and effects measured by in vivo and in vitro assays, *Environmental Toxicology and Chemistry* 30 (2) (2011): 413–426.
34. W. Brack, H. Segner, M. Möder, et al., Fixed-effect-level toxicity equivalents: A suitable parameter for assessing ethoxyresorufin-o-deethylase induction potency in complex environmental samples, *Environmental Toxicology and Chemistry* 19 (2000): 2493–2501.
35. M. Engwall, D. Broman, et al. Toxic potencies of lipophilic extracts from sediments and settling particulate matter (SPM) collected in a PCB-contaminated river system, *Environmental Toxicology and Chemistry* 15 (1999): 213–222.

36. J. B. Walker, L. Seddon, E. McMullen, et al., Organochlorine levels in maternal and umbilical cord blood plasma in Arctic Canada, *Science of the Total Environment* 302 (2003): 27–52.
37. A. W. Glynn, A. Wolk, M. Aune, et al., Serum concentrations of organochlorines in men: A search for markers of exposure, *Science of the Total Environment* 263 (2000): 197–208.
38. T. J. Schrader and G. M. Cooke, Effects of aroclors and individual PCB congeners on activation of the human androgen receptor in vitro, *Reproductive Toxicology* 17 (2003): 15–23.
39. J. P. Giesy and K. Kannan, Dioxin-like and non-dioxin-like toxic effects of polychlorinated biphenyls (PCBs): Implications for risk assessment, *Critical Reviews in Toxicology* 28 (1998): 511–569.
40. J. Suh, J. S. Kang, K. H. Yang, et al., Antagonism of aryl hydrocarbon receptor-dependent induction of CYP1A1 and inhibition of IgM expression by di-ortho-substituted polychlorinated biphenyls, *Toxicology and Applied Pharmacology* 187 (2003): 11–21.
41. A. O. Cheek, K. Kow, J. Chen, et al., Potential mechanisms of thyroid disruption in humans: Interaction of organochlorine compounds with thyroid receptor, transthyretin, and thyroid-binding globulin, *Environmental Health Perspectives* 107 (1999): 273–278.
42. H. R. Andersen, A. M. Andersson, S. F. Arnold, et al., Comparison of short-term estrogenicity tests for identification of hormone-disrupting chemicals, *Environmental Health Perspectives* 107 (1999): 89–108.
43. C. Schmutzler, I. Gotthardt, P. J. Hofmann, et al., Endocrine disruptors and the thyroid gland—a combined in vitro and in vivo analysis of potential new biomarkers, *Environmental Health Perspectives* 115 (Supp. 1) (2007): 77–83.
44. R. Waring, Dysregulation of endogenous steroid metabolism potentially alters neuronal and reproductive system development: Effects of environmental plasticisers, 2006, http://ec.europa.eu/research/quality-of-life/ka4/pdf/report_endomet_en.pdf.
45. B. Deutch, H. S. Pedersen, G. Asmund, et al., Contaminants, diet, plasma fatty acids and smoking in Greenland 1999–2005, *Science of the Total Environment* 372 (2007): 486–496.
46. M. Pliskova, J. Vondracek, R. F. Canton, et al., Impact of polychlorinated biphenyls contamination on estrogenic activity in human male serum, *Environmental Health Perspectives* 113 (2005): 1277–1284.

47. H. R. Andersen, F. Nielsen, J. B. Nielsen, et al., Xeno-oestrogenic activity in serum as marker of occupational pesticide exposure, *Occupational and Environmental Medicine* 64 (2007): 708–714.
48. J. P. Bonde, G. Toft, L. Rylander, et al., Fertility and markers of male reproductive function in Inuit and European populations spanning large contrasts in blood levels of persistent organochlorines, *Environmental Health Perspectives* 116 (2008): 269–277.
49. A. Stronati, G. C. Manicardi, M. Cecati, et al., Relationships between sperm DNA fragmentation, sperm apoptotic markers and serum levels of CB-153 and p,p'-DDE in European and Inuit populations, *Reproduction* 132 (2006): 949–958.
50. T. Kruger, M. Spano, M. Long, et al., Xenobiotic activity in serum and sperm chromatin integrity in European and Inuit populations, *Molecular Reproduction & Development* 75 (2008): 669–680.
51. M. Long, A. Stronati, D. Bizzaro, et al., Relation between serum xenobiotic-induced receptor activities and sperm DNA damage and sperm apoptotic markers in European and Inuit populations, *Reproduction* 133 (2007): 517–530.

14

Environmental Health Risk Assessment

Ryan G. Sinclair, Kristen Gunther, and Rhonda Spencer-Hwang

Introduction

Every day of our lives, we make decisions based on perceived risk and probability. Since ancient times human civilizations have engaged in a form of risk analysis to support the decision-making process.[1] The Asipu people, who lived near the Tigris River around 3200 BC, had persons living within their community who were consulted as professionals to weigh the risks and benefits and help with making decisions. Today our risk analysis decisions are often made unconsciously about a variety of topics, including health, finances, sports, environment, and many more. We usually complete an informal risk analysis without even knowing that this is the process we used to arrive at a decision. Our unconscious decisions or *intuition* is often incomplete and biased and may not provide the best assessment. For this reason, the process of risk analysis is formalized by many groups, with many different meanings: Wall Street analysts calculate financial risks; governmental health agencies calculate health risks from factors such as industrial emissions or pathogens in drinking water. What all risk analysis activities have in common is the use of a measurable probability to assess the likelihood of an event.[2] In the case of health, we estimate the likelihood of injury, disease, or death resulting from human exposure to a potential environmental hazard.

In the context of environment, *health risk assessment* is defined as the process used to characterize the nature and magnitude of potential risks to public health by combining the results of an exposure assessment with the results of a probability analysis for the contaminant in question, over a specified time interval. For example, we might be exposed to a pathogen in the drinking water distribution system and need to decide whether or not the municipality should order a boil-water warning. In another case, we might estimate the prevalence of cancer in a community exposed to diesel exhaust for the past twenty years.

Health risk assessment developed as a discipline in the 1980s in response to the nuclear industry and the availability of better detection technologies for contaminants in our environment and their relationship to our health. The U.S. Environmental Protection Agency (EPA) was involved in the process of conducting risk assessments in the 1970s; however, it wasn't until the 1980s that the process was formalized (www.epa.gov/risk assessment/history.htm). The congressional mandate of the Clean Air Act established, and the Safe Drinking Water Act required, an improved estimate of potential hazards made for risk management purposes. The U.S. government chartered an independent scientific group to consider the different areas of risk assessment and develop a formalized process. The National Research Council published *Risk Assessment in the Federal Government: Managing the Process*, commonly referred to as the "Red Book," in 1983,[3] which first officially recognized the field of risk assessment. The EPA's 1986 publication of the *Carcinogenic Risk Assessment Guidelines* established the risk assessment areas of hazard identification, exposure assessment, dose-response assessment, and risk characterization.[4] The first risk assessment framework considered laboratory toxicology data and human health data for chemical risk assessment. In the past decade, the federal government used a similar process to develop microbial risk assessment, which considers additional uncertainties. Examples of common risks in the United States are shown in table 14.1. In the 1980s the EPA also developed and released the Integrated Risk Information System (IRIS) database, which contains information on the human health effects of exposure to various substances in the environment. For chemical and microbial health risk assessment, the four steps are hazard identification, exposure assessment, dose response, and risk characterization. These are described briefly in Table 14.2 and in more detail below.

TABLE 14.1
Comparative Risks of Death for Some Common Activities in the United States

Activity	Risk	Number of Deaths/Year	Lifetime Risk
Driving	Motor vehicle accidents	46,000	1/65
Staying home	Home accidents	25,000	1/130
Smoking cigarettes	Lung cancer (one pack of cigarettes per day)	80,000	1/4

Source: Based on data from C. Gerba, Risk assessment, in *Environmental and pollution science*, edited by I. Pepper, C. Gerba, and M. Brusseau (Burlington, MA: Elsevier, 2006), 212–232.

TABLE 14.2
The Four Components of Risk Assessment

Hazard Identification	Defines and describes the hazard's anticipated effect on human health by systematically reviewing all related data exhibiting an acute or chronic health effect.
Exposure Assessment	Determines the route of exposure, amount, and duration of exposure to an agent. Determines the size and nature of population exposed.
Dose-response	Characterizes the relationship between the dose of a hazard received and the incidence of an adverse health effect in exposed populations.
Risk Characterization	The exposure assessment and dose response are integrated together to estimate the magnitude of the public health problem and evaluate variability and uncertainty.

Source: Based on data from U.S. Environmental Protection Agency, *Guidelines for carcinogen risk assessment*, EPA/630/R-00/004 (Washington, DC: Risk Assessment Forum, U.S. Environmental Protection Agency, 1986).

Other Risk Analysis Components

The process of health risk analysis includes risk assessment as the formal procedure to describe the risk. Without additional effort, this science-based outcome can often be misinterpreted or not properly emphasized by educators and the media. More effort needs to accompany risk assessment for management and communication needs. *Risk management* is defined as the judgment and analysis that combine the scientific results of a risk assessment with economic issues, political factors, legal needs, social factors, and anticipated engineering challenges to produce a decision about an environmental action. *Risk communication* is an interactive next step, which communicates the risk assessment/management information to the public, government officials, and the media. Risk communication includes the participatory process of information exchange between individuals, groups, and other stakeholders. A strong risk communication process will always have a two-way information exchange between risk assessors and affected communities. Designing the appropriate message for risk communication is an important step that must be taken carefully, because this is when media and other stakeholders are likely to misinterpret the scientific outcomes from the risk assessment. The relationship among risk assessment, risk management, and risk communication is illustrated in Figure 14.1.

Risk Assessment Steps

Hazard Identification

This step defines and describes the hazard's anticipated effect on human health. The hazard is defined by systematically reviewing all data exhibiting an acute or chronic health effect. The process will also define the route of exposure and attempt to determine if toxic or infectious effects in one setting will likely occur in other settings. An important component of hazard identification is selection of the key studies to be included in the assessment: the studies that will provide accurate and timely information on the substance of interest that poses a potential health hazard.

Hazard identification may rely on data from several types of studies. Typically human epidemiological data may provide the strongest evidence for determining whether or not a substance is a hazard to human health. Epidemiological studies usually fall into one of the following categories: cohort (prospective) study, case-control (retrospective) study, cross-sectional design, or clinical trial. The different study designs each have their own strengths and weaknesses. Cohort data comprise information collected from study participants who are free from disease at baseline and are followed over a designated period of time to determine whether

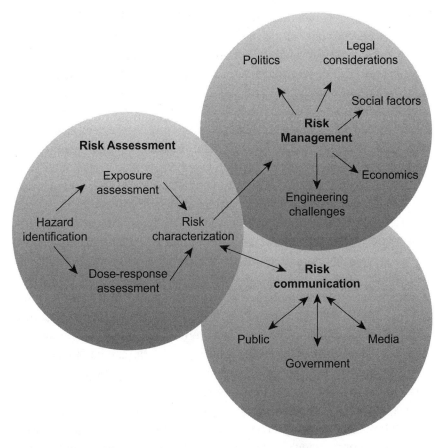

Figure 14.1 Relationship among risk assessment, risk management, and risk communication. (Authors, based on data from [2], [4].)

or not they develop disease. In the simplest type of cohort, study subjects are divided into two groups, exposed and nonexposed. An advantage of a cohort study is that the time sequence of exposure to disease is in agreement with real time; however, this type of study may take a long time to complete and may be costly. Case-control data consist of information from a past event wherein a population or individual had a response to some specific hazardous agent. Case-control data are often less expensive than cohort data and often will yield results in a more timely manner. However, case-control data are subject to a number of biases, such as an interview bias or nonresponse bias. A cross-sectional study involves the comparison of groups rather than individuals to identify any association. A cross-sectional study allows the researcher to more quickly obtain available data. This type of study allows room for a greater range of exposure assessments. These studies can be more expensive and time consuming if the data are not readily available. Clinical trials are typically considered the gold standard, because in this type of study design subjects are exposed to a substance while in a controlled laboratory setting. Researchers are better able to determine the direct biological impact on human health of the substance in question, while preventing other confounding factors from affecting the results. However, it is highly unethical to expose human subjects to many substances to determine their health effects. In the worst-case scenario, there are *no data* available for the hazard in question. When there are inadequate data or there is no evidence available from a human model, it is possible to consider animal data.

If quality human data are available, animal data are not preferable, because animals are not humans. For new chemicals, animal data are often the only choice, because no humans have been exposed to the chemical in question. Animal data are also very useful in providing insight into the biological mechanism of the chemical impacting human health. Animals used for these studies are mice and rats for toxicity data, because they have similar liver, heart, and skin to humans; pigs and guinea pigs for skin studies; and white rabbits for teratological effects (i.e., birth defects) and eye studies. Primates are rarely used for chemical studies, but are used more often in infectious disease studies. Advantages of using animal data are many, especially when no other data are available. We can control the population in animal studies, where we know that certain species will have certain predictable sensitivities and care requirements. Animals also have a rapid age progression. When a rat is two years old, it is equal in some ways to a seventy-year-old human. Rats are inexpensive, so animal studies can be affordable. The ethical procedures for use of animals in toxicology studies are considerably less restrictive than the procedures to consider when using human subjects. Animals also have similar organs; rats have more similarities with humans than dissimilarities. The major disadvantages of animal studies are that many animals have a low variability of response, where the only measurable outcome is death and not cancer formation. This low variability may be common among the entire species, especially in laboratory strains, which are inbred and sometimes

TABLE 14.3
EPA Weight of Evidence Categories for Carcinogenic Groups

Class	Description	Data Type
A	Human carcinogen	Human epidemiological data
B1	Probable carcinogen	Animal data and limited human data
B2	Probable carcinogen	Animal data only
C	Possible carcinogen	Limited animal data
D	Not classifiable as to carcinogenicity	Without adequate animal data
E	No evidence	Animal or human

Source: Adapted from U.S. Environmental Protection Agency, *Guidelines for carcinogen risk assessment*, EPA/630/R-00/004 (Washington, DC: Risk Assessment Forum, U.S. Environmental Protection Agency, 1986).

considerably less robust. Another difficulty in using animal data is physiologic variability. An example of physiologic variability is di(2-ethylhexyl) phthalate (DEHP). DEHP causes cancer in rodents, but has been confirmed as noncarcinogenic to humans by the International Agency for Research on Cancer (IARC), the World Health Organization's (WHO) authority on cancer.[5]

The hazard identification step weighs all evidence and sometimes combines data from human toxicity studies and animal studies. If a strong animal study is available, its data may be preferred over a human study that has only limited data availability. Information on animal and human studies is described in a scheme to assess Weight-Of-Evidence (WOE) for carcinogenicity developed by the U.S. Environmental Protection Agency (EPA).[4] Table 14.3 describes this scheme, in which more weight is given to human than to animal evidence.

Exposure Assessment

This step determines the size and nature of the population exposed: the route, amount, and duration of the exposure to an agent. Exposure assessment also can be used to estimate hypothetical exposures that might arise from release of new microbes or chemicals into the environment. An exposure pathway is defined as the journey that a hazardous chemical or microbe takes from a source to a human receptor.

Exposure assessment determines the extent and frequency of the human exposure by asking questions such as "How much?", "How often?", and "How certain?" It is also necessary to quantify the number of people exposed and the degree of absorption by various routes of exposure and to consider the effects of the hazard on high-risk groups (i.e., pregnant women, infants, the elderly, the immune compromised, etc.). Routes of exposure to consider are ingestion through drinking and eating, skin absorption from liquid and soil exposures, and inhalation. The total dose exposed is equal to the sum of the doses from all the routes of exposure. The resulting exposure calculation will be based on the average response among typical individuals. A confident understanding of the initial exposure or initial study is needed for a successful risk assessment.

Exposure levels are estimated using direct, reconstructive, or predictive measures. The preferred direct measure is an assessment of the actual concentrations of chemical or microbe contaminants and the actual doses to the affected individuals. If a direct measurement is not available, it is possible to use biomarkers to reconstruct the levels of chemicals or infection in the body. Biomarkers allow use of previously published toxicology studies to calculate the hypothetical dose to the individuals. Cotinine concentration in blood serum is an example of a useful biological marker for determining the extent of cigarette smoking. If previous studies are not available, it is possible to predict exposure through measuring the suspect hazard in the environment and then estimate human contact and the resulting dose to the average person.

In determining the exposure of persons to chemical agents, one must fully consider the exposure pathway. Exposure assessment should consider the environmental medium and assess the route by which exposure might occur. Assessment should also include examining how conditions might change over time and how this change might influence human exposure. An example is a contaminant in an open, unused land area where the area is sited to be developed into a soccer field. Assessment should include the environmental medium in which the substance has been identified, such as air, groundwater, surface soil contamination, sediment, surface water or groundwater, and how the substance may be transformed or transported in any one of these mediums. Also, assessment should include identification of potential human exposures in each of the potential media the substance may come in contact with. If contaminated groundwater is pumped to a local household residence, the occupants of the house may be exposed via ingesting a glass of contaminated water and breathe in or have dermal exposure to the contaminant substance while taking a shower.

Dose-Response Assessment

This step characterizes the relationship between the dose of a hazard received (or administered) and the incidence of an adverse health effect in exposed populations. The dose-response process considers intensity of exposure, age pattern of exposure, gender, lifestyle, and other modifying factors. This assessment extrapolates from high to low doses and from animals to humans. The dose response evaluation is performed to estimate the incidence of the adverse effect as a function of the magnitude of human exposure to a chemical or microbe.

The dose and the human response to that dose are only one step in the risk assessment process. Because there is so much uncertainty associated with this step, the entire risk assessment process is dependent on a reasonable estimation of the agent's ability to cause harm. In a dose-response investigation, the study subjects are exposed to increasing doses of the toxicant or microbe in question. The result of this type of experiment is the *dose-response relationship*, which describes the exposed species' response to the hazardous agent in question. In chemical and microbial risk assessment, the dose is indicated as milligrams of substance or pathogen ingested, inhaled, or absorbed through the skin per kilogram of body weight per day (mg kg^{-1} day^{-1}).[2]

The dose-response relationship is based on human or animal studies. Human studies include case reports and epidemiological studies, which come in the form of cohort, case-control, cross-sectional, and clinical trial studies. When a human study is not available, animal studies are used, wherein the agent in question is evaluated for four types of responses: acute, subacute, subchronic, or chronic. Acute studies look at immediate responses in animals to a set of doses administered over a short time interval. Chronic studies assess an animal's response to the dose of an agent given over its entire lifetime. Chronic studies are good for cancer assessments, while acute studies are often used for systemic or immediate effects. Animal studies are further characterized by additional specialized

TABLE 14.4
Types of Studies for Microbial or Chemical Dose-Response

Human Studies	Animal Studies	
Case reports	General studies	
Epidemiological studies	• Acute	• Subchronic
• Cross-sectional	• Subacute	• Chronic
• Cohort	Specialized studies	
• Case-control	• Teratology—physiologic abnormalities	
• Clinical trial	• Genotoxicity—deleterious to genetic material	
	• Developmental	
	• Male/female reproductive	
	• Cancer	
	• Endocrine disruption	

Source: Adapted from U.S. Environmental Protection Agency, *Guidelines for carcinogen risk assessment*, EPA/630/R-00/004 (Washington, DC: Risk Assessment Forum, U.S. Environmental Protection Agency, 1986).

efforts mentioned in Table 14.4. The types of studies in Table 14.4 can have a wide variety of responses, which may include no observable effect, temporary effects, permanent damaging effects, chronic functional impairment, or death. In the case of microbial risk assessment, studies also need to consider the risk of the infectious pathogen infecting and causing a recognizable symptom.

Dose-Response Curves

Data resulting from experimental dose-response studies are usually presented in a dose-response curve. The curve in Figure 14.2 has an x-axis, which describes the dose of the hazardous agent in question, and a y-axis, which describes the risk of some adverse health effect or infection. When the dose of the hazardous agent increases, the risk of adverse health effects also increases.

Cancer vs. Noncancer

For chemicals that are noncarcinogenic, it is believed that a *threshold* exists. Below this threshold point there is no toxic response to the chemical exposure. When the hazardous agent is noncarcinogenic, the

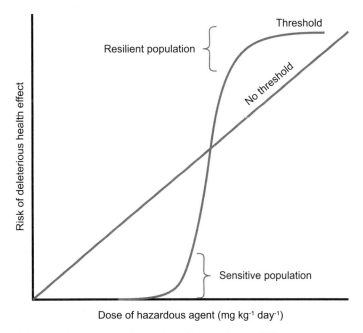

Figure 14.2 Dose-response relationship. (Adapted from [9].)

dose-response curve will be an S shape, as seen in Figure 14.2, representing the varying degrees of risk. A person can be exposed to low levels of noncarcinogens without experiencing adverse health effects. In contrast, carcinogens are considered *nonthreshold*, which means that even a small exposure to the chemical can put the person at risk of developing cancer. Therefore, the only safe level of a carcinogen is zero. The dose-response curve for carcinogens will be linear, or almost linear, and always pass through the origin (see Figure 14.2).

Dose-Response Models

Dose-response curves are often obtained from animal data and cannot be interpreted without understanding some assumptions. Animal studies are often expensive, so researchers are limited in the number of test animals they use. This limits a study's ability to show adverse effects at low doses, because the low-dose studies will often take much longer and many more animals than the high-dose response studies. Because of this expense, researchers are forced to extrapolate low-dose response data from their

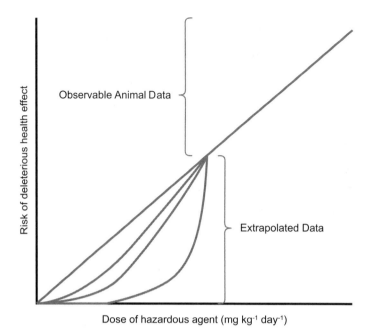

Figure 14.3 Dose-response curves based on extrapolated data. (Adapted from [9].)

high-dose studies. Scientists and researchers use the available animal data to interpret a reasonable low dose for human exposure. The dose-response curves shown in Figure 14.3 are derived from extrapolated animal data. Because humans are more often exposed to low doses of adverse agents than to high doses, the extrapolated animal data from high-dose response studies must be interpreted with reservations. This explains the EPA's WOE classification system, which assigns a greater worth to human studies than animal studies (Table 14.3).

In order to extrapolate data from low-dose studies, several mathematical models are used. These models are often controversial, because no model can be proved or disproved from the data, and no model can be shown to be more accurate than the other. Thus, the model used depends on the particular agency using it and the policies implemented within that agency. The following models are frequently used: one-hit, multistage, multihit, and linear multistage.[2]

The *one-hit model* for carcinogenesis is the simplest, with two assumptions made in determining dose-response. The first is that a single chemical exposure, or "hit," is capable of creating a malignant change. For example, a single encounter with a carcinogen will damage the DNA structure, leading to the development of a tumor. Once this malignant change occurs, tumor development will continue regardless of the dose the person was exposed to. The second assumption is that this change will occur in a single stage.

The *multistage model* is more scientifically based and considers the development of tumors to be the result of multiple biological stages rather than one single occurrence. The chemical or hazardous agent must pass through a series of biological stages without being altered before a tumor can develop. For example, the chemical must go through stages such as metabolism, covalent bonding, and DNA repair, while maintaining its structure, before a tumor can develop. The dose received during exposure determines how fast the chemical passes through these stages and how fast tumor development begins. This model produces a linear relationship between the dose and the risk.

The *multihit model* assumes that many exposures or hits from the chemical are needed to develop a response leading to tumor development. In this model all hits must be a direct result of the dose and cannot occur spontaneously, as in the multistage model. Therefore, multihit models often produce flatter curves at low doses and predict lower risk than the multistage model would.

The *linear multistage model* is similar to the multistage model; however, it assumes that there are many stages to tumor development, involving multiple carcinogens, co-carcinogens, and promoters (carcinogens that activate cancer-forming genes). This model involves a series of mathematical steps to determine risk, which are discussed in the next section. Figure 14.4 shows the data points involved in generating a linear multistage model.

Dose-Response Concepts

The EPA uses the linear multistage model to determine risk, because this model tends to be more cautious and overemphasizes risk. As mentioned previously, the linear multistage model uses mathematical equations to determine the lifetime risk from exposure. The slope of the dose-response curve produced by this model is called the *potency factor* (PF) or q_1^*. The potency factor is the inverse of the concentration of chemical measured in milligrams per kilogram of animal body weight per day, or $1/(\text{mg kg}^{-1}$

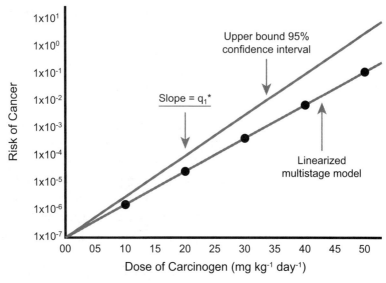

FIGURE 14.4 Datapoints generated using a linear multistage model. (Adapted from [9].)

day^{-1}). A list of potency factors for common chemicals can be found in the EPA's Integrated Risk Information System (IRIS).[6] The potency factor is then multiplied by the lifetime average dose (AD) of 1 mg kg^{-1}. Equation 1 shows the lifetime risk of a carcinogen.

$$\text{Lifetime risk} = \text{AD} \times \text{PF} \qquad \text{Eq. 1}$$

The *chronic daily intake* is what allows us to determine the risk of getting cancer. This determines the probability of getting cancer from a certain dose taken over a typical seventy-year human lifetime.

When determining the dose-response for noncarcinogens, thresholds exist from the observable effects of the toxin. One threshold is used when no observable toxic effects are seen below a certain dose or exposure to a chemical; this threshold is known as the no-observed-adverse-effect-level (NOAEL). Below the NOAEL, the body is able to naturally remove the chemical without creating harm to itself. The LOAEL, or lowest observed adverse effect level, may be used when the NOAEL is not available. The LOAEL threshold represents the lowest dose of a substance observed that causes an adverse health effect. When the LOAEL is used, an additional uncertainty factor must be applied to account for the adverse effects seen.

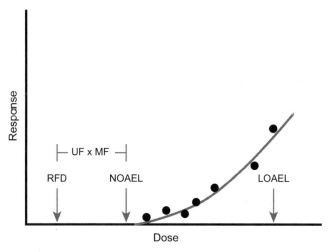

Figure 14.5 Dose-response showing NOAEL, LOAEL, and RfD. (Adapted from [9].)

The *reference dose* (RfD) of a substance demonstrates the intake or dose of the substance per unit of body weight per day (mg kg^{-1} day^{-1}) that is tolerable without producing adverse health effects to human populations, including more sensitive groups such as children or pregnant women. Figure 14.5 shows the dose-response plot for carcinogens with regard to the NOAEL and LOAEL.

Certain substances have higher potency factors and lower reference doses, often leading to higher toxicity. In these cases, safety factors or uncertainty factors are used to account for the variability that is seen as a result of these differences. Intraspecies variability is used to describe populations in the same species that may be more sensitive to exposure than the normal adult population, such as pregnant women, children, and the elderly. To account for this variability, the NOAEL is divided by a factor of 10. Interspecies variability is seen when the evidence of harmful effects from an exposure is based on animal data. In this case an additional factor of 10 is used to determine the reference dose of the substance in humans. Another uncertainty factor is used when the duration of the study is less than the lifetime of the subject. For example, if a skin test is performed on an animal to determine the effect of a personal care product, but the study is completed before the end of the animal's life, a less than lifetime uncertainty factor would be used to determine the reference dose. Finally, an

TABLE 14.5
Summary of Uncertainty Factors and Reference Dose Concepts

Uncertainty Factor (UF)
    Intraspecies variability (10)
    Interspecies variability (10)
    Less than lifetime study duration (10)
    Lack of study NOAEL (10)
    Modifying factor:
        Extra UF based on severity of effect (1–10)

Reference Dose Concepts
    No-Observed-Adverse-Effect-Level (NOAEL)
    Uncertainty Factor (UF)
    Modifying Factor (MF)
    Reference Dose (RfD) = $\dfrac{\text{NOAEL}}{\text{UF} \times \text{MF}}$

Source: Adapted from C. Gerba, Risk assessment, in *Environmental and pollution science*, edited by I. Pepper, C. Gerba, and M. Brusseau (Burlington, MA: Elsevier, 2006), 212–232. With permission from Peter M. Sandman.

uncertainty factor of 10 is often applied when limited or questionable data are available from both human and animal studies. This is called the lack of study NOAEL. A summarization of these uncertainty factors is provided in Table 14.5. Modifying factors are another safety level based on professional judgments regarding uncertainties that may not be explained by the use of typical uncertainty factors. Modifying factors also account for the severity of the hazardous effect.

To determine the RfD for a substance, the NOAEL is divided by the product of all uncertainty factors (UF) and modifying factors (MF). Equation 2 shows the general formula used to obtain the RfD. The RfD can also be used to determine quantitative risk of a substance by using the relationship between the potency factor and chronic daily intake:

$$\text{RfD} = \frac{\text{NOAEL}}{\text{UF}_1 \times \text{UF}_2 \ldots \times \text{UF}_n \times \text{MF}} \qquad \text{Eq. 2}$$

$$\text{Risk} = \text{Potency Factor} \,(\text{Chronic Daily Intake} - \text{RfD}) \qquad \text{Eq. 3}$$

Environmental Health Risk Assessment 297

Although equation 3 is not often used, the RfD can be a simple indicator of potential risk by comparing it to the chronic daily intake. If the chronic daily intake is below the RfD, then it is assumed that there is minimal risk to the average population.

Risk Characterization

The last step in the risk assessment process is risk characterization. This process estimates the incidence of health effects under the various conditions described in the risk assessment. The exposure assessment and dose-response are integrated together in order to estimate the magnitude of the public health problem and to evaluate variability and uncertainty. The outcome of the risk characterization process is a movement toward risk management and risk communication.

When determining the risk of developing cancer resulting from a certain chemical exposure, a framework called weight-of-evidence (WOE) is used. Within this framework previously gathered evidence is used to make a scientific judgment on the possible risk of developing cancer based on the quality, accuracy, and consistency of the data. The first step in the WOE framework is using evidence from human and animal studies individually to determine the risk of cancer development. The second step is combining the risk of cancer development from human and animal studies to determine the overall weight-of-evidence for human risk. The last step in the framework is to evaluate all of the evidence and determine if it should be modified or if it is a reasonable calculation of risk for the population. See Table 14.3 for the EPA's weight-of-evidence scale. Various approaches are used to conduct cancer and noncancer risk characterizations, explained below.

Cancer Risk Analysis

To determine the overall lifetime risk of developing cancer, the chronic daily intake of the chemical is determined and multiplied by the potency factor of that chemical, determined from the EPA's IRIS database. To find the chronic daily intake of the carcinogen, the following components are needed: the average concentration during exposure; the amount of contaminated medium, typically measured in L/day or m^3/day; the frequency and duration of exposure; the body weight of the person exposed; and the time period of exposure. Equation 4 represents how to determine the chronic daily intake:

$$\text{Chronic Daily Intake} = \frac{\text{Exposure concentration} \times \text{Contact rate} \times \text{Exposure frequency and duration}}{\text{Body weight} \times \text{Average time of exposure}} \qquad \text{Eq. 4}$$

Once the chronic daily intake is determined using previously gathered data, it can be multiplied by the potency factor to determine lifetime risk of developing cancer. This risk is expressed as a number. Often people are exposed to chemicals through drinking water, contaminated food, and the air we breathe. The method of determining the chronic daily intake of a chemical can be used for all of these exposure routes; however, be cautious about the units used when performing calculations. For example, for waterborne chemicals, make sure to express the exposure amount in L/day compared to mg/day for ingestion of chemicals through food.

Noncancer Risk Analysis

When determining the risk associated with noncarcinogenic substances, hazard quotient and hazard indexes are used. A hazard quotient is used for a single substance exposure and can be determined by dividing the average daily dose during the exposure period (in mg kg^{-1} day^{-1}) by the reference dose (in mg kg^{-1} day^{-1}). If the hazard quotient is less than 1.0, it is not considered to be toxic and does not pose significant risks. If the hazard quotient is above 1.0, there is a potential risk associated with the substance, but further investigation is needed to certify this. The hazard index is used when there is exposure to more than one chemical. The hazard index is simply the sum of the hazard quotient of each chemical.

The information gathered from the risk assessment process resulting in a numeric risk characterization can be used by agencies to develop guidelines to alert the public about a potential hazard and reduce the risk of harmful effects. However, it is important to remember that the numerical estimates of risk cannot stand alone. Consideration of the uncertainties and limitation in calculation must be taken into account, leading to effective scientific interpretation.

Risk Management and Communication

After the completion of the risk assessment process, risk characterization can be used to alert appropriate authorities and the public about the hazards of chemical exposure. This process is often participatory, with multiple stakeholders involved in the analysis and dissemination of information. The process of risk management takes the risk characterization and relates it to legal factors, socioeconomic factors, and public concern. Members of the risk management team ultimately decide, based on evidence and other situational factors, whether the chemical exposure causes a significant harm to the surrounding population and how to control this substance. Often concepts of benefit identification and cost-effectiveness will impact the risk management team's actions in regard to controlling the substance. The team then determines how this information should be relayed to the public through the process of risk communication.

Risk communication is very important to ensure public awareness of the hazard, methods for resolution, and public acceptance of these efforts. This process often involves principles of public health and community engagement. While large agencies and organizations must be notified of the possible effects of chemical exposure, community members must also be informed. This communication process can be through media outlets, including news reports and public service announcements, as well as using the assistance of community members to communicate risk and resolution methods. By empowering community members with information on harmful substances as well as how to prevent exposure to these chemicals, they can spread the word to their peers. One approach is through the use of health promoters, who are knowledgeable, respected community members. They are first informed of the health information through participatory learning, then they transfer this information to their peers and neighbors through informal communication and community presentations. This is only one example of the participatory process of risk communication; there are many other ways to engage multiple stakeholders in the information process to prevent exposure and adverse health effects. Risk management and risk communication will ideally lead to the control of hazardous chemicals to protect the public's health. In certain cases where no action is taken to control these chemicals, the risk assessment process can be used to provide sufficient evidence for advocacy by interested organizations and

TABLE 14.6
The 25 Key Crisis Communication Recommendations

Don't over-reassure.	Establish your own humanity.
Put reassuring information in subordinate clauses.	Tell people what to expect.
Err on the alarming side.	Offer people things to do.
Acknowledge uncertainty.	Let people choose their own actions.
Share dilemmas.	Ask more of people.
Acknowledge opinion diversity.	Acknowledge errors, deficiencies, and misbehaviors.
Be willing to speculate.	Apologize often for errors, deficiencies, and misbehaviors.
Don't overdiagnose or overplan for panic.	Be explicit about "anchoring frames."
Don't aim for zero fear.	Be explicit about changes in official opinion, prediction, and policy.
Don't forget emotions other than fear.	Don't lie, and don't tell half-truths.
Don't ridicule the public's emotions.	Aim for total candor and transparency.
Legitimize people's fears.	Be careful with risk comparisons.
Tolerate early over-reactions.	

Source: Adapted from P. Sandman and J. Lanard, *Crisis communication: Guidelines for action; planning what to say when terrorists, epidemics, or other emergencies strike* (The American Industrial Hygiene Association, 2004), http://www.psandman.com/handouts/AIHA-DVD.htm (accessed October 1, 2010), with permission of Peter M. Sandman.

community members. Peter Sandman is a leading author and expert in the field of risk communication.[7] Table 14.6 presents his twenty-five key crisis communication recommendations for communicating risks.

Risk Perception

Each person has a certain set of standards, whether conscious or unconscious, that he or she uses to determine the risk of various situations. This process is called *risk perception* and often involves our natural feeling of intuition. Psychologists and researchers have found that the general public and experts perceive risk differently. The average person is more likely to base the riskiness of a situation on his or her beliefs and intuitions rather than on scientific fact. A person's upbringing, social, and cultural factors all impact what he or she

Environmental Health Risk Assessment 301

sees as a hazard and how he or she will respond to these hazards. Typically the public will perceive more risk in situations that invoke dread, in unfamiliar situations, and if there is a potential for a catastrophe. For example, people perceive more risk from nuclear power plants than they do from driving in a motor vehicle. In this case, nuclear power is unfamiliar and is perceived to have the potential to hurt more people than driving in a motor vehicle. People are also more likely to perceive lower risks with activities that they enjoy, that they are familiar with, or that they perceive have greater benefits than risks, such as riding a motorcycle or skiing. In contrast, experts perceive greater risk in situations that have been scientifically associated with increased harm. Experts usually correlate risk with the number of annual fatalities related to a certain activity. Regardless, there is a certain degree of risk perceived with each activity that a person performs, which varies depending on the activity or hazard and the person judging the risk.

Risk mapping is a structure used to estimate the perceived risk among a population. Activities and hazards are mapped out based on two factors: 1) whether or not the risk is observable and 2) whether or not the risk is controllable (see Figure 14.6a). If the obvious risks of an activity or

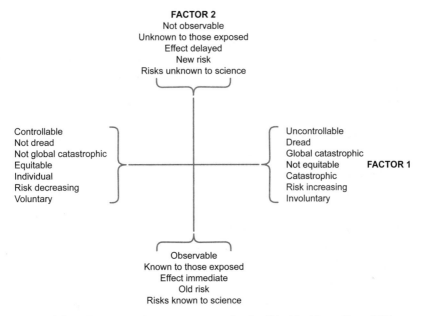

Figure 14.6a Factors used to map perceived risks. (Modified from Slovic [8].)

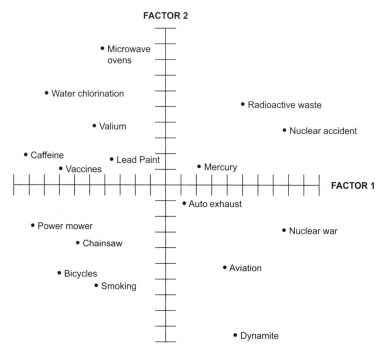

Figure 14.6b Location of hazards on a risk map. (Modified from Slovic [8].)

hazard are unobservable, the perceived risk is typically higher and will be shown higher up on the risk map. If the hazard is uncontrollable, has the potential to invoke dread, and can be catastrophic, the perceived risk is higher and will be shown farther to the right on the risk map. Certain unknown hazards such as radioactive wastes and nuclear power have greater perceived risk than do using a lawn mower or drinking caffeine, as seen in Figure 14.6b.

What the public perceives as a risk can greatly influence their response to a hazard, an educational campaign, or a media report on that particular hazard. Therefore, risk perception and risk mapping are particularly important when environmental health professionals need to communicate a potential hazard to the public. If the public does not see the activity as a potential risk, they will be more unlikely to comply with suggestions or regulations set forth by experts. On the other hand, if the public sees a hazard as posing a large risk, they may feel undermined when experts say that there is no potential risk. By understanding these perceptions, professionals and experts can more effectively communicate the risk characterization of a hazard and increase the community's understanding of that hazard.

Conclusion

Risk perception and the process of risk assessment are commonly performed subconsciously by people every day. Government agencies have developed a formalized process to determine the risk of a hazard based on these perceptions. This formalized process is often used by health departments and agencies; it impacts the public through regulations and suggestions related to hazard exposure. The determination of the risks we face from environmental exposures is important to reduce adverse health effects and control the presence of the hazardous exposure.

To recap, a four-step process is used to determine the risk of a hazard, leading to both risk management and communication. The first step of hazard identification describes the chemical substance and its potential effect on humans. The use of scientific studies and animal data is helpful in understanding the potential effects that a hazard can have on humans; however, caution must be used when equating animal studies to human reactions. Exposure assessment identifies the way that people are exposed to the chemical, how much of the chemical they are receiving, and how long they have been exposed to the chemical. Dose-response identifies the pattern through which the chemical will affect human health. This step is important to gain a broader perspective on whether a small dose will create harmful effects, as in the case of carcinogens, or if a person can receive a certain dose without experiencing adverse health effects, such as noncarcinogens. Last, risk characterization uses all the gathered data to determine the overall risk that the chemical poses to human populations.

The information gathered from this risk assessment process is then used to determine management strategies to reduce the adverse health effects from the chemical or hazard. Communication of these risks and mitigation strategies is then provided to the local population. Both risk management and risk communication must include all stakeholders to determine the most appropriate and effective ways of reducing adverse health effects. All of the processes in risk assessment, management, and communication are intended to reduce the harmful effects a substance may have on a population and improve the quality of life among that population.

Notes

1. V. Covello and J. Mumpower, Risk analysis and risk management: An historical perspective, *Risk Analysis* 5 (2) (1985): 103–120.
2. C. Gerba, Risk assessment, in *Environmental and pollution science*, edited by I. Pepper, C. Gerba, and M. Brusseau (Burlington, MA: Elsevier, 2006), 212–232.
3. U.S. National Research Council, *Risk assessment in the federal government: Managing the process* (Washington, DC: National Academy Press, 1983).
4. U.S. Environmental Protection Agency, *Guidelines for carcinogen risk assessment*, EPA/630/R-00/004 (Washington, DC: Risk Assessment Forum, U.S. Environmental Protection Agency, 1986).
5. C. Brody, J. DiGangi, T. Easthope, et al., IARC downgrading of DEHP, *International Journal of Occupational and Environmental Health* 9 (4) (2003): 399–400.
6. U.S. Environmental Protection Agency, *Risk assessment* (2010), http://www.epa.gov/risk_assessment (accessed October 1, 2010).
7. P. Sandman and J. Lanard, *Crisis communication: Guidelines for action; planning what to say when terrorists, epidemics, or other emergencies strike* (The American Industrial Hygiene Association, 2004), http://www.psandman.com/handouts/AIHA-DVD.htm (accessed October 1, 2010).
8. P. Slovic, Perception of risk, *Science* 236 (4799) (1987): 280–285.
9. W. Hartley, Lecture notes from health risk assessment class, Tulane University School of Public Health and Tropical Medicine, New Orleans, LA, 2003.

15

The Environmental Health Policy-Making Process

Barry L. Johnson

Introduction

Protection of human health and well-being from the adverse consequences of hazards in the environment is the essence of environmental health. Humankind has learned over the ages how to take certain precautions to prevent adverse effects of environmental hazards. For instance, our primordial ancestors surely had to learn how to avoid contact with those feral animals whose intent was to act upon their carnivorous instincts and also to learn which poisonous plants and creatures to avoid. Over eons of time, people learned the importance of clean water and air, sanitary management of human wastes, and the imperative of safe food to consume. These learning experiences were coincident with evolution of social structures, as humans gathered themselves into societies from tribes to megacities. With these social changes came the advent of environmental health policy making, the subject of this chapter.

It is important at this juncture to define two key terms, *policy* and *politics*, because making policy for any of life's endeavors is important to our social development and individual well-being. *Policy* is a definite course or method of action selected from among alternatives and in light of given conditions to guide and determine present and future directions.[1] More to the point of this chapter, policy can also be defined as a plan that embraces the general goals and acceptable procedures in governmental action.[1] As implied by the definition, effective policy making requires making choices among alternatives and will be based on conditions at hand. Environmental health policy therefore consists of actions selected from alternatives used to guide and determine present and future directions that are focused on environmental hazards and conditions.

Politics is a) the art of science of government, b) political affairs or business, and c) the total complex of relations between people living in society.[1] Although most people associate politics with politicians and government, in fact, politics occurs within families, businesses, civic organizations, schools, and other societal structures. In all these examples, politics must incorporate dialogue, debate, negotiation, and ultimately, compromise among the interested parties.

Politics permeates the development and execution of environmental health policy. Some people may have a negative opinion of politics and politicians, because the practice of politics necessarily involves negotiation and compromise, and some politicians have been poor examples of ethical behavior. Thus, to associate a somewhat unwholesome opinion of politics with an altruistic image of environmental health might seem contradictory. Moreover, if environmental health is about preventing the adverse consequences of environmental hazards, should not something so important "be above" politics? The answer, of course, is no. Politics involves relationships among people, and the core of public health is all about people. How health and environmental organizations reach out to the public is a matter of politics, involving communication, negotiation, and compromise. Further, government organizations must compete with other government programs for budgets, personnel allocations, and authorities—all of which requires political acumen and wisdom.

Developing environmental health policy, according to our chosen definition, must involve the identification of alternatives that might be applied to specific situations. From the alternatives, policy makers (e.g., a legislative body, tribal council, or parent) determine the best (applying stated criteria) alternative, communicate their decision to interested parties, and apply the policy when future circumstances arise the response to which must be based on policy.

Having now a working definition of *environmental health policy*, we can proceed to discuss who makes environmental health policy, how environmental policy is developed, how the public's policy expectations are to be met, and examples of some key environmental health policies.

Who Makes Environmental Health Policy?

It is not overly simplistic to say that we are all environmental health policy makers. We make decisions each day on such matters as whether to drive to work or take public transportation, thereby contributing to an increased air pollution burden. We decide on how to manage our household and business wastes, thus adding to a waste stream that can potentially adversely affect persons residing near waste sites. And we decide whether or not to participate in public health programs of vaccination against infectious diseases. So one group of environmental health policy makers are individuals. But there are other groups who make environmental health policies.

Government at all levels is a significant environmental health policy maker. At the federal government level, policies to protect the public against environmental hazards are enacted by the U.S. Congress into specific laws that address basic human needs: potable water, clean air, safe food, sanitary management of wastes, and protection against chemical, physical, and biological hazards. In the United States, examples of such laws include the Clean Water Act, Clean Air Act, Food, Drug and Cosmetic Act, Solid Waste Act, Toxic Substances Control Act, and the Public Health Service Act.[2] State governments, often in compliance with federal environmental laws, enact laws through state legislatures that effectuate provisions of federal laws and in addition sometimes address specific environmental conditions that are unique to a state. Both at the federal level and in most states, environmental health policies are structured around a system of dual responsibilities: environmental protection agencies and public health departments. For example, at the federal level, the U.S. Environmental Protection Agency (USEPA) develops policies in compliance with specific federal environmental statutes (e.g., Clean Air Act) that are intended to protect human and ecosystems health, whereas the U.S. Department of Health and Human Services has components (e.g., National Institutes of Health) that conduct research on the health impacts of various environmental hazards and provide services to the states and others in the administration of public health programs that protect against environmental hazards, e.g., infectious diseases.

Significant environmental health policies are made by local government, such as city, county, and regional governments. Indeed, a citizen who would like to promote a desired environmental health policy can often achieve implementation of that policy through local government action. This is because local government policy makers are the elected officials who are closest to the public and thereby, in theory, are more amenable to public pressure. Subsequent material in this chapter will discuss how policy can be made or influenced. Examples of environmental health policies enacted by local government include bans on tobacco smoking in public places, inspections of food service operations, and programs of vaccination against infectious diseases.

In addition to government, important environmental health policies are developed by entities in the private sector. Large corporations develop environmental health policies for various reasons that comport with their corporate values and strategies. For instance, some large companies support programs of car-pooling and work-at-home, programs that reduce air pollution emissions from vehicles. Other business enterprises make environmental health policies that stipulate their purchase of "green commerce" products, thereby lessening the pollution load on the environment. Other organizations, such as environmental advocacy groups like the Sierra Club, develop environmental health policies that are intended to influence their members and others to take environmentally responsible actions. Advocacy groups often use their policies as a means to influence government policy making. This occurs through lobbying of elected officials and other government policy makers.

It merits mentioning that international organizations such as the United Nations and the European Union make environmental health policies that in turn affect national governments and eventually individuals within a society. Regarding the United Nations, its World Health Organization makes policies on such environmental matters as identifying pandemic diseases, providing services to nations in need of protection against an environmental hazard, e.g., famine, and training health professionals to deal with specific environmental hazards. Regarding the European Union (EU), policies and directives are made by the European Parliament for purpose of controlling specific environmental hazards and threats to human health. Examples of such policies are policies to control the shipment of hazardous wastes between member countries of the EU and the requirement that chemical producers provide the EU with information about the toxicity of specific chemicals that have the potential for release into the environment.

How Is Environmental Health Policy Made?

There is no cookbook for how to make environmental health policy, nor is there a *Policy Making for Dummies* book. Making specific environmental policies will vary in approach according to the conditions and purposes of each situation. However, there are general principles of policy making that apply to the specific situation of environmental health policy making. These general principles will be described in this section, wherein experiential advice is offered, followed by a useful, simplified model of policy making in general.

As prior narrative has stated or inferred, setting environmental health policies is a difficult, complex undertaking. It is a process that is thoroughly political—and should be, because one definition of politics is the "the total complex of relations between people living in society."[1] Politics forces issues into the arena of discussion and debate, whether in the U.S. Congress or in one's family. Inevitably, all politics must include discussion, negotiation, and, ultimately, compromise.

Advocacy Is a Must

Making environmental health policy requires advocacy, even if the policy maker is one's self. Few persons or organizations are able to make environmental health policy themselves. More commonly, policy is made by persons with the access to authority to make policy, e.g., a legislative or executive government body or a corporate office. Typically, legislatures or similar bodies contain committees that are entrusted to develop environmental policies. To have a committee or similar body adopt an environmental health policy requires that a member of the committee become an advocate for the desired policy. In other words, the desired policy, e.g., anti-smoking policy, must be on a committee member's agenda.[3]

To convince a policy maker, especially an elected official, to adopt an environmental health policy as an advocate is usually best broached through the official's staff. An official's staff are relied upon to gauge proposed policies and other matters that are likely to be political in nature and impact. Avoidance of an official's staff is a mistake. Most staff are quite competent and can provide good advice on whether a particular proposed policy needs revision in order to make it politically palatable.

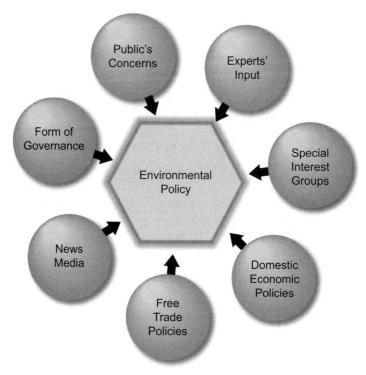

Figure 15.1 Factors that can influence environmental health policy. (Johnson BL [1]:44. Used with the permission of CRC Press.)

Influences on How Environmental Health Policy Is Made

Those desirous of making environmental health policy should be aware that many influences can and usually will come to bear on how the policy is made. Some such influences are shown in Figure 15.1.[2] The elements in this figure were derived from discussions with public and private sector policy makers. These influences are generally listed in what is perceived to be their descending order of impact on policy makers.

The Public's Concerns

To establish environmental health policy through new legislation or refinement of existing regulations or ordinances requires concern expressed by the public—not all of the public, of course, but groups or individuals with special interests or specific concerns. These interests can take the form of an environmental organization, a citizens group, business association, public health agency, or similar groups with a specific concern about an environmental hazard. Setting environmental health policy is seldom initiated by individual elected officials; few elected officials are willing to lead legislative efforts that they may perceive as politically risky. On the other hand, elected officials are generally responsive to concerns from the public—if the pressure is great enough and the political risks are relatively low.

Special Interest Groups

A necessary component of setting environmental health policy is the presence of special interest groups. These are organizations that have specific points of view relative to a particular environmental hazard or proposed policy. To be more specific, examples of special interest groups include national environmental organizations (e.g., World Wildlife Fund), business advocacy organizations (e.g., U.S. Chamber of Commerce), and public health associations (e.g., American Public Health Association). As a particular environmental policy becomes of interest to a legislative body (e.g., U.S. Congress, state legislature, county commission), special interest groups will emerge to present their point of view and attempt to persuade legislators to support their view. This interaction occurs through presentations made by special interest groups in public meetings (e.g., congressional hearings) and private meetings with members of a legislative body and their staffs. Special interest groups are important for assuring a full range of debate and discussion on a proposed environmental health policy, but can be an impediment to democratic processes if they wield undue influence on policy makers.

Newsmedia, Internet, and Social Network Communications

Newsmedia include newspapers, television, radio, cable, and other commercial and public outlets of news and information. Internet communications include Web sites, e-mails, and blogs. Social networks include Facebook, Twitter, YouTube, and similar entities. All these resources serve a vital role in informing the public and shaping opinion. Much of what the

U.S. public knows about environmental hazards is based on information from the news media, particularly from television sources. Readers and viewers are presented with words and images that depict the presence and consequences of specific environmental hazards. A recent example is the news coverage of the Gulf of Mexico oil spill disaster.

Few communication resources have proliferated as quickly and widely as the Internet. It began in the early 1960s at the Massachusetts Institute of Technology. University staff prepared concept papers that laid out the ideas of networking digital computers for purpose of exchanging packets of information. The World Wide Web technology was built upon the Internet infrastructure, allowing users to communicate globally.[5] More recently, the social networks have proliferated. With both the Internet and the social networks, one now has the capability of sending general information and personal communiqués around the globe, almost instantaneously. It is now expected that any group that seeks public interest in its products, services, and agenda must have a Web site on the Internet or a place on one or more of the social networks. This, of course, includes environmental and public health groups and organizations. From such resources can be obtained information on specific environmental hazards, copies of environmental laws and regulations, government policy positions, and special interest groups' stances. All this body of information can inform the public and thereby be used to protect an individual's health or become material for personal and group advocacy, e.g., letters to elected officials. However, there is a caution that must accompany use of information from the Internet and the social networks. The credibility of a Web site or network source can vary, particularly on matters of science, because there are relatively few constraints placed on them. On matters of science, the normal expectations of independent peer review of research reports and their interpretations are seldom followed. The consequence can be unsupportable scientific assertions that are intended to advocate a particular point of policy or other course of action.

In a policy context, news media, Internet, and social network resources can bring environmental problems to the attention of the public, leading to concerns in special interest groups and potentially affected members of the public. These concerns are soon brought to the attention of elected officials and policy makers, who in turn are expected by a concerned public to take action.

Experts' Input

Environmental health policy can be influenced by the findings and recommendations from individual experts or groups of experts. Experts can be representatives from government agencies, universities, corporations, science councils, and such. Legislators and policy makers are often challenged to personally understand or appreciate the seriousness of a particular environmental hazard. In response to this challenge, legislative bodies will often turn to expert groups for their analysis and recommendations about a particular environmental hazard or issue. As an example of experts' input, the U.S. National Academy of Sciences has often been requested by the U.S. Congress and federal government agencies to provide advice on matters of environmental health policy, for example, on risk assessment of environmental hazards. At the local level, university faculty can provide similar advice on matters of local interest such as integrated pest management strategies.

Domestic Economic Policies

Federal environmental health policy, and to a lesser extent, state and local policies, are influenced by domestic economic policies. During times of weak national economies, it is difficult for elected officials and policy makers to impose laws and regulations that could result in further economic hardship. For instance, a particular proposed environmental policy that could harm international trade would be difficult to enact in times of economic austerity. Special interest business groups would argue that the contemplated environmental policy would result in job reductions, lessened corporate profits, stock devaluation, and such. Few elected officials would be willing to favor an environmental policy that might exacerbate a period of economic fragility.

Trade Policies

Free trade policies remove trade barriers between nations and eliminate tariffs on goods. The European Union has no trade barriers between EU Member States. In another example, Canada, Mexico, and the U.S. entered into the North American Free Trade Agreement (NAFTA) on January 1,

1994, a treaty among the three countries that removed barriers to trade across their borders. Some persons and groups argue that free trade translates into lowered environmental and public health protections. They assert that transnational corporations will relocate polluting and injurious (to workers) industries from countries with stringent environmental control to countries without such controls. A particular environmental health policy may founder on arguments that it would complicate free trade and economic well-being.

Form of Governance

How a country, region, state/province, tribal nation, or locality chooses to govern itself can influence how environmental health policy is implemented. In the U.S., the three branches of federal government (legislative, executive, judicial) are mirrored at the state and local government levels. Any federal or state environmental statute, regulation, or local government ordinance is subject to judicial processes if litigation is brought by a party that disagrees with some aspect of the statute, regulation, or ordinance. In the U.S., the method of governance is a democratic republic, which means representatives are democratically elected and authorized to act for other persons who reside in a specific geographic area (e.g., a congressional district). As a democratic republic, serviced by elected officials, considerable input from the public on matters of environmental policy is both possible and desirable.

Environmental health policy established within parliamentary systems is largely set by the relevant ministries of environment and health, operating under broad authorities of parliament. The amount of public input varies from one country to another, but is generally less than in the U.S. system of government.

The Public's Policy Expectations

Policy makers who shape or administer environmental health policies need to be aware that the persons affected by their policies have expectations of what these policies should contain.[2] These expectations can be written into a policy or made part of how the policy is administered. A short alphabetized list of such expectations follows.

Accountability

The notion of accountability is rooted in both ethics and law. For the former, human experience has evolved through religious teachings and secular wisdom to hold a person accountable for his or her actions. For the latter, holding government and corporations accountable for their actions is a relatively recent public policy in the U.S. At the federal level, the U.S. Congress enacted the Government Performance and Results Act (GPRA) in 1993, in part to "improve Federal program effectiveness and public accountability by promoting a new focus on results, service quality, and customer satisfaction."[6] This act was meant to hold executive branches of government more accountable to the public. To the extent that the act is meeting its goal is currently unknown. On a corporate level, the financial irregularities associated with the U.S. economic recession of 2008 onward contributed to public skepticism and demands for more controls on how banks and insurance companies manage their financial accountability to the public. In regard to environmental health policy, the public expects environmental and public health agencies to respond to their concerns and to be accountable for protecting environmental quality and human health.

Cost-Benefit Analysis

Although the U.S. public looks skeptically at personal health decisions that are based on their financial cost—consider the negative reactions of many persons to health maintenance organizations, where costs allegedly drive decisions on health care—the emergence of cost-benefit analyses and hazard management decisions have become policy within government agencies and business operations. Cost-benefit policy has occurred in government because of legislative directives and federal court decisions and within business operations that must have a sense of costs associated with their products, working conditions, and consumer affairs.

Environmental Justice

How environmental hazards are experienced by people of color became a matter of social justice in the 1970s in the U.S. Minority communities and persons of low income expressed their belief that toxic chemicals, in particular, were being deliberately released into their communities from hazardous waste dumps, incinerators, and pollution from industry. The resulting expressions of concern led to the establishment of environmental justice, defined by USEPA as "[t]he fair treatment and meaningful involvement of all people regardless of race, color, national origin, or income with respect to the development, implementation, and enforcement of environmental laws, regulations, and policies."[7] Federal government policies emerged in the early 1990s that were intended to address a number of environmental justice concerns. State governments have emulated the federal example by creating offices that investigate environmental justice allegations and recommend corrective actions. While the effectiveness of federal, state, and some private sector actions to prevent environmental injustices continues to be debated, the public policy to prevent environmental injustices has become a cornerstone in the foundation of U.S. social justice.

Federalism

The sharing of power and authorities between the U.S. federal government and its territories, the states, and local governments is a public policy that crosscuts almost all social programs in this country, including environmental statutes. As an example pertaining to environmental health policy, the states have the primary responsibility for enforcing air pollution regulations developed by the federal government (i.e., USEPA), with overall responsibility for the development of air quality standards vested with the federal government.

Polluter Pays for Consequences of Pollution

In the 1970s, environmental groups adopted the strategy that those who cause pollution should pay the costs for its effects on the environment and remediation. The first federal expression of this thesis is found in the Comprehensive Environmental Response, Compensation, and Liability Act of 1980. In particular, companies and others that created hazardous waste dumps became liable for the costs of cleanup of the dumps and associated effects on human health and natural resources. A similar legal philosophy has been adopted by the European Union and elsewhere.

Prevention Is Preferred to Remediation

Common sense tells us that avoiding a problem is preferable to having to fix its consequences. Prevention of disease and disability is the cornerstone of public health. This cornerstone is policy at all levels of public health, from federal to state to local health departments and their programs. Disease and disability reduce a person's quality of life, can lead to costly health care, and lessen societal strength by eliminating or reducing ability to work and contribute to society. Although not always evident, environmental statutes in general are expressions of the prevention policy. All such statutes are predicated on protection of human health and environmental quality. This protection is addressed through control of environmental hazards, as pursued through regulations and standards and their enforcement.

Product Safety

With consumerism has come public policy specific to product safety. Consumers do not want products that could harm them or their children. The result has been federal resources directed to the prevention of harmful consumer products' entry into commerce. The Consumer Product Safety Commission (CPSC), in particular, identifies commercial products (e.g., children's toys) that can be hazardous. As an example, the CPSC has proposed banning the sale in the U.S. of baby cribs that have side rails that can be raised or lowered because of concern that some young children have suffocated when the rails collapsed on their head.

The Public's Right to Know

It has become public policy in the U.S. that individuals have the right to know of conditions that may be hazardous to their health and well-being. The public's right to know is not absolute. For example, matters of national security and confidential business secrets are excluded from public view, unless ordered by a court. While the importance of right to know may seem self-evident, it has not always been the case, especially in the realm of information available to workers and consumers of commercial products. Not until the passage of the U.S. Occupational Safety and Health Act in 1970 were there general requirements to inform U.S.

workers of on-the-job hazards (e.g., hazardous chemicals). Similarly, producers of consumer products had no requirement to inform U.S. consumers that flaws in product design or manufacture could be hazardous to them until enactment of the federal Consumer Product Safety Act of 1972.

Risk Assessment and Risk Management

Risk assessment and risk management have become particularly dominant in the area of environmental health, beginning circa 1980.[8] Federal regulatory agencies, primarily USEPA and the U.S. Occupational Safety and Health Administration (OSHA), have implemented formal risk assessment procedures to estimate the degree of risk posed by individual environmental hazards. The impetus for this development came from federal court decisions which found that USEPA and OSHA regulations had failed to establish the degree of risk in certain proposed regulatory actions. In effect, the courts ruled that risk must be consequential, not trifling or unsupported, in any regulatory action.

Social Support

Cultures have over eons of evolution learned that their survival depended on social support systems. These included hunting in groups for food, banding together to defend territory, and traveling in groups on trading expeditions. Living in close proximity, not as isolated individuals, gave protection to groups, eventually forming villages, cities, and nation states. Nations have developed social support structures that are intended to enhance the survival of both the state and the individual. Assistance to persons who need food, shelter, and education is commonplace, generally worldwide. In the U.S., assistance programs include public education, subsidized housing, health services to the elderly and the indigent, and monetary subsidies to farmers, among many others. These are examples of public policy meant to provide social support to meet basic needs of members of a society. Unmet basic needs can lead to societal disruptions (e.g., political turmoil, chaos) that reduce a society's ability to protect itself, foster economic gains, and enhance quality of life for individuals. How much of a society's resources should be devoted to social support systems is, and will remain, a legitimate debate.

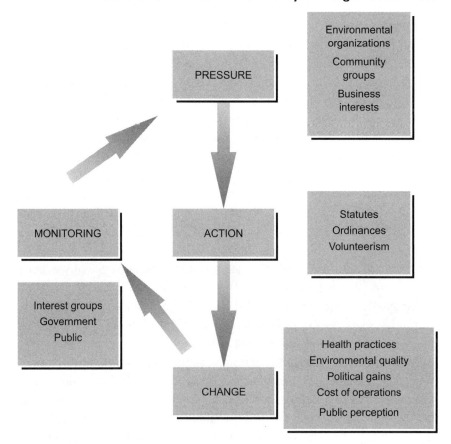

Figure 15.2 Simplified flowchart of environmental policy making. (Johnson BL [1]:52. Used with the permission of CRC Press.)

The PACM Model of Policy Making

Establishing any policy involves a political dimension, and the accompanying policy making and practice of politics are intertwined. Recognition of this reality leads to a simplified diagram of policy making. Policy making in general can be viewed as the process shown in Figure 15.2, which is referred to here as the PACM model, where P represents Pressure, A refers to Action, C stands for Change, and M represents Monitoring.[2] The PACM model is a classic feedback loop model, wherein adjustments are taken on the basis of information obtained from sources within the process.

Pressure. Because setting any policy is a political event, and recognizing that all political systems have a certain amount of inertia, putting pressure on the system is required. This is because elected officials are often slow to support a proposed policy initiative until they have calculated the political implications of their support. The need to bring pressure on elected officials and other policy makers is particularly important on matters of environmental health policy, where economic impacts of proposed policies (e.g., regulating the siting and operating of incinerators) often become controversial.

Bringing pressure on political systems to set federal environmental health policies occurs through lobbying by environmental organizations (e.g., Environmental Defense, Natural Resources Defense Council), business associations (e.g., U.S. Chamber of Commerce), trade associations (e.g., American Petroleum Institute), professional societies (e.g., American Public Health Association), state governments, and specialty experts (e.g., National Academy of Sciences). The pressure is exerted through meetings with policy makers and their staffs, testimonies given at public hearings, reports in the newsmedia, and other outlets that can influence public opinion and motivate elected officials.

Pressure can be brought to bear on business enterprises to achieve environmental goals and changes in business practices. Like pressure directed at government policies and practices, meetings between environmental groups and policy makers in business can bring about debate, negotiation, and compromise. An example is the successful pressure applied to the McDonald's food chain to effect changes in how food was packaged. Specifically, environmental groups sought to replace plastic containers with those fabricated of paper and cardboard. Environmentalists were concerned about the relatively lack of biodegradability of plastic materials, resulting in their longevity in municipal landfills much longer than their paper counterparts.

Action. When sufficient pressure is brought to bear on a political system, action can occur. Action means simply that something occurs, gets done, or moves. Actions can take the form of enacted legislation by federal or state legislatures or ordinances promulgated by county and local governments. Action can also be new or revised federal or state environmental regulations or new policies.

Change. Change, simply put, means that a different course of action will occur from what was previously done. For example, whereas the Air Pollution Control Act of 1955 was limited primarily to research and technical and economic assistance to state and local governments, the Clean Air Act Amendments of 1963 enlarged USEPA's responsibilities for controlling air pollution. At the same time, the act continued congressional intent that "[t]he prevention and control of air pollution at the source is the primary responsibility of state and local governments."[4] Change can also occur when pressure is directed to Executive Branch agencies and departments. Such pressure is more narrowly focused and not always as visible to the public as pressure directed to legislative bodies.

Monitoring. Just because a change in a policy has occurred does not always mean that those who initiated the change will be satisfied. The changes may be perceived as being ineffective, too costly, or misdirected. Monitoring (or surveillance) of the changes is a common means to assess the impact of a change in policy. Monitoring data can, in turn, be used to refine policies or even get them withdrawn.

In summary, an understanding of the PACM model is fundamental in order to effect policy development, be it a national environmental policy or a family budget. Application of this simple model must be tempered with the wisdom that politics will be a consideration at each step. No policy gets established without the art of debate, negotiation, and compromise.

Conclusion

As discussed in this chapter, making environmental health policies can be influenced by many factors. For instance, making environmental health policies involves both politics and determination. Politics must be factored into the policy-making process, since policy makers are often elected officials and they work within a political framework. Determination is required because environmental health policies often require that policy makers be convinced of the worth of a proposed policy. Moreover, the public has policy expectations, as described in this chapter, that must be factored into the adoption of environmental health policies. To facilitate the development of environmental health policies, a simple PACM model is described that can help focus attention on the actions necessary to develop and implement a particular environmental health policy. In the end, the effort involved in making environmental health policy is often extensive, but the outcomes are worth the effort, since what is at stake is protection of human health and the environment.

Notes

1. *Webster's ninth new collegiate dictionary* (Springfield, MA: Merriam-Webster Publishers, 1986).
2. B. L. Johnson, *Environmental policy and public health* (Boca Raton, FL: CRC Press, 2007).
3. W. A. Rosenbaum, *Environmental politics and policy*, 4th ed. (Washington, DC: Congressional Quarterly, 1999), 179, 184.
4. J. Fromson, A history of federal air pollution control, *Environmental Law Review—1970* (1970): 214.
5. Internet Society, *A brief history of the Internet*, http://www.isoc.org/internet/history/brief.shtml (accessed January 30, 2011).
6. U.S. Congress, *Government performance and results act of 1993* (Washington, DC: U.S. Government Printing Office, 1993).
7. U.S. Environmental Protection Agency, *Environmental justice*, http://www.epa.gov/environmentaljustice/ (accessed January 30, 2011).
8. National Research Council, *Risk assessment in the federal government: Managing the process* (Washington, DC: National Academy Press, 1983).

16

Significant Environmental Health Statutes and Key Regulations

Barry L. Johnson

Introduction

Over time the U.S. federal government has become the nation's major environmental health authority. This is not to say that the federal government acts alone; indeed, much of environmental protection and practice is conducted in cooperation with the states, territories, and tribal nations. This kind of cooperative relationship is often the product of specific mandates in federal statutes. The purpose of this chapter is to describe the more significant federal environmental health statutes and their key regulations. The chapter is organized around a public health perspective, focused on both protection of the environment as well as promotion of human welfare. More specifically, a healthful environment can be characterized by clean air, potable water, safe food, sanitary management of wastes, safe consumer products, and workplaces that do not present unsafe and unhealthful working conditions. These essentials of environmental health are used as the organizational structure for this chapter.

It is important to understand the process by which environmental statutes and regulations are created in the United States. To create a statute, a member of Congress has to propose a bill. Sometimes bills have several cosponsors. Bills are assigned to pertinent congressional committees for deliberation and action. Bills that receive committee approval are voted on by all members of the House of Representatives and the Senate. If both houses of Congress approve a bill, it goes to the president, who has the option to either approve it or veto it. If approved by the president or by Congress overriding a presidential veto, the new statute is called an *act*, and the text of the act is known as a *public statute*. For the purposes of this chapter, *statute* is used as a synonym for *law*. Once an act is passed, the House of Representatives standardizes the text of the statute and publishes it in the *United States Code*, which is the official record of all federal statutes.

Statutes often do not include all the details of Congress's intent. In order to make the statutes work on a day-to-day level, Congress authorizes specific federal agencies to create regulations. *Regulations* set specific rules about what is legal and what isn't. The process of making federal regulations is described in the following sections.

Federal Regulations

Prior to discussing individual federal environmental health statutes, it is important to observe that much of the effectiveness of the statutes derives from regulations that effectuate the statutes' implementation. That is not to say that environmental regulations are the only means by which to protect the environment and human health. Indeed, there is a host of nonregulatory actions that contribute to environmental protection and mitigation of environmental hazards. Voluntary actions taken by individuals, such as water conservation, are an example. However, the core of the substantive actions taken for environmental health purposes stems from regulatory actions authorized by specific environmental statutes.

Once a regulation is completed and has been printed in the *Federal Register* as a final rule, it is "codified" by being published in the *Code of Federal Regulations* (CFR). The CFR is the official record of all regulations created by the federal government. It is divided into fifty volumes, called titles, each of which focuses on a particular area. (The CFR can be accessed at http://ecfr.gpoaccess.gov.) Almost all environmental regulations appear in Title 40.

The following sections describe in brief the federal government's regulatory programs, the rulemaking process, and one key regulatory strategy, called standards and regulations.

The Federal Government's Regulatory Programs

Federal agencies such as the U.S. Environmental Protection Agency (EPA), the Occupational Safety and Health Administration (OSHA), and the Food

Significant Environmental Health Statutes and Key Regulations 325

and Drug Administration (FDA), and at least fifty others are called *regulatory agencies* because they are authorized by specific federal statutes enacted by Congress to develop, promulgate, and enforce rules (i.e., regulations) that carry the full force of law, which means they bring enforcement. Most federal government regulations fall into one or more of five categories: *Process requirements* are regulations that control emissions from sources of pollution and usually require the use of performance standards for pollution control technologies. *Product controls* cover certain commercial products, including a product's design and potential uses. *Notification requirements* mandate notification of government regulatory agencies and/or the public about a company's actions. *Response requirements* require that a specific federal regulatory agency be notified when emergency conditions occur. *Compensation requirements* require an individual or business entity to reimburse the federal government for actions detrimental to the public's welfare.[1]

The Federal Government's Rulemaking

The federal government's process of making regulations, called *rulemaking*, proceeds along the following course. 1) Congress must enact legislation, signed into law by the president (or by way of Congress's overriding a presidential veto), that requires a federal agency to develop and promulgate rules (i.e., regulations) specific to a particular congressional intent. Such legislation is called *authorizing legislation* or *enabling legislation*. 2) The designated executive branch agency (e.g., EPA) develops proposed regulations following the requirements of the Administrative Procedures Act, which requires that the agency publish them in the *Federal Register*, the federal government's journal of daily announcements, in compliance with a specific statute. Meetings held during the rulemaking process must generally be open to the public. The public has the opportunity to submit data to the rulemaking agency during the rulemaking process. Draft and final rules must be published in the *Federal Register* and on the agency's Web site. 3) Final rules take effect upon publication in the *Federal Register*, but often provide a time schedule for compliance by the regulated community.[2]

Standards and Regulations

Before describing the major federal environmental health statutes, it is important to mention that the primary environmental health regulatory policy is called *standards and regulation* (also sometimes called *command and control*).[3] This policy is operative in many of the statutes discussed in this chapter. In particular, the distinction between quality and emission standards needs elaboration. Q*uality standards* are the maximum concentrations of contaminants permitted in specific environmental media, such as air, water, and food. They are established by regulatory agencies through a process of review of scientific literature, risk estimation, and health impact. In distinction, *emission standards* prescribe the amount of contaminant discharges that can be released within a prescribed length of time from significant emission sources, such as industrial facilities, municipal discharges into water, and landfills. It is important to understand the meanings of both kinds of standards and their essential interrelation.

Significant Federal Environmental Health Statutes

The U.S. federal government plays a most significant role in protecting the American people against the health and ecological consequences of environmental hazards. This has not always been the case. In the early years of the United States, states and local governments, but especially individual persons, bore the responsibility for protections against such environmental hazards as unsafe food, contaminated water, and unwise sanitation practices. One of the first expressions of federal environmental relevance occurred in 1798, with the enactment by Congress of An Act for the Relief of Sick and Disabled Seamen.[4] This statute provided health care for American mariners returning from seafaring duties. This same act also addressed the problem of shipboard vermin, such as rodents, that arrived with incoming ships.

But it was not until the twentieth century that the federal government became the principal authority in protecting the American public against such environmental health hazards as unsafe food, contaminated water, foul air, dangerous consumer products, unsanitary disposal of wastes, and hazardous workplace conditions. Much of this seminal legislation was adopted during the 1960s and 1970s, a period characterized by considerable social unrest due to such societal changes as African Americans' fight for civil rights and students' protests against U.S. involvement in the Vietnam War. Against this backdrop of changes in direction and national policy, an environmental movement arose that championed the enactment of environmental protection statutes. The major federal statutes that flowed from the American public's gradual embrace of federal leadership in environmental and public health protection are described below. Each statute's description is accompanied by a brief statement of the statute's key regulations.

Air Pollution Statutes

Pollution of the air we breathe for our very existence is a problem likely dating from antiquity, perhaps from the time when humans first came into contact with smoke from fires used for warmth and warding off predators. One source cites an event in the year 1306, when citizens of London petitioned their government to take action to reduce levels of smoke in ambient air.[5] In response, King Edward I issued a royal proclamation to prohibit artificers (i.e., craftsmen) from burning sea coal (i.e., low-quality coal gathered from sea shores), as distinguished from cleaner-burning charcoal, in their furnaces. This is an early example of government taking action against the effects of air pollution, which can be defined as the contamination of the atmosphere by gaseous, liquid, or solid wastes. Given this fourteenth-century example of one government's attempts to improve citizens' air quality, it is not surprising to learn that in the twentieth century the U.S. public's concern about air pollution also led to government action.

The Clean Air Act
The federal government began trying to control air pollution in the 1950s with the passage of the Air Pollution Control Act of 1955.[6] This was the first federal air pollution law, and it mandated federal research programs to investigate the health and welfare effects of air pollution. It also authorized the federal government to provide technical assistance to states. In 1963 the Clean Air Act (CAA) replaced the Air Pollution Control Act, with the purpose of empowering the Secretary of Health, Education, and Welfare to define air quality criteria based on scientific studies and to provide grants to state and local air pollution control agencies. In 1965 the Motor Vehicle Air Pollution Control Act directed the EPA to establish automobile emission standards. The Federal Air Quality Act was enacted in 1967 and established a framework for defining "air quality control regions" based on meteorological and topographical factors of air pollution.

The 1970 CAA amendments required the EPA to develop National Ambient Air Quality Standards (NAAQS) for individual air contaminants and then placed most of the responsibility on the states to achieve compliance with the standards. The 1977 CAA amendments added provisions for geographic areas with air cleaner than national standards in order to prevent their deterioration in air quality, and provisions were added pertaining to *nonattainment areas*, that is, geographic areas that had failed to meet national air quality standards. Under the 1977 amendments, states were required to develop State Implementation Plans (SIPs) that would meet air quality standards.

The 1990 CAA amendments added comprehensive provisions to regulate emissions of air toxicants, acid rain, and substances thought to be a threat to the earth's ozone layer. In addition, the 1990 amendments added a comprehensive air emissions permit program and markedly strengthened enforcement provisions and requirements for mobile sources of emissions, content of automobile fuels, and geographic areas that fail to meet air quality standards (i.e., nonattainment areas). These amendments finally banned the sale in the United States of gasoline that contained lead additives, ending one of the twentieth century's worst environmental health missteps, since lead in ambient air became a significant source of children's exposure to this toxicant. The 1990 amendments also changed the way hazardous air pollutants are regulated. The CAA, as now amended, in effect recognizes two kinds of ambient air pollutants: the six Criteria Air Pollutants and, basically, everything else. The CAPs consist of oxides of sulfur, oxides of nitrogen, ground-level ozone, particulate matter, lead, and carbon monoxide. They are associated with emissions from internal combustion engines. The second category of air pollutants comprises what are called Hazardous Air Pollutants (HAPs), which are pollutants primarily released from industrial operations.

Key Clean Air Act Regulations
40 CFR part 50 requires the EPA to set National Ambient Air Quality Standards for pollutants considered harmful to public health and the environment. The Clean Air Act established two types of national air quality standards. *Primary standards* set limits to protect public health, including the health of sensitive populations such as asthmatics, children, and the

elderly. *Secondary standards* set limits to protect public welfare, including protection against decreased visibility and damage to animals, crops, vegetation, and buildings. Both primary and secondary air quality standards are legally enforceable.

40 CFR part 63 contains national emission standards for hazardous air pollutants (NESHAP) established pursuant to section 112 of the act as amended November 15, 1990. These standards regulate specific categories of stationary sources that emit (or have the potential to emit) one or more hazardous air pollutants listed in this part.

Water Quality Statutes

Water suitable for human consumption and other uses has historically been of importance to public health. Indeed, around 400 BC Hippocrates, the father of medicine, emphasized the importance of boiling and straining water for health purposes. Modern programs of water quality protection can be dated from the late nineteenth century, when chlorine was found to be an effective water disinfectant when added in low concentrations to drinking water supplies. In 1902 Belgium became the first country to make continuous use of chlorine as an additive to drinking water supplies. Chlorination of public drinking water supplies in the United States dates to 1908, when the Boonton reservoir supply, Jersey City, New Jersey, was chlorinated, triggering a series of lawsuits that were ultimately decided by courts in favor of water chlorination as a means of water purification.

Prior to the enactment of federal water quality statutes, states bore the responsibility for dealing with issues of sanitation and drinking water quality. The federal government's involvement with water pollution control dates to the turn of the twentieth century, when water pollution control regulations were included in the Rivers and Harbors Act of 1899. This act authorized the regulation of industrial discharges of pollution into waters that might cause navigation problems. Later the Public Health Service Act of 1912 expressed the first federal policy on the disposal of human wastes, which authorized the Public Health Service to provide technical advice and assistance to communities and for federal research on sanitary waste disposal methods. Over time more comprehensive, focused federal water quality legislation was enacted by Congress, as described herein.

The Clean Water Act

The principal federal statute now governing pollution of the nation's waterways is the Federal Water Pollution Control Act, more commonly called the Clean Water Act (CWA).[7] The original purpose of the act was to establish a federal program to award grants to states for construction of sewage treatment plants. Although originally enacted in 1948, the act was completely revised by amendments in 1972, giving the CWA most of its current shape. The 1972 legislation declared as its objective the restoration and maintenance of the chemical, physical, and biological integrity of U.S. waters.

The CWA, as amended, contains a number of complex elements of overall water quality management. Foremost is the requirement that states must establish ambient water quality standards for water bodies, consisting of the designated use or uses of a water body (e.g., recreational, public water supply, or industrial water supply) and the water quality criteria that are necessary to protect water's uses. Through the issuance of water discharge permits, states or the EPA impose wastewater discharge limits on individual industrial and municipal facilities in order to ensure that water quality standards are attained. Under the CWA, states must identify lakes, rivers, and streams for which wastewater discharge limits are not stringent enough to achieve established water quality standards, after implementation of technology-based controls by industrial and municipal dischargers. For each of these water bodies, a state is required to set a *total maximum daily load* (TMDL) of pollutants at a level that ensures that applicable water quality standards can be attained and maintained. A TMDL sets the maximum amount of pollution a waterbed can receive without violating water quality standards, including a margin of safety.

The 1977 CWA amendments focused on toxic pollutants, establishing the basic structure for regulating discharges of pollutants into waters of the United States. Further, the amendments established a program to regulate the discharge of dredged and fill material into U.S. waters, including wetlands. In 1987 the CWA was reauthorized and again focused on toxic substances. Prior to 1987 the CWA only regulated pollutants discharged to surface waters from *point sources* (i.e., pipes, ditches, and similar conveyances of pollutants), unless a water discharge permit was obtained under provisions in the act. For the first time, *nonpoint sources* (e.g., stormwater runoff from agricultural lands) were covered by the 1987 amendments. In 2000 the CWA was amended when Congress enacted the Beaches

Significant Environmental Health Statutes and Key Regulations 331

Environmental Assessment and Coastal Health (BEACH) Act, which provides the EPA with additional authority to regulate the quality of water used for recreation at beaches.

Key Clean Water Act Regulations

40 CFR part 122 implements the National Pollutant Discharge Elimination System (NPDES) program. The NPDES program requires that water discharge permits be issued for the discharge of pollutants from any point source into waters of the United States. The EPA can delegate this program to individual states.

40 CFR part 131 establishes the authority for setting water quality standards and defines the water quality goals of a water body, or portion thereof, by designating the use or uses to be made of the water and by setting criteria necessary to protect the uses.

The Safe Drinking Water Act

The Safe Drinking Water Act (SDWA) of 1974 was enacted to protect the quality of drinking water in the United States.[8] This statute focuses on all waters actually or potentially designed for drinking use, whether from surface or underground sources. The SDWA, as amended in 1986 and 1996, gives the EPA the authority to set drinking water standards. *Drinking water standards* are regulations that the EPA sets to control the level of contaminants in the nation's drinking water. There are two categories of drinking water standards.

A National Primary Drinking Water Regulation (NPDWR or primary standard) is a legally enforceable standard that applies to public water systems. Primary standards protect drinking water quality by limiting the levels of specific contaminants that can adversely affect public health and are known or anticipated to occur in water. They take the form of Maximum Contaminant Levels or Treatment Techniques. A National Secondary Drinking Water Regulation (NSDWR or secondary standard) is a nonenforceable guideline regarding contaminants that may cause cosmetic effects (such as skin or tooth discoloration) or aesthetic effects (such as taste, odor, or color) in drinking water. The EPA recommends secondary standards for water systems but does not require systems to comply with them. However, states may choose to adopt them as enforceable standards.

In 1996 Congress made sweeping changes to the SDWA. Originally the SDWA focused primarily on treatment as the means of providing safe drinking water at the tap. The 1996 amendments modified the existing statute by recognizing source water protection, operator training, funding for water system improvements, and public information as important components of safe drinking water programs. A cost-benefit analysis and a risk assessment are required before a drinking water standard can be set.

Key Safe Drinking Water Act Regulations

40 CFR part 141 establishes national primary drinking water standards. These regulations control chemical and biological contaminants in drinking water that can affect human health.

40 CFR part 142 specifies that the states have the primary enforcement responsibility for public water systems in the state during any period for which the EPA administrator determines that the state has adopted drinking water regulations that are no less stringent than the national NPDWRs in effect under part 141 of this chapter and has adopted and is implementing adequate procedures for the enforcement of such state regulations.

40 CFR part 143 establishes NSDWRs. These regulations control contaminants in drinking water that primarily affect the aesthetic qualities relating to the public acceptance of drinking water. The regulations are not federally enforceable but are intended as guidelines for the states.

Food Safety Statutes

Americans in the twenty-first century expect not to be harmed by the food and medicinal drugs and therapeutic devices with which they come into contact. That expectation is the product of personal experience (e.g., few of us have had serious illnesses from eating impure food or from taking contaminated legal drugs) and general trust in public health systems (e.g., restaurant inspections). Although episodes of illness occur as the result of impure food (e.g., undercooked meat in hamburgers), the current situation is vastly different from that of our ancestors.

The Food, Drug, and Cosmetic Act

In the nineteenth and early twentieth centuries, any government control of food and drugs was the responsibility of the states. State statutes, if extant, varied greatly. In that era, use of chemical preservatives and toxic colors added to food was virtually uncontrolled. During the 1870s the grassroots Pure Food Movement arose and soon became the principal source of political support for federal food and drugs legislation. In 1903 Dr. Harvey W. Wiley became the director of the U.S. Department of Agriculture's (USDA's) Division of Chemistry and soon thereafter helped arouse public opinion against impure consumer products that he and his staff had identified.[9]

The Food and Drugs Act of 1906 prohibited the manufacture and interstate shipment of adulterated and mislabeled foods and drugs.[10] President Theodore Roosevelt was an active advocate for this legislation. The statute enabled the federal government to initiate litigation against the producers of alleged illegal products, but lacked affirmative requirements to guide compliance with the statute. The 1906 statute also lacked key provisions necessary to make it effective in identifying harmful food and drug products. The USDA's Division of Chemistry enforced the provisions of the Food and Drugs Act of 1906 until 1931, when the Food and Drug Administration (FDA) was created.

In 1938, following the deaths of more than 100 adults and children who had taken an untested medicine, the Federal Food, Drug, and Cosmetic Act (FDCA) of 1938 was enacted and was signed into law by President Franklin D. Roosevelt, giving the FDA the authority to require that drugs be proven safe before they could be marketed, establish "tolerances" for certain harmful substances, and set standards of identity and quality for drugs. This act and its subsequent amendments have provided the FDA with the authority to regulate food, drugs, cosmetics, and medical devices, and as such impacts daily the lives of all Americans. In 1940 the FDA was transferred from the USDA to what later became the U.S. Department of Health and Human Services.

In more recent times, two significant statutes have expanded the FDA's regulatory and policy authorities. One statute, the Family Smoking Prevention and Tobacco Control Act (FSPTCA), was signed into law by President Barack Obama on June 22, 2009. The law provides the FDA with new responsibilities for controlling the public health carnage caused by use of tobacco products. Tobacco use is the leading cause of early deaths in the United States, with more than 400,000 persons dying annually from cigarette smoking and exposure to tobacco smoke. The FSPTCA, among other provisions, a) requires companies who manufacture or import tobacco products to provide the FDA with a listing of the amounts of all ingredients in the products; b) gives the FDA the authority to require companies to provide information about the amount of nicotine in their products; c) gives the FDA the authority to promulgate standards for nicotine yields and for the reduction of other harmful substances in tobacco products; d) requires warnings on tobacco products to cover 50 percent of the front and back panels of the package and be in large and legible type; and e) requires the FDA to issue regulations concerning the advertising of, and access to, tobacco products.

A second significant change in the FDA's public health authority occurred on January 4, 2011, when President Barack Obama signed into law the Food Safety Modernization Act. The act had been enacted by Congress following reports of deaths of persons in the United States who had consumed unsafe foods. The modernization act's general provisions require companies that produce food to develop and implement written food safety plans; in addition, the provisions give the FDA the authority to better respond to complaints and require food recalls, and further require the agency to better ensure that imported foods are as safe for consumption as are those produced in the United States. More specifically, a) food facilities must have a written preventive controls plan that specifies the possible problems that could affect the safety of their products; b) the FDA is required to increase the frequency of inspections of food production facilities, with high-risk U.S. facilities receiving an initial inspection within the first five years of the law's implementation; c) the FDA must establish science-based standards for the safe production and harvesting of fruits and vegetables; and d) the FDA is given authority to mandate a recall of unsafe food if the producing company fails to do so voluntarily.

Key FDCA Regulations: 21 CFR part 7 specifies procedures for removing or correcting consumer products that are in violation of statutes administered by the FDA. Recall is a voluntary action taken by manufacturers and distributors. A failure to recall can lead to the FDA's seeking court action to remove the product.

21 CFR part 101 applies to food labeling, including such requirements as labeling of food ingredients, nutrition information, health claims, and any food warnings, among other requirements.

21 CFR part 129 specifies criteria for the processing of bottled drinking water. Regulations pertain to assuring that bottled drinking water is

safe and has been processed, bottled, held, and transported under sanitary conditions.

21 CFR part 314 sets forth procedures and requirements for the submission to, and review by, the FDA of applications and abbreviated applications to market a new drug.

Federal Meat Inspection Act

Consumption of meat and meat products has long been part of the human diet, although debate continues about the ethics of raising animals as a food source. Our ancestors—whether indigenous people or colonists—hunted the forests, plains, and bodies of water for food sources such as birds, mammals, fish, and shellfish. With the passage of time, rural Americans grew their own food in gardens and processed domesticated animals into meat and meat products. A family's meat quality and personal health protection were therefore at the mercy of a farmer's skill and resources in food preservation. The federal government had little role to play in what were essentially personal matters of family diet and health.

Until 1906 states had the primary responsibility for protecting the public against impure food, including meat products. Needless to say, food inspection programs varied considerably among the states. The public's ignorance of conditions in the meatpacking industry began to change in the early years of the twentieth century. In particular, a major influence on public opinion was Upton Sinclair's 1906 book, *The Jungle*, which graphically described unsanitary conditions and inhumane slaughter of animals in the Chicago meatpacking industry. On June 30 Congress enacted the Federal Meat Inspection Act of 1906 (FMIA).[11] The act was substantially amended by the Wholesome Meat Act of 1967. The FMIA was the first federal government involvement in the food safety of meat and meat products. The provisions of the FMIA are administered by the USDA.

The primary goals of the FMIA, as amended, are to prevent adulterated or misbranded livestock and products from being sold as food and to ensure that meat and meat products are slaughtered and processed under humane and sanitary conditions. These requirements apply to animals and their products produced and sold within states as well as to imports, which must be inspected under equivalent foreign standards. The key provisions of the FMIA, as amended, are as follows: inspection by USDA inspectors of animals prior to slaughter; humane methods of slaughter; and post-mortem inspection and labeling of approved carcasses and meat products. Inspections include both visual examinations of carcasses as well as tests for microbacterial contamination. The Poultry Products Inspection Act and the Egg Products Inspection Act provide the USDA with parallel authority similar to that under the FMIA.

Key FMIA regulations: 9 CFR part 416 stipulates that meat processing establishments must be operated and maintained in a manner sufficient to prevent the creation of unsanitary conditions and to ensure that products are not adulterated.

9 CFR part 500 states the conditions under which the USDA may take a regulatory control action against a meat processor because of such conditions as unsanitary practices, product adulteration, and inhumane handling or slaughtering of livestock.

Waste Management Statutes

How humankind manages its waste is a vital issue of environmental health. Our ancestors learned from experience that water and food that were contaminated with unsanitary waste could cause diseases such as cholera and dysentery. Regarding federal statutes, the Clean Water Act, previously described, provides resources and regulations that address water treatment facilities. More recently we have learned that mismanagement of hazardous waste can also cause adverse health effects to human and ecological systems. This section addresses federal legislation that deals with the management of solid waste, both hazardous and nonhazardous.

The Resource Conservation and Recovery Act

The Resource Conservation and Recovery Act (RCRA) of 1976 established the federal program that regulates the management of solid and hazardous wastes. The RCRA amends earlier legislation, the Solid Waste Disposal Act of 1965, but the amendments of 1976 were so comprehensive that the act is commonly called RCRA rather than by its official title.

As background, the Solid Waste Disposal Act of 1965 focused on research, demonstrations, and training. It provided for sharing with the states the costs of making surveys of waste disposal practices and problems and of developing waste management plans. The Resource Recovery

Act of 1970 changed the whole tone of the legislation, from efficiency of disposal to concern with the reclamation of energy and materials from solid waste. The RCRA of 1976 instituted the first federal permit program for hazardous waste management and prohibited open dumps. The RCRA amendments of 1984 prohibited land disposal of untreated hazardous wastes; set liner and leachate collection requirements for land disposal facilities; set deadlines for closure of facilities not meeting standards; and established a corrective action program for facilities that store, treat, or dispose of hazardous waste.[12]

For conceptualization purposes, the RCRA's regulatory authorities can be divided into the categories of hazardous waste and nonhazardous solid waste. For the former, the EPA regulates commercial enterprises as well as federal, state, and local government facilities that generate, transport, treat, store, or dispose of hazardous waste. For nonhazardous solid waste, the EPA establishes minimum criteria that landfills must meet in order to manage solid waste. These landfills are predominantly regulated by state and local governments, but must meet EPA standards.

Key RCRA Regulations: 40 CFR part 256 provides guidelines for development and implementation of states' solid waste management facilities. These guidelines contain methods for achieving the objectives of environmentally sound management and disposal of solid and hazardous waste, resource conservation, and maximum utilization of valuable resources. Plans must prohibit the establishment of new open dumps within the states.

40 CFR part 264 establishes minimum national standards that define the acceptable management of hazardous waste. The standards in this part apply to owners and operators of all facilities that treat, store, or dispose of hazardous waste.

The Comprehensive Environmental Response, Compensation, and Liability Act

The Comprehensive Environmental Response, Compensation, and Liability Act (CERCLA, or Superfund) was enacted in 1980 and was reauthorized by the Superfund Amendments and Reauthorization Act of 1986.[13] The statute was a direct consequence of the discoveries of releases of hazardous substances from abandoned landfills into community residences. In particular, the community of Love Canal, a suburb of Niagara Falls,

New York, was evacuated following the discovery that it overlay an abandoned chemical dump. Love Canal captured the American public's attention because of intense news media coverage.

The CERCLA was therefore the product of great public concern that toxic materials could invade private homes and cause harm to children and future generations. The intent of the law is stated to be, "To provide for liability, compensation, and emergency response for hazardous substances released into the environment and the cleanup of inactive hazardous waste disposal sites."[13] The CERCLA's basic purposes are to provide funding and enforcement authority for remediating (i.e., cleaning up) uncontrolled hazardous waste sites, responding to hazardous substance emergencies, reacting to public health concerns, identifying "potential responsible parties" to pay for remediation costs, and removing emergency chemical spills.

The Emergency Planning and Community Right-to-Know Act of 1986 was enacted as Title III of the 1986 CERCLA amendments. Under Title III, state and local governments are required to develop emergency plans for responding to unanticipated environmental releases of acutely toxic materials. In addition, businesses covered by Title III must notify state and local emergency planning agencies of the presence and amounts in inventory of hazardous materials on their premises and to notify federal, state, and local authorities of planned and uncontrolled environmental releases of those substances. All this information is made available to the public.

Key CERCLA Regulations: 40 CFR part 300 references the National Oil and Hazardous Substances Pollution Contingency Plan (NCP), which provides the organizational structure and procedures for preparing for and responding to discharges of oil and releases of hazardous substances, pollutants, and contaminants.

40 CFR part 370 establishes reporting requirements for providing the public with important information on the hazardous chemicals in their communities. Reporting raises community awareness of chemical hazards and aids in the development of state and local emergency response plans.

Product Safety Statutes

We all come into contact with commercial products that we accept for specific purposes. Such products include the food we eat, the clothes we wear, the medications we use, the vehicles we utilize for transportation, the cosmetics we apply, and the household and industrial chemicals we use for pest control, gardening, and other purposes. We expect these products to be beneficial and to provide some enhancement of our lives. But this is not always the case. Some products, especially if used improperly, can adversely affect our health and well-being. Some pesticides, food additives, and children's toys are such examples.

This section describes three major federal statutes that address products that can have safety implications if used improperly. One might argue that the Food, Drug, and Cosmetic Act should be included here, but the importance of food safety merits a separate section of its own.

The Federal Insecticide, Fungicide, and Rodenticide Act

Although federal pesticide legislation was first enacted in 1910, its aim was to reduce economic exploitation of farmers by manufacturers and distributors of adulterated or ineffective pesticides. Congress did not address the potential risks to human health posed by pesticide products until it enacted the 1947 version of the Federal Insecticide, Fungicide, and Rodenticide Act (FIFRA).[14] The USDA became responsible for administering the pesticide statutes during this period. However, responsibility was transferred to the EPA when that agency was created in 1970. Broader congressional concerns about long- and short-term toxic effects of pesticide exposure on pesticide applicators, wildlife, nontarget insects and birds, and food consumers subsequently led to a complete revision of the FIFRA in 1972. The 1972 statute, as amended, is the basis of current federal policy.

The FIFRA, as amended, requires the EPA to regulate the sale and use of pesticides in the United States through registration and labeling of the estimated 21,000 pesticide products currently in use. The act directs the EPA to restrict the use of pesticides as necessary in order to prevent unreasonable adverse effects on humans and the environment, taking into account the costs and benefits of various pesticide uses. The FIFRA prohibits sale of any pesticide in the United States unless it is registered and labeled for approved uses and restrictions. The EPA registers each pesticide for specific approved uses, for example, to control boll weevils on cotton. In addition, the FIFRA requires the EPA to reregister older pesticides based on new data that meet current regulatory and scientific standards. Establishments that manufacture or sell pesticide products must be registered by the EPA. Facility managers are required to keep certain records and to allow inspections by the EPA or state regulatory representatives.

Most pesticides currently registered in the United States are older pesticides and were not subject to modern safety reviews. Amendments in 1972 to the FIFRA directed the EPA to re-register approximately 35,000 older products, thereby assessing their safety in light of current knowledge. Many of the 35,000 products will not be reviewed, and their registrations will be canceled because registrants do not wish to support re-registration.

In 1996 the Food Quality Protection Act (FQPA) of 1996 revised both the FIFRA and the FDCA. The overall purpose of the FQPA is to protect the public from pesticide residues found in the processed and unprocessed foods they eat. Essentially, the FQPA amended the FIFRA and the FDCA so that a single health-based standard would be issued to alleviate problems concerning the inconsistencies between the two statutes. The health-based standard is to be based on a "reasonable certainty of no harm." The EPA is required to review all pesticide tolerances within ten years, giving particular attention to protection of young children exposed to pesticide residues.

Key FIFRA Regulations: 40 CFR part 152 sets forth procedures, requirements, and criteria concerning the registration of pesticide products.

40 CFR part 156 regulates the labeling of pesticide products, specifying that such labels must contain the product's name, name of the producer, the registration number, direction for use, hazard statements, and other information.

40 CFR part 171 establishes standards for certification of commercial pesticide applicators.

The Toxic Substances Control Act

The enactment of the Toxic Substances Control Act (TSCA) of 1976 directs the EPA to execute the following key actions: 1) require chemical manufacturers and processors to conduct tests of existing chemicals; 2) prevent future risks through premarket screening and regulatory tracking of new chemical products; 3) control unreasonable risks already known or as they are discovered for existing chemicals; and 4) gather and disseminate information about chemical production, use, and possible adverse effects to human health and the environment.[15]

Significant Environmental Health Statutes and Key Regulations 341

At the time of the TSCA's enactment, the statute allowed continued production of the 62,000 chemicals already in commercial use, which were called *existing chemicals*. Another 18,000 chemicals have been introduced into commerce since 1976, known as *new chemicals*. In sum, approximately 80,000 chemicals potentially fall under the regulatory provisions of the TSCA. The TSCA authorizes the EPA to screen existing and new chemicals used in commerce in order to identify potentially dangerous products or uses that should be subject to federal control. The EPA may require manufacturers and processors of chemicals to conduct and report the results of tests to determine the effects of potentially dangerous chemicals on living organisms. Based on test results and other information, the EPA may regulate the manufacture, importation, processing, distribution, use, and/or disposal of any chemical that presents an unreasonable risk of injury to human health or the environment. A variety of regulatory tools are available to the EPA under the TSCA, ranging in severity from a total ban on production, import, and use to a requirement that a product bear a warning label at the point of sale.

Key TSCA regulations: 40 CFR part 710 requires persons who manufacture, import, or process chemical substances for commercial purposes to report information to the EPA, which is compiled into an inventory of chemical substances manufactured or processed for a commercial purpose.

40 CFR part 720 requires any person who intends to manufacture a new chemical substance in the United States for commercial purposes to submit a notice, unless the substance is excluded under TSCA provisions. The person must specify the identity of the substance, the total amount produced, and the basic technology for the manufacturing process.

The Consumer Product Safety Act

Of the major federal regulatory agencies of relevance to environmental health, the smallest in resources and perhaps least known to the American public is the Consumer Product Safety Commission (CPSC). The CPSC was created by Congress under the Consumer Product Safety Act of 1972.[16] The purposes of the act are 1) to protect the public against unreasonable risks of injury associated with consumer products; 2) to assist consumers in evaluating the comparative safety of consumer products; 3) to develop uniform safety standards for consumer products and to minimize conflicting state and local regulations; and 4) to promote research and investigation into the causes and prevention of product-related deaths, illnesses, and injuries.

Although the CPSC has statutory authority to develop safety standards for products under their jurisdiction, voluntary standards are mandated by the act, which states, "The Commission shall rely upon voluntary consumer product standards rather than promulgate a consumer product safety standard prescribing requirements [w]henever compliance with such voluntary standards would eliminate or adequately reduce the risk of injury addressed and it is likely that there will be substantial compliance with such voluntary standards."[16] According to the Commission, since 1990 their cooperative work with industry has resulted in 214 voluntary standards. During the same period, 35 mandatory rules were issued. By law, the CPSC can issue a mandatory standard only when a voluntary standard has been determined by the Commission not to have eliminated or adequately reduced the risk of injury or death or if a voluntary standard is unlikely to be met with substantial compliance.

Fueled by reports of imported toys that contained lead and other toxic materials, a major upgrade of the Consumer Product Safety Act resulted from congressional legislation enacted in 2008 and signed into law by President George W. Bush on August 14, 2008. The Consumer Product Safety Improvement Act of 2008 provides the CPSC with additional authorities and resources for addressing product safety issues. In particular, the act requires that toys and infant products be tested by manufacturers before they are sold and bans the presence of lead and phthalates in toys. The act also creates the first comprehensive publicly accessible consumer complaint database and further provides the CPSC with new resources, increases civil penalties that CPSC can impose on violators of the Commission's regulations, and provides protection for whistle-blowers who report product safety defects.

Key CPSC Regulations: 16 CFR part 1508 sets forth the requirements that must be met by manufacturers of full-size baby cribs.

16 CFR 1508 establishes regulations that pertain to the manufacture and sale of bicycles, including regulations for a bicycle's braking system, steering system, pedals, drive chain, tires, and other components.

Significant Environmental Health Statutes and Key Regulations 343

Workplace Safety and Health Statutes

Where we work can be considered an environment. With this understanding, there are two primary environmental health statutes. One is the Occupational Safety and Health Act of 1970. The other is the Mine Safety and Health Act of 1977. The former covers general workplace conditions in the United States; the latter covers a specific kind of workplace, surface and underground mines. Only the former act is described in this chapter because of its more general relevance to environmental health concerns and policies.

The Occupational Safety and Health Act

The Occupational Safety and Health Act of 1970 (OSHA) was the first U.S. federal legislation to deal comprehensively with health and safety problems in the workplace. The declared purpose and policy of the statute is "[t]o assure so far as possible every working man and woman in the Nation safe and healthful working conditions and to preserve our human resources."[17] To effectuate this goal, the act created three new government agencies. The Occupational Safety and Health Administration (OSHA), which is located in the U.S. Department of Labor, is directed to develop workplace safety and health standards, conduct workplace inspections, enforce regulatory actions developed by the agency, help set up state occupational safety and health programs and monitor their effectiveness, and conduct education and training programs for safety and health professionals. The National Institute for Occupational Safety and Health (NIOSH), which is a part of the U.S. Department of Health and Human Services, is tasked with conducting scientific research, performing workplace health evaluations, and developing criteria documents that contain recommendations for safety and health exposure conditions. The third agency is the Occupational Safety and Health Review Commission, which was established to adjudicate disputes arising from enforcement of the act.

The most consequential OSHA responsibility under the act is the development and promulgation of safety and health standards. This is done through regulatory action, taking into account all relevant scientific information and public concerns. OSHA standards affect most workplaces, large and small. Employers convicted of willful violation of OSHA standards may face civil or criminal penalties, depending on the seriousness of the infraction. OSHA's inspectors conduct inspections of workplaces to assess compliance with OSHA safety and health standards. Where no specific standards exist for a workplace condition, the act directs each employer to provide "[a] place of employment which is free from recognized hazards that are causing harm to employees" (p. 370).[17] This "general duty" clause is used by OSHA to control workplace hazards that are obvious and for which no specific standard exists.

Key OSHA Regulations: 29 CFR part 1910 promulgates occupational safety and health standards that have been found to be national consensus standards or established federal standards.

29 CFR part 1910.1200 requires that the hazards of all chemicals produced or imported be evaluated and that information about these hazards be transmitted to employers and employees. This transmittal of information is to be accomplished by means of comprehensive hazard communication programs, which are to include container labeling and other forms of warning, material safety data sheets, and employee training.

Conclusion

Over time, a significant set of federal environmental health statutes and attendant regulations have evolved in the United States. Some of these statutes date back to the beginning of the twentieth century, whereas others are the product of mid-twentieth-century social conditions, which included the American public's embrace of environmental protection. All these statutes are intended in one way or another to protect the American public against the adverse health effects of various environmental hazards. These hazards include air pollutants, water pollutants, unsafe food, hazardous consumer products, and unsafe working conditions. Described in this chapter are those statutes and illustrative key regulations that are intended by the federal government to protect the American public against these kinds of environmental hazards.

Notes

1. Office of Management and Budget, *Report to Congress on the costs and benefits of federal regulations*, http://www.whitehouse.gov/omb/inforeg/chap1.html (accessed January 30, 2011).

Significant Environmental Health Statutes and Key Regulations 345

2. B. L. Johnson, *Environmental policy and public health* (Boca Raton, FL: CRC Press, 2007).
3. W. A. Rosenbaum, *Environmental politics and policy*, 4th ed. (Washington, DC: Congressional Quarterly, 1998).
4. F. Mullan, *Plagues and politics: The story of the United States public health service* (New York: Basic Books, 1987).
5. J. A. Fromson, A history of federal air pollution control, *Environmental Law Review—1970* (1970): 214.
6. J. E. McCarthy, L. B. Parker, L. Schierow, et al., *Clean air act: Summaries of environmental laws administered by EPA* (Washington, DC: Congressional Research Service, 1999).
7. C. Copeland, *Clean water act: Summaries of environmental laws administered by EPA* (Washington, DC: Congressional Research Service, 1999).
8. M. Tiemann, *Safe drinking water act: Summaries of environmental laws administered by EPA* (Washington, DC: Congressional Research Service, 1999).
9. F. Mullan, *Plagues and politics: The story of the United States public health service* (New York: Basic Books, 1987).
10. Center for Food Safety and Nutrition, FDA (CFSAN), *The story of the laws behind the labels: Part I. 1906—The federal food, drug, and cosmetic act* (Washington, DC: U.S. Food and Drug Administration, 1981).
11. House Agriculture Committee, Federal meat inspection act of 1906, http://www.fsis.usda.gov/Help/glossary-f/index.asp (accessed January 30, 2011).
12. J. E. McCarthy and M. Tiemann, *Solid waste disposal act/resource conservation and recovery act: Summaries of environmental laws administered by EPA* (Washington, DC: Congressional Research Service, 1999).
13. M. Reisch, *Superfund: Summaries of environmental laws administered by EPA* (Washington, DC: Congressional Research Service, 1999).
14. L. Schierow, *Federal insecticide, fungicide, and rodenticide act: Summaries of environmental laws administered by EPA* (Washington, DC: Congressional Research Service, 1999).
15. L. Schierow, *Toxic substances control act: Summaries of environmental laws administered by EPA*, National Library for the Environment,

http://ncseonline.org/nle/crsreports/briefingbooks/laws/k.cfm (accessed January 30, 2011).
16. Consumer Product Safety Act (CPS Act), U.S. code collection, http://www.thecre.com/fedlaw/legal5c/uscode15-2051.htm (accessed January 30, 2011).
17. R. Moran, Occupational safety and health act, in *Environmental law handbook*, edited by J. G. Arbuckle (Rockville, MD: Government Institutes, 1991).

17

The Role of State and Local Public Health Departments in Environmental Enforcement

Mark B. Horton, Alison F. Dabney, Cindy A. Forbes, Patrick Kennelly, Richard A. Kreutzer, Gregg W. Langlois, Carl Lischeske, Robert Schlag, and Gary H. Yamamoto

Introduction

Environmental enforcement is "the set of actions that governments or others take to achieve compliance within the regulated community and to correct or halt situations that endanger the environment or public health. Enforcement by the government usually includes inspections, negotiations, and legal action. It may also include compliance promotion."[1]

The goal of environmental enforcement is to protect the natural environment and promote healthy places that "improve the quality of life for all people who live, work, worship, learn, and play within their borders—where every person is free to make choices amid a variety of healthy, available, accessible, and affordable options."[2]

Environmental enforcement addresses a wide range of issues, such as ensuring the safety of public drinking water systems and the food supply; ensuring that drugs, medical devices, and certain other consumer products are safe and effective and free from adulteration, misbranding, and false advertising; ensuring that radiologic materials, medical waste, and sewage are handled and disposed of safely; preventing, monitoring, and controlling vector-borne diseases; tracking and preventing pesticide-related illness; ensuring the quality of indoor and outdoor air; and promoting and ensuring the safety of beach and other recreational waters.

Public health plays a critical role in protecting and promoting healthy places by engaging in the three core functions—assessment, policy development, and assurance—and the ten essential services of public health (see Table 17.1). Environmental enforcement agencies perform the core functions and essential services through activities as varied as conducting environmental surveillance and biomonitoring; diagnosing source conditions; training and educating industries, workers, the media, and the public; forming multilayer partnerships; developing regulatory policies; enforcing laws; training and maintaining a competent environmental public health workforce; evaluating policy effectiveness; and researching new ideas and alternatives. And when policies and enforcement activities fail to perform at the desired level, or unforeseen environmental events expose people or the environment to harmful agents, protective or treatment modalities are identified to avoid or mitigate the negative health impact. For example, environmental enforcement does not fully protect all children from lead exposure, but childhood lead prevention programs link exposed children to healthcare providers for testing, evaluation, and treatment.

Effective environmental enforcement integrates regulatory activity at the federal, state, and local levels. Increasingly, those governmental regulatory activities are supplemented by citizens acting in the public interest, educating people about environmental hazards, advocating for policy and regulatory change, and taking legal action to ensure enforcement of environmental health laws and regulations. For example, the California Safe Drinking Water and Toxic Enforcement Act of 1986 (known as Proposition 65) specifically authorizes lawsuits by private parties "acting in the public interest" as a means of enforcement. Table 17.2 shows the respective roles of federal, state, and local governments in selected areas of environmental enforcement in California.

This chapter briefly describes four areas of environmental enforcement to illustrate the breadth and scope of activities addressing the wide range of environmental threats mentioned above. Short California case studies in each area demonstrate the variation in the vertical integration of federal, state, and local public health activities and the horizontal integration of public health activities with the environmental protection activities of other government agencies. The case studies also illustrate how providing the ten essential services of public health ensures comprehensive and

TABLE 17.1

Environmental Enforcement and the Ten Essential Services of Public Health

	Examples of Public Health Environmental Enforcement Activities			
Ten Essential Services of Public Health	Drinking Water	Food Safety	Shellfish	Medical Waste
1. *Monitoring* health status to identify community health problems, including health disparities.	X	X		
2. *Detecting and investigating* health problems and health hazards in the community.	X	X	X	X
3. *Informing, educating, and empowering* people and organizations to adopt healthy behaviors to enhance health status.	X	X	X	X
4. *Partnering* with communities and organizations to identify and solve health problems and to respond to public health emergencies.	X	X	X	X
5. Developing and implementing public health *interventions* and best practices that support individual and community health efforts and increase healthy outcomes.	X	X		X
6. *Enforcing* laws and regulations that protect health and ensure safety.	X	X	X	X
7. *Linking* people to needed personal health services and ensuring the provision of population-based health services.				
8. Assuring a competent public health *workforce* and effective public health leadership.	X	X		X
9. *Evaluating* effectiveness, accessibility, and quality of public health services, strategies, and programs.	X	X		
10. *Researching* for insights and innovative solutions to public health problems.	X	X	X	X

TABLE 17.2
Roles of Federal, State, and Local Government in Selected Areas of Environmental Enforcement in California

	Federal Role	State Role	Local Role
Drinking Water	• Promulgate Clean Water Act • Promulgate Safe Drinking Water Act • Develop laboratory standards and methods • Perform research on water treatment methods • Conduct risk assessment for contaminants • Conduct risk management of contaminants	• Conduct risk assessment for contaminants • Conduct risk management of contaminants • Monitor water quality • Monitor water wells • Inspect underground fuel storage tanks • Approve and monitor recycled water projects that replace drinking water supplies	• Monitor water wells • Track abandoned wells • Monitor beaches • Inspect underground fuel storage tanks • Seal unused wells
Food Safety	• Promulgate food, drug, cosmetic, and consumer product laws • Conduct foodborne illness investigations	• Conduct foodborne illness investigations • Provide laboratory support for food outbreak investigations • Conduct food inspections • Issue food and consumer product embargoes and recalls • Train and certify food service workers	• Conduct retail food vendor inspections • Investigate food vendor complaints
Shellfish	• Conduct National Shellfish Sanitation Program (NSSP) • Promulgate NSSP Guide for the Control of Molluscan Shellfish (Model Ordinance) • Interstate Shellfish Sanitation Conference	• Certify commercial growing areas • Conduct sanitary surveys and pollution source evaluation • Assess water quality • Monitor for marine neurotoxins • Prevent illegal harvesting	• Regulate small wastewater treatment systems and onsite sewage disposal that could impact growing areas • Collect samples for marine toxin analysis • Post information on biotoxin quarantines
Medical Waste	• None	• Inspect health care facilities	• Inspect health care facilities

(continued)

TABLE 17.2
(Continued)

Federal Role	State Role	Local Role
	• Register medical waste transporters • Permit transfer and treatment facilities • Register trauma scene waste practitioners • Review alternative treatment technologies • Maintain local enforcement agency consistency • Register home-generated sharps waste consolidation points • Investigate incidents and complaints	• Register home-generated sharps waste consolidation points • Investigate incidents and complaints

effective environmental enforcement. Our case studies highlight drinking water, food and drug, medical waste, and shellfish activities in California.

Drinking Water

In the summer of 1854, London experienced the two most severe cholera outbreaks in the city's history. Dr. John Snow, perhaps the first epidemiologist, plotted the deaths on a map. The result resembled a target, with the greatest concentration of deaths around a well on Broad Street. Upon inspection of the well site, Snow concluded that a cesspool on the property was the cause of the outbreak and ordered the well's pump handle removed so no one could use the well water. Though it would be several decades before the germ theory of disease was developed, Dr. Snow was already using several tools of environmental health enforcement that are used today.[3]

The federal Interstate Quarantine Act of 1893 gave the surgeon general of the U.S. Public Health Service (PHS) the authority "to make and enforce such regulations as are necessary to prevent the introduction, transmission, or spread of communicable disease from foreign countries into the states of possessions, or from one state or possession into any other state of possession."[4] The first water-related regulation was adopted in 1912. PHS adopted the first drinking water standards in 1914; revised drinking water standards were issued in 1925, 1942, 1946, and 1962.[5]

In 1974 the Public Health Service Act was amended by the federal Safe Drinking Water Act (SDWA) and gave the U.S. Environmental Protection Agency (EPA) jurisdiction over public water systems. A public water system is a system that provides water for human consumption through pipes or other constructed conveyances that has fifteen or more service connections or regularly serves at least twenty-five individuals daily at least sixty days out of the year.

EPA jurisdiction includes authority to establish and enforce drinking water standards called maximum contaminant levels (MCLs) and power to delegate this authority to state health agencies via "primacy" agreements. Primacy states must adopt laws that conform to federal drinking water laws, take enforcement actions against public water systems that violate these laws, perform routine inspections of public water systems, and provide federal and state funding to public water systems to make necessary improvements. In 1996 the SDWA was amended to include better oversight of all water systems, especially small water systems, and establish the Safe Drinking Water State Revolving Fund, which provides funding to water systems to help them comply with drinking water standards.

Although the EPA is responsible for developing national standards for drinking water contaminants, many states have acted independently to establish drinking water standards to protect consumers from emerging contaminants of concern to the state, such as perchlorate and methyl tertiary butyl ether (MTBE).

The following case studies describe waterborne illness outbreaks and drinking water compliance actions.

Case Study:
Recreational Vehicle Park Waterborne Disease Outbreak

In July 2000 an outbreak of gastroenteritis affected 147 users of a private recreational vehicle park located on the shores of a rural lake. The California Department of Public Health and local investigators found that the leach system for an on-site sewage system had failed. The leach field was located sixty-five feet from the primary well that supplied the park. Investigators immediately ordered the camp operator to post boil-water notices on all water fixtures in the park. Bacterial analysis of samples from the well and distribution system confirmed that the water supply was contaminated with coliform bacteria. Subsequent samples indicated the presence of *Escherichia coli* (*E. coli*) bacteria and Calicivirus in the well. Calicivirus was also found in stool specimens collected from ill campers. The local enforcement agency ordered the park operator to disinfect the distribution system, disconnect the contaminated well from the system and use the backup well, provide fail-safe chlorination for the backup well, and construct a new well and leach field. The operator complied with these directives, and no further illness occurred.

Case Study:
Uncovered Drinking Water Reservoirs

In some communities a significant portion of drinking water is stored in uncovered distribution reservoirs. Water in these surface sources can be contaminated by a variety of sources, including surface water runoff, bird and animal wastes, human activity, algal growth, insects, fish, and airborne deposition of particulates. As a result, uncovered reservoirs have been subject to increasing regulation at both the state and federal levels. For example, four reservoirs of a large public water system receive surface runoff and were subject to state and federal regulations requiring surface water treatment. In July 1993 the large public water system entered into an agreement with the California Department of Public Health to bring these four reservoirs into compliance with state regulations. The compliance agreement required the water system to implement interim operations plans to mitigate the public health risk of operating the reservoirs, provide ongoing public notification regarding the compliance status of the reservoirs, and comply with state regulations at each reservoir according to a state-approved schedule. Six other uncovered reservoirs remain in this large water system, including a 3.3-billion-gallon main reservoir. To comply with current federal surface water treatment rules, the water system must remove these reservoirs from service or treat the discharge in accordance with a state-approved schedule. The water system submitted a plan and schedule to install floating covers on two of these uncovered reservoirs and to replace the other four with covered storage or buried tanks.

Case Study:
Methyl Tertiary Butyl Ether (MTBE)

In California the risk assessment and risk management components of drinking water regulation occur in two separate state entities. First, the California Environmental Protection Agency's Office of Environmental Health Hazard Assessment performs a risk assessment to set a public health goal (PHG) for a drinking water contaminant of concern. A PHG is the concentration of a drinking water contaminant that poses no significant health risk if consumed for a lifetime, based on current risk assessment principles, practices, and methods. After the Office of Environmental Health Hazard Assessment establishes a PHG for a contaminant, the California Department of Public Health develops a maximum contaminant level (MCL) for that contaminant. State law requires the MCL to be as close to the PHG as is technically and economically feasible, placing primary emphasis on the protection of public health. In addition to MCLs intended to protect the public against adverse health effects, California law and the federal SDWA allow for establishing secondary MCLs to address contaminants that affect aesthetic quality such as taste, odor, and appearance. Although secondary MCLs established under federal law are not enforceable, California's law requires public water systems to comply with secondary MCLs adopted by the state unless a majority of the customers agree that they are not willing to pay for the treatment technology necessary to comply.

An example of the process by which California chooses drinking water contaminants to regulate and how the final regulation is established is the case of the chemical MTBE. MTBE is a colorless, liquid hydrocarbon that has been used as an octane booster in gasoline since the 1970s.

In 1991 MTBE was detected in a groundwater source being considered for drinking water use. In response, the California Department of Public Health established an action level (a nonregulatory guidance level) of 35 micrograms per liter (ug/l) for MTBE in drinking water. Over the next several years, MTBE was detected in twenty-six drinking water sources; the most significant contamination in groundwater was 610 ug/l in one city. The contamination affected 50 percent of that city's drinking water supply.

The Role of State and Local Public Health Departments 355

As a result of these findings, in 1997 the California Department of Public Health required public water systems to monitor for MTBE. Water systems detected MTBE in 1.4 percent (67 of 4,889) of the drinking water sources tested. Significant public criticism over the continued use of MTBE as a gasoline additive and demands for action to protect drinking water sources prompted legislation that required the California Department of Public Health to establish both a primary MCL and a secondary MCL for MTBE.

In response to the statutory mandate to establish a primary and secondary MCL for MTBE, the California Department of Public Health asked the Office of Environmental Health Hazard Assessment to establish a PHG for MTBE. Based on MTBE's potential carcinogenic effects, the Office of Environmental Health Hazard Assessment established a PHG of 13 ug/l, assuming a *de minimus* theoretical excess individual cancer risk of 10^{-6} (one in a million) from exposure to MTBE. A plaintiff challenged the adoption of the PHG in California's Superior Court, arguing that the PHG was a regulation and therefore had to be adopted through the normal regulatory process defined in California law. The court rejected the argument, and subsequent law explicitly states that PHGs are not regulations.

The California Department of Public Health reviewed taste and odor studies and established a secondary MCL for MTBE of 5 ug/l, as the studies indicated the majority of consumers would not experience objectionable tastes and odors at that level. The Department established a primary MCL for MTBE of 13 ug/l (the same level as the PHG). California is one of at least eleven states that have adopted a primary drinking water MCL for MTBE. As of 2010, a national drinking water standard for MTBE had not been established.

Food Safety

Food safety regulations in the United States emerged during the early colonial era, and regulations to protect consumers continue to evolve. In 1862 President Abraham Lincoln established the Bureau of Chemistry, the predecessor of the current U.S. Food and Drug Administration (FDA). After revelations of scandalous and unsanitary conditions in the food, drug, and meatpacking industries, in 1906 Congress passed the Pure Food and Drug

Act and the Meat Inspection Act. By prohibiting misbranded and adulterated foods, drinks, and drugs; controlling the conditions in processing plants; and banning the use of poisonous preservatives and dyes in food, these two laws set the foundation for ensuring food and drug safety nationwide. In 1938 Congress passed the federal Food, Drug, and Cosmetic Safety Act. In 1949 the FDA published guidance for the food processing industry, Procedures for the Appraisal of the Toxicity of Chemicals in Food. From 1950 to 1952, the congressional Delaney Committee investigated the safety of chemicals in foods and cosmetics. In 1958 the Food Additives Amendment required manufacturers to demonstrate the safety of food additives. In the following decades, the federal government continued to enact legislation and regulations to address specific food safety issues, such as the Infant Formula Act, Toxicological Principles for the Safety Assessment of Direct Food Additives and Color Additives Used in Food, the Nutritional Labeling and Education Act, the Dietary Supplement Health and Education Act, the Current Good Manufacturing Practice regulations, and the Food Allergy Labeling and Consumer Protection Act.[6]

Food safety laws and regulations are enforced through a close partnership among federal, state, and local environmental health professionals. At the state level, food safety enforcement activities include risk-based food safety inspections and investigations of food manufacturers and distributors. These activities help prevent foodborne illness and verify compliance with applicable statutes and regulations, including regulations related to good manufacturing practices, standards of identity, labeling, and food and color additives. Failure of process controls in food manufacturing or distribution systems can put consumers at risk of severe illness or even death. Many of these events require extensive investigations of complex food manufacturing processes. These investigations include detailed reviews of each point of the food production continuum, from the growing field to the point of sale, to determine how the product became contaminated. Enforcement activities also include responding to complaints, natural disasters, fires, product tampering, and other events that can compromise the safety of the food supply.

If necessary, the state takes enforcement action to ensure consumers are not exposed to adulterated or misbranded foods. The state may levy embargoes against adulterated, misbranded, and falsely advertised products. Businesses in violation of the law can face civil, criminal, and/or administrative penalties including incarceration in jail, fines, and loss of operating licenses.

In the wake of a number of foodborne illness outbreaks attributed to California products, the California Department of Public Health formed the California Food Emergency Response Team (CalFERT). CalFERT comprises state and federal investigators and scientists who jointly investigate outbreaks. CalFERT has investigated outbreaks such as the 2006 *E. coli* O157:H7 outbreak, attributed to bagged spinach. CalFERT also works with manufacturers on product recalls to ensure that adulterated foods are removed quickly from the marketplace.

Case Study:
CalFERT Response to *Salmonella* Rissen Outbreak

In 2009 CalFERT investigated a multistate outbreak of *Salmonella* Rissen, a rare strain of salmonella. A total of eighty-seven individuals in five states (California, sixty-five; Nevada, ten; Oregon, eight; Washington, three; and Idaho, one) became ill with *Salmonella* Rissen associated with consuming white pepper. Illness onset dates ranged from December 9, 2008, to April 29, 2009.

Public health officials in the affected states collected and analyzed data to determine a common source that may have led to these illnesses. Initially they were unable to find any common food exposure that would explain a majority of the cases. Several illness clusters among people who ate at buffet style restaurants made the variety and complexity of potential food sources quite large. On March 27, 2009, the Oregon Public Health Division isolated *Salmonella* Rissen from pepper samples collected from a restaurant in Oregon where a sickened individual had eaten. The samples came from opened containers of ground white pepper and ground black pepper from the same manufacturer. These samples were subsequently matched to the outbreak strain of *Salmonella* using pulse field gel electrophoresis (PFGE), which allows the comparison of the DNA fingerprint of the bacteria found in food items to those found in case patients. Additional pepper samples from the same manufacturer were also collected from restaurants in California, Nevada, and Washington.

On March 27, 2009, CalFERT initiated an investigation at the pepper manufacturing facility. Investigators collected environmental and product samples and firm manufacturing information, and reviewed product processing and shipping documentation. After a thorough inspection of the facility and its processing methods, CalFERT embargoed all raw materials and in-process and finished products at the facility because of the potential for cross-contamination from the pepper grinding operations.

CalFERT collected 391 product and environmental samples at the manufacturing facility. Sixty-three (16 percent) of these samples were positive for *Salmonella*. Twenty-five of the sixty-three samples (40 percent) were further tested and found positive for *Salmonella* Rissen that matched the outbreak strain. This 100 percent match of the further tested samples made it unnecessary to perform additional testing on the other thirty-eight samples positive for *Salmonella*. Ground white pepper collected from closed and in-process containers was the only product collected from the facility that was a PFGE match to the *Salmonella* Rissen outbreak strain.

CalFERT worked with the local district attorney to obtain a preliminary injunction strictly controlling the manufacturer and preventing it from further operations until the California Department of Public Health was satisfied with the remediation efforts at the facility. The department required the manufacturer to establish protocols to ensure that raw materials were adequately inspected for potential contamination and establish process controls to prevent contamination or cross-contamination of products produced or repackaged at the facility. The manufacturer resumed operations at the facility in approximately four months. The district court subsequently issued a permanent injunction to ensure continued compliance with the California Department of Public Health requirements to ensure food products are unadulterated.

Case Study:
Reducing Illness and Death from Eating Raw Oysters
Although state health departments work closely with their federal counterparts, states have the authority to take regulatory action to ensure the safety of the food supply when federal agencies do not act. The National Shellfish Sanitation Program is the federal/state cooperative program recognized by the U.S. Food and Drug Administration

The Role of State and Local Public Health Departments 359

and the Interstate Shellfish Sanitation Conference for the sanitary control of shellfish produced and sold for human consumption.[7] In cooperation with the Interstate Shellfish Sanitation Conference, the National Shellfish Sanitation Program issues guidelines for ensuring the safety of shellfish for human consumption.

Despite the guidelines, throughout the 1990s significant numbers of illnesses and deaths each year were related to *Vibrio vulnificus infections* following the consumption of raw oysters from the Gulf of Mexico. *Vibrio vulnificus* bacteria are naturally occurring marine pathogens that can cause serious infections and death in immunocompromised persons or those with liver diseases who consume contaminated raw oysters. Nationally, from 1989 to 2000 there were 275 foodborne *Vibrio vulnificus* illnesses (resulting in 143 deaths) in California and the Gulf Coast states (Florida, Louisiana, Texas, Alabama, and Mississippi). Of those, California had 44 illnesses and 29 deaths.

In March 1991 the California Department of Public Health adopted emergency regulations requiring retailers selling Gulf Coast oysters in California to warn at-risk persons of the danger of eating raw oysters. *Vibrio vulnificus* illnesses continued, particularly among Spanish speakers, and in October 1996 the department revised the regulation to require the warning in both English and Spanish. However, this educational warning strategy did not reduce illnesses and death. In 2003 the California Department of Public Health promulgated emergency regulations that restricted the sale of raw Gulf Coast oysters in California from April through October unless (1) the oysters were completely shucked, packed in jars or containers, and labeled for cooking only or (2) treated with a scientifically validated process that can reduce *Vibrio vulnificus* to nondetectable levels. Since the promulgation of the regulation, California has not identified any deaths attributable to the consumption of raw Gulf Coast oysters.

Pre-Harvest Shellfish Protection and Marine Biotoxin Monitoring

California established the Pre-Harvest Shellfish Protection and Marine Biotoxin Monitoring Program in accordance with the National Shellfish Sanitation Program, which sets sanitary requirements for shellfish growing waters and regulates all aspects of the commercial growing and harvesting of shellfish for human consumption. State and local regulatory agencies cooperatively implement elements of this program. Surveillance activities include frequent water quality monitoring in commercial growing areas and routine shellfish and phytoplankton monitoring of recreational areas and commercial aquaculture sites to protect the public from potentially deadly naturally occurring toxins responsible for paralytic shellfish poisoning (PSP) and amnesic shellfish poisoning (ASP) caused by domoic acid. The monitoring is conducted by over 100 individual participants from federal, state, and local agencies and citizen volunteers. Enforcement activities include imposing harvest closures when a growing area is affected by marine neurotoxins or pollution, imposing quarantines or health advisories on species of sport-harvested shellfish due to dangerous toxin levels, and patrolling closed growing areas to prevent illegal harvesting.

During the past five years the state program has partnered with university researchers in a project funded by the National Oceanographic and Atmospheric Administration's Monitoring and Event Response for Harmful Algal Blooms. One of the goals of this research is to improve detection of potentially toxic phytoplankton blooms by using new technologies for rapid species identification and toxin detection in the field. Improving these capabilities will inform sampling and monitoring efforts and improve the ability to alert the public about unsafe conditions.

Case Study:
Domoic Acid

During 2007 a domoic acid event occurred at a site in a coastal pilot study to evaluate field-based monitoring methods. California Department of Public Health staff observed a strong positive relationship among the qualitative field data (phytoplankton observations and shellfish toxin assays) and the quantitative laboratory data (gene probes for toxic species and high performance liquid chromatography analysis of shellfish), allowing early detection of a toxic bloom. Based on this success, the field monitoring methods were implemented by a commercial shellfish grower operating in a region of the Pacific Coast that has been a hot spot for domoic acid.

The commercial grower was trained in the collection and field identification of toxic phytoplankton and the use of the field test kit for domoic acid in shellfish. The commercial grower has now conducted weekly sampling, field observations, and toxin testing for three years. During each year, this commercial grower has been the first to detect the presence of domoic acid in the region. This awareness allowed the company to avoid distributing potentially toxic shellfish until public health regulatory laboratory test results were available. The toxin concentration reported by the laboratory was well above the federal alert level (20 ppm), resulting in closure of the growing area. The early field detection resulted in improved public health protection, prevented the need for a product recall, allowed the state department of public health to focus additional sampling in the affected region, and helped use limited laboratory resources efficiently by prioritizing samples from the high-risk region.

The success of this effort led to transfer of the field methodology to two additional commercial shellfish companies in another coastal community where domoic acid is a potential threat. A similar protocol was also developed for the field detection of the PSP toxins and implemented by a commercial grower operating in a region that experiences at least annual increases in these toxins. In all cases, the commercial growers have found that the methods are quick and easy and allow them to better manage their harvest activities based on trends in toxin occurrence. Their field observations and toxin assays provide the state regulatory agency with advance warning of potential toxic events, the first stage in protecting consumers.

Case Study:
Paralytic Shellfish Poisoning

In July 2009 routine weekly monitoring of commercial shellfish beds in a California coastal estuary revealed detectable concentrations of PSP toxins at sentinel mussel stations. This region is the focal point of PSP activity along the Pacific Coast and has experienced the greatest number of human illnesses and deaths due to consuming contaminated shellfish. PSP toxin concentrations can increase exponentially, going from nondetectable levels to well above the federal alert level (80 ug/100 g tissue) within days. Although the measured concentrations were below the federal alert level and not yet present in the oyster beds, the risk of a rapid increase in toxicity prompted the California Department of Public Health to order the shellfish bed operator to increase sampling frequency to twice per week. The department also worked closely with the oyster bed operator to monitor water samples for toxin-producing phytoplankton as an early warning tool. Detectable toxin levels persisted at the sentinel mussel stations through mid-August but remained undetectable in oyster harvesting locations. In late August 2009 the toxin levels in oysters from the harvest beds in the outer estuary increased fivefold, well above the federal alert level. The California Department of Public Health immediately closed all the outer beds to harvest and ensured that no contaminated product was distributed. The department also notified the local health department, shellfish companies in a nearby growing area, the National Park Service, the California Department of Fish and Game, and the U.S. Food and Drug Administration. Participants in the state's Marine Biotoxin Program provided additional coastal shellfish and phytoplankton samples to help determine the spatial extent of the bloom. Fortunately the state's annual mussel quarantine was already in effect as of May 1 to protect the public. The annual quarantine alerts the public to avoid recreational harvest of mussels for human consumption from May to 1 through October 31 due to the likelihood of dangerous levels of PSP toxins in coastal mussels.

Despite the high toxin concentrations in the oyster beds of the outer estuary, the oyster beds in the innermost reaches of the estuary remained at nondetectable levels for the PSP toxins. The threat of a sudden increase in toxicity within the inner estuary prompted shellfish program staff to require each batch of harvested oysters from this area to be tested for PSP toxins by the California Department of Public Health laboratory to determine if the product was safe for human consumption. During this batch harvest testing, the oysters from the inner estuary were found to be safe, allowing the company to maintain a safe operation while ensuring that contaminated product from the closed outer estuary was not harvested. Throughout this PSP event, the California Department of Public Health worked with the oyster harvester to track the spatial and temporal trends in toxicity throughout the estuary. The batch harvest restrictions remained in place until mid-September, when toxin levels declined to safe or undetectable levels throughout the estuary for consecutive weeks and the absence of toxin-producing phytoplankton was confirmed. The department removed harvest restrictions, and operations returned to normal for the remainder of the year.

Medical Waste Management

California is one of a handful of states with laws to address medical waste handling, disposal, and treatment issues. In 1990, in response to public concerns over possible HIV and hepatitis B transmission from needles and syringes found on California beaches, the California legislature enacted the Medical Waste Management Act. The federal government does not have a comparable requirement for managing medical waste.

To protect the public and workers from exposure to infectious disease-causing agents, the California Department of Public Health and thirty-five local enforcement agencies regulate the generation, management, and disposal of medical waste. The state regulatory program registers, permits, and inspects healthcare facilities, treatment facilities, transfer stations, trauma scene waste management practitioners, and transporters. In some counties and cities, local enforcement agencies, rather than the state, register and inspect healthcare facilities. State officials work with local agencies, healthcare facilities, and the medical waste management industry to ensure compliance with state laws and regulations.

Case Study:
Improper Disposal of Medical Waste
On November 6, 2006, a regional medical center in Southern California notified the California Department of Public Health of improper disposal of ninety-seven bags of untreated medical waste at a local solid waste landfill three days earlier. Landfill staff identified the medical waste during the disposal process, reported it to the facility, and removed it to a proper disposal site. The California Department of Public Health investigated the incident and determined that the medical waste originated from the surgery area of the medical center. The waste had been placed in black trash bags instead of the required red bags designed to handle medical wastes. As a result of the investigation, the department issued a notice of violation and administrative penalty order to the medical center. The medical center paid a fine for violating the Medical Waste Management Act.

Conclusion

The complexity and dynamics of the global market, local environmental justice issues, emerging environmental threats, conflicts at the industrial-agricultural-urban interface, and advancing biomedical and chemical technologies challenge environmental public health professionals to use innovative tools that ensure effective environmental enforcement. Environmental public health must deal with new challenges not only in familiar roles of ensuring water, food, and consumer product safety, but also in new roles addressing water scarcity, food security, product design, and environmental compatibility.

To keep pace with emerging environmental enforcement challenges, public health must strengthen its monitoring capacity, employ new tools for assessing environmental and public health risks, and increase its technological competence and agility. Emerging environmental enforcement issues include responding to climate change, regulating pharmaceutical disposal, monitoring carcinogens and reproductive toxicants in personal care products, and ensuring the quality and appropriate use of recycled water.

Environmental enforcement activities must move farther "upstream," addressing issues before regulatory problems arise. Upstream activities include influencing community development toward better balance with the natural environment, such as smaller impacts on other species' habitats; reducing greenhouse gas production; and promoting sustainable use of water and other renewable resources and sustainable food production. These activities rely on tools such as health impact assessments that involve communities in important decisions, evaluate policy options, and support decisions that maximize public health benefits while minimizing environmental harm.

Notes

1. U.S. Environmental Protection Agency, *Principles of environmental enforcement*, part I, introduction (July 15, 1992). http://www.inece.org/princips/glossary.pdf. Accessed September 1, 2010.
2. Centers for Disease Control and Prevention, *Designing and building healthy places*, http://www.cdc.gov/healthyplaces. Accessed September 1, 2010.

3. F. W. Pontius, *Drinking water regulation and health* (New York: John Wiley & Sons, 2003), 6–9.
4. Pontius, *Drinking water regulation and health*, 14.
5. James M. Montgomery, Consulting Engineers, Inc., *Water treatment principles and design* (New York: John Wiley & Sons, 1985), 66–67.
6. U.S. Food and Drug Administration, *Milestones in food and drug law history*, http://www.fda.gov/AboutFDA/WhatWeDo/History/Milestones/ucm081229.htm (accessed September 1, 2010).
7. U.S. Food and Drug Administration, National Shellfish Sanitation Program, http://www.fda.gov/Food/FoodSafety/Product-SpecificInformation/Seafood/FederalStatePrograms/NationalShellfishSanitationProgram/default.htm (accessed September 1, 2010).

18

Risk Communication and Environmental Health: Principles, Strategies, Tools, and Techniques

Vincent T. Covello

Introduction

The U.S. National Academy of Sciences has defined risk communication as "an interactive process of exchange of information and opinion among individuals, groups, and institutions."[1] Risk communication is the two-way exchange of information about risks, including environmental health risks associated with hazardous waste, water contamination, air pollution, and radiation.

Numerous studies have highlighted the importance of effective risk communication in enabling people to make informed choices and participate in deciding how risks should be managed. Effective risk communication provides people with timely, accurate, clear, objective, consistent, and complete risk information. It is the starting point for creating an informed population that is involved, interested, reasonable, thoughtful, solution-oriented, cooperative, and appropriately concerned about the risk and more likely to engage in appropriate behaviors.

Effective risk communication is especially critical during an emergency. For example, under normal circumstances the elaborate infrastructures and mechanisms that protect public health and the environment generally go unnoticed. In the middle of an emergency, however, such as the BP oil spill in 2010, there will be intense interest.

The primary objectives of effective risk communication are to build, strengthen, or repair trust; educate and inform people about risks; and encourage people to take appropriate actions.

These objectives apply to all four major types of risk communication:

1. Information and education
2. Behavior change and protective action
3. Disaster warning and emergency notification
4. Joint problem solving and conflict resolution

Risk communication will directly influence environmental health policy and public perceptions of environmental health issues. Poor risk communication about environmental health issues can fan emotions and undermine public trust and confidence. Poor risk communication can also create stress and conflict and precipitate a crisis. Good risk communication about environmental health issues can rally support, calm a nervous public, provide needed information, encourage cooperative behaviors, and potentially help save lives. A spokesperson who communicates badly may be perceived as incompetent, uncaring, or dishonest, thus losing trust. One who communicates well, however, can reach large numbers of people with clear and credible environmental health messages. Well constructed, practiced, and delivered risk communication messages about environmental health issues will inform the public, reduce misinformation, and provide a valuable foundation for informed decision making.

The principles of risk communication are supported by a large body of behavioral and social science research (see Figure 18.1). Over the past thirty years, thousands of articles on risk communication have been published in peer-reviewed scientific journals. Several reviews of the literature have been published by major scientific organizations, such as the National Academy of Sciences in the United States and the Royal Academy of Sciences in Great Britain.

One of the main principles of risk communication indicates that when people are highly upset, they often have difficulty hearing, understanding, and remembering information. Research shows that the mental stress caused by exposure to real or perceived risks can significantly reduce a person's ability to process information. Factors that cause the highest

FIGURE 18.1
Selected Risk Communication References

Auf der Heide, E. 2004. Common misconceptions about disasters: Panic, the "disaster syndrome" and looting. In *The First 72 Hours: A Community Approach to Disaster Preparedness*, edited by M. O'Leary, pp. 340–80. Lincoln, NE: iUniverse Publishing.

Becker, S. M. 2007. Communicating risk to the public after radiological incidents. *British Medical Journal* 335 (7630): 1106–7.

Bennett, P., and K. Calman, eds. 1999. *Risk communication and public health*. New York: Oxford University Press.

Bennett, P., D. Coles, and A. McDonald. 1999. Risk communication as a decision process. In *Risk communication and public health*, edited by P. Bennett and K. Calman. New York: Oxford University Press.

Blendon, R. J., J. M. Benson, C. M. DesRoches, E. Raleigh, and K. Taylor-Clark. 2004. The public's response to Severe Acute Respiratory Syndrome in Toronto and the United States. *Clinical Infectious Diseases* 38: 925–31.

Brunk, D. 2003. Top 10 lessons learned from Toronto SARS outbreak: A model for preparedness. *Internal Medicine News* 36(21): 4.

Centers for Disease Control and Prevention. 2002. *Emergency and risk communication*. Atlanta, GA: CDC.

Centers for Disease Control and Prevention, National Center for Health Marketing. 2007. *Plain English thesaurus for health communications*. Atlanta, GA: CDC. www.nphic.org/files/editor/file/thesaurus_1007.pdf.

Chess, C., B. J. Hance, and P. M. Sandman. 1986. *Planning dialogue with communities: A risk communication workbook*. New Brunswick, NJ: Rutgers University, Cook College, Environmental Media Communication Research Program.

Covello, V. 1992. Risk communication: An emerging area of health communication research. In *Communication yearbook 15*, edited by S. Deetz, pp. 359–73. Newbury Park and London: Sage Publications.

Covello, V. T. 2003. Best practice in public health risk and crisis communication. *Journal of Health Communication* 8 (supp. 1, June): 5–8.

Covello, V. T. 2006. Risk communication and message mapping : A new tool for communicating effectively in public health emergencies and disasters. *Journal of Emergency Management* 4 (3): 25–40.

Covello, V. T. 2011. Guidance on developing effective radiological risk communication messages: Effective message mapping and risk communication with the public in nuclear plant emergency planning zones. NUREG/CR-7033. Washington, DC: Nuclear Regulatory Commission.

Covello, V. T. 2011. Guidance on developing an emergency risk communication (ERC)/joint information center (JIC) plan for a radiological emergency. NUREG/CR-7032. Washington, DC: Nuclear Regulatory Commission.

(continued)

FIGURE 18.1
(Continued)

Covello, V. T., D. B. McCallum, and M. T. Pavlova, eds. 1989. *Effective risk communication: The role and responsibility of government and nongovernment organizations.* New York: Plenum Press.

Covello, V. T., K. Clayton, and S. Minamyer. 2007. Effective risk and crisis communication during water security emergencies: Summary report of EPA sponsored message mapping workshops. EPA Report No. EPA600/R-07/027. Cincinnati, OH: National Homeland Security Research Center, Environmental Protection Agency.

Covello, V. T., and P. Sandman. 2001. Risk communication: Evolution and revolution. In *Solutions to an environment in peril,* edited by A. Wolbarst, pp. 164–78. Baltimore, MD: Johns Hopkins University Press.

Covello, V. T., P. Slovic, and D. von Winterfeldt. 1986. Risk communication: A review of the literature. *Risk Abstracts* 3 (4): 171–82.

Covello, V. T., R. Peters, J. Wojtecki, and R. Hyde. 2001. Risk communication, the West Nile Virus epidemic, and bio-terrorism: Responding to the communication challenges posed by the intentional or unintentional release of a pathogen in an urban setting. *Journal of Urban Health* 78 (2, June): 382–91.

Davies C. J., V. T. Covello, and F. W. Allen, eds. 1987. *Risk communication: Proceedings of the national conference on risk communication.* Washington, DC: The Conservation Foundation.

Douglas, M., and A. Wildavsky. 1982. *Risk and culture: An essay on the selections of technological and environmental dangers.* Berkeley: University of California Press.

Environmental Protection Agency (US). 2007 Communicating radiation risks: Crisis communication for emergency responders. EPA-402-F-07-008. Washington, DC: United States Environmental Protection Agency, Office of Radiation and Indoor Air, July.

Embrey, M., and R. Parkin. 2002. Risk communication. In *Handbook of CCL microbes in drinking water,* by M. Embrey et al. Denver, CO: American Water Works Association.

Fischhoff, B. 1995. Risk perception and communication unplugged: Twenty years of progress. *Risk Analysis* 15 (2): 137–45.

Hance, B. J., C. Chess, and P. M. Sandman. 1990. *Industry risk communication manual.* Boca Raton, FL: CRC Press/Lewis Publishers.

Heath, R., and D. O'Hair, eds. 2010. *Handbook of risk and crisis communication.* New York: Routledge.

Hyer, R., and V. Y. Covello. 2007. *Effective media communication during public health emergencies: A World Health Organization handbook.* Geneva: World Health Organization.

Johnson, B. B., and V. Covello. 1987. *The social and cultural construction of risk: Essays on risk selection and perception.* Dordrecht, Holland: D. Reidel Publishing.

(continued)

FIGURE 18.1
(Continued)

Kahneman, D., and A. Tversky. 1979. Prospect theory: An analysis of decision under risk. *Econometrica* 47 (2): 263–91.

Kahneman, D., P. Slovic, and A. Tversky, eds. 1982. *Judgment under uncertainty: Heuristics and biases.* New York: Cambridge University Press.

Kasperson, R. E., O. Renn, P. Slovic, H. S. Brown, J. Emel, R. Goble, J. X. Kasperson, and S. Ratick. 1987. The social amplification of risk: A conceptual framework. *Risk Analysis* 8: 177–87.

Krimsky, S., and A. Plough. 1988. *Environmental hazards: Communicating risks as a social process.* Dover, MA: Auburn House.

Lindell, M. K., and V. E. Barnes. 1986. Protective response to technological emergency: Risk perception and behavioral intention. *Nuclear Safety* 27(4, October–December).

Lofstedt, R. E., and O. Renn. 1997. The Brent Spar controversy: An example of risk communication gone wrong. *Risk Analysis* 17 (2): 131–35.

Lundgren, R., and A. McKakin. 2004. *Risk communication: A handbook for communicating environmental, safety, and health risks.* 3rd ed. Columbus, OH: Batelle Press.

McKechnie, S., and S. Davies. 1999. Consumers and risk. In *Risk communication and public health*, edited by P. Bennett, p. 170. Oxford: Oxford University Press.

Mileti, D. S., and L. Peek. 2000. The social psychology of public response to warnings of a nuclear power plant accident. *Journal of Hazardous Materials.* 75 (2): 181–94.

Mileti, D. S., and S. Beck. 1975. Communication in crisis: Explaining evacuations symbolically. *Communication Research* 2 (1, January).

Morgan, M. G., B. Fischhoff, A. Bostrom, and C. J. Atman. 2001. *Risk communication: A mental models approach.* Cambridge, UK: Cambridge University Press.

Morgan, M. G., B. Fischhoff, A. Bostrom, L. Lave, and C. J. Atman. 1992. Communicating risk to the public. *Environmental Science and Technology* 26 (11): 2048–56.

National Academy of Sciences. 1996. *Understanding risk: Informing decisions in a democratic society.* Washington, DC: National Academy Press.

Peters, R., D. McCallum, and V. T. Covello. 1997. The determinants of trust and credibility in environmental risk communication: An empirical study. *Risk Analysis* 17(1): 43–54.

Powell, D., and W. Leiss. 1997. *Mad cows and mother's milk: The perils of poor risk communication.* Montreal: McGill-Queen's University Press.

Renn, O., W. J. Bums, J. X. Kasperson, R. E. Kasperson, and P. Slovic. 1992. The social amplification of risk: Theoretical foundations and empirical applications. *Journal of Social Science Issues* 48: 137–60.

Sandman, P. M. 1989. Hazard versus outrage in the public perception of risk. In Effective Risk Communication: The Role and Responsibility of Government and Non-government Organizations, edited by V. T. Covello, D. B. McCallum, and M. T. Pavlova, pp. 45–49. New York: Plenum Press.

(continued)

FIGURE 18.1
(Continued)

Slovic, P. 1987. Perception of risk. *Science* 236: 280–85.

Slovic, P., ed. 2000. *The perception of risk*. London: Earthscan Publication, Ltd.

Slovic, P., B. Fischhoff, and S. Lichtenstein. 2000. Facts and fears: Understanding perceived risk. In *The perception of risk*, edited by P. Slovic, pp. 137–53. London: Earthscan Publications Ltd.

Stallen, P. J. M., and A. Tomas. 1988. Public concerns about industrial hazards. *Risk Analysis* 8: 235–45.

Weinstein, N. D. 1987. *Taking care: Understanding and encouraging self-protective behavior*. New York: Cambridge University Press.

---

levels of worry, anxiety, and mental stress include, but are not limited to, perceptions that

- the risk is under the control of others, especially those who are not trusted;
- the risk is involuntary;
- the risk is inescapable;
- the risk is of human origin versus natural origin;
- the risk involves a type of risk that is unfamiliar or exotic;
- the risk threatens a form of injury or death that is dreaded;
- the risk is characterized by a great deal of uncertainty; and
- the risk is likely to cause injury or death to children, pregnant women, or other vulnerable populations.

The challenge for risk communicators is to overcome the communication barriers created by such anxiety-provoking factors.

Seven Cardinal Rules for Effective Risk Communication

There are seven cardinal rules for effective risk communication.[2]

1. **Allow people to have a voice and participate in decisions that affect their lives.**
2. **Plan and tailor risk communication strategies.** Different goals, audiences, and communication channels require different risk communication strategies.

3. **Listen to your audience.** People are usually more concerned about psychological factors, such as trust, credibility, control, voluntariness, dread, familiarity, uncertainty, ethics, responsiveness, fairness, caring, and compassion, than about the technical details of a risk. To identify real concerns, a risk communicator must be willing to listen carefully to and understand the audience.

4. **Be honest and transparent.** Honesty and transparency are critical for establishing trust and credibility. Trust and credibility are among the most valuable assets of a risk communicator.

5. **Coordinate and collaborate with credible sources of information and trusted voices.** Communications about risks are enhanced when accompanied by referrals to credible, neutral sources of information. Few things hurt credibility more than conflicts and disagreements among information sources.

6. **Plan for media influence.** The media play a major role in transmitting risk information. It is critical to know what messages the media deliver and how to deliver risk messages effectively through the media.

7. **Speak clearly and with compassion.** Technical language and jargon are major barriers to effective risk communication. Abstract and unfeeling language often offends people. Acknowledging emotions, such as fear, anger, and helplessness, is typically far more effective.

Risk Communication Models

Effective risk communication is based on several models that describe how risk information is processed, how risk perceptions are formed, and how risk decisions are made. Together, these models provide the intellectual and theoretical foundation for effective risk communication.

The Risk Perception Model

One of the most important paradoxes identified in the risk perception literature is that the risks that kill or harm people, and the risks that alarm and upset people, are often very different. For example, there is virtually no correlation between the ranking of hazards according to statistics on expected annual mortality and the ranking of the same hazards by how upsetting they are to people. There are many risks that make people worried and upset many people, but cause little harm. At the same time, there are risks that kill or harm many people, but do not make people worried or upset.

This paradox is explained in part by the factors that affect how risks are perceived. Several of the most important are trust, voluntariness, controllability, familiarity, fairness, benefits, dread, and effects on children. These factors, together with actual risk numbers, determine a person's emotional response to risk information. For example, they affect levels of public fear, worry, anxiety, anger, and outrage. Levels of these emotions tend to be greatest and most intense when a risk is perceived to be involuntary, unfair, not beneficial, not under one's personal control, and managed by untrustworthy individuals or organizations.

The Mental Noise Model

The mental noise model focuses on how people process information under stress. Mental noise is caused by the stress and strong emotions associated with exposures to risks.

When people are stressed and upset, their ability to process information can become severely impaired. In high-stress situations, people typically display a substantially reduced ability to process information. Exposure to risks associated with negative psychological attributes (e.g., risks perceived to be involuntary, not under one's control, low in benefits, unfair, or dreaded) contributes greatly to mental noise.

People under stress typically

- have difficulty hearing, understanding, and remembering information;
- focus most on the first and last things they hear;
- focus on the negative more than the positive;
- process information at several levels below their educational level;
- attend to no more than three messages at a time;
- focus intensely on issues of trust, benefits, fairness, and control; and
- want to know that you care before they care what you know.

The Negative Dominance Model

The negative dominance model describes the processing of negative and positive information in high-concern and emotionally charged situations.

In general, the relationship between negative and positive information is asymmetrical, with negative information receiving significantly greater weight.

The negative dominance model is consistent with the concept of "loss aversion," a central theorem of modern psychology. According to the concept of "loss aversion," people put greater value on losses (negative outcomes) than on gains (positive outcomes). When people face uncertainty, they do not typically evaluate the information carefully or compute the risks. Instead, they base their risk decisions and judgments on a brief list of emotions, instincts, and mental shortcuts. As Joshua Lehrer points out, "These shortcuts aren't a faster way of doing the math; they're a way of skipping the math altogether" (pp. 76–77).[3] People assign a much higher weight to the pain of loss than to the pleasure of gain. In human decision making, losses are feared more than gains are anticipated. Negatives loom larger than positives.

One practical implication of the negative dominance model is that it takes several positive or solution-oriented messages to counterbalance one negative message. On average, in high-concern or emotionally charged situations, it takes three or more positive messages to counterbalance a negative message. Another practical implication of negative dominance theory is that communications that contain negatives—for example, words such as *no*, *not*, *never*, *nothing*, *none*, and other words with negative connotations—tend to receive closer attention, are remembered longer, and have greater impact than messages with positive words. As a result, the use of unnecessary negatives in high-concern or emotionally charged situations can have the unintended effect of drowning out positive or solution-oriented information. Risk communications are often most effective when they focus on positive, constructive actions; on what is being done, rather than on what is not being done.

The Trust Determination Model

A central theme in the risk communication literature is the importance of trust in effective risk communications. Trust is generally recognized as the single most important factor determining perceptions of risk. Only when trust has been established can other risk communication goals, such as consensus building and dialogue, be achieved.

Trust is typically built over long periods of time. Trust is easily lost. Once lost, it is difficult to regain.

Because of the importance of trust in effective risk communication, a significant part of the risk communication literature focuses on the determinants of trust. Research indicates that the most important trust determination factors are (1) listening, caring, empathy, and compassion; (2) competence, expertise, and knowledge; and (3) honesty, openness, and transparency. Other factors in trust determination are accountability, perseverance, dedication, commitment, responsiveness, objectivity, fairness, and consistency. Trust determinations are often made in nine to thirty seconds. Initial trust impressions are often lasting trust impressions.

Trust is created in part by a proven track record of caring, honesty, and competence. It can be substantially enhanced by endorsements from trustworthy sources.

Trust in individuals varies greatly depending on their perceived attributes and their verbal and nonverbal communication skills. Trust in organizations also varies greatly. For example, surveys indicate that among the most trustworthy individuals and organizations in many risk controversies are (in no priority) informed citizen advisory panels, educators, firefighters, safety professionals, doctors, pharmacists, meteorologists, nurses, and faith leaders

Challenges to Effective Risk Communication

These four models are the backdrop for two of the most important challenges to effective risk communication:

- Selectivity and bias in media reporting about risk
- Psychological, sociological, and cultural factors that create public misperceptions and misunderstandings about risks

Each challenge is briefly discussed below.

Selectivity and Bias in Media Reporting about Risks

The media play a critical role in the delivery of risk information. However, journalists are often highly selective in their reporting about risks. For example, they often focus their attention on

- controversy;
- conflict;
- events with high personal drama;
- failures;
- negligence;
- scandals and wrongdoing;
- risks or threats to children; and
- stories about villains, victims and heroes.

Much of this selectivity stems from a host of professional and organizational factors, including deadlines and ratings.

Psychological, Sociological, and Cultural Factors That Create Public Misperceptions and Misunderstandings about Risks

People typically use only a small amount of available information to make risk decisions. Several factors contribute to this phenomenon.

The first factor is information "availability." The availability of an event (one that is accessible or easily remembered) often leads to overestimation of its frequency. Because of availability, people tend to assign greater probability to events of which they are frequently reminded (e.g., in the news media, scientific literature, or discussions among friends or colleagues) or to events that are easy to recall or imagine through concrete examples or dramatic images.

A second factor is conformity. This is the tendency on the part of people to behave in a particular way because "everyone else is doing it," or because everyone else believes something.

A third factor is overconfidence in one's ability to avoid harm. A majority of people, for example, consider themselves less likely than average to get cancer, get fired from their jobs, or get mugged. Overconfidence is most prevalent when high levels of perceived personal control lead to reduced feelings of susceptibility. Many people fail to use seat belts, for example, because of the unfounded belief that they are better or safer than the average driver. In a similar vein, many teenagers often engage in high-risk behaviors (e.g., drinking and driving, smoking, unprotected sex) because of perceptions, supported by peers, of invulnerability and overconfidence in their ability to avoid harm.

A fourth factor is "confirmatory bias." Confirmatory bias is the tendency of people to

1. seek out and accept information that is consistent with their beliefs or biases,
2. ignore information that is not consistent with their beliefs or biases, and
3. interpret information to support or confirm their beliefs or biases.

Once a belief about a risk is formed, new evidence is generally made to fit, contrary information is filtered out, ambiguous data are interpreted as confirmation, and consistent information is seen as "proof."

A fifth factor is the public's aversion to uncertainty. This aversion often translates into a marked preference and demand by the pubic for statements of fact over statements of probability: the language of risk assessment. Despite statements by experts that precise information is seldom available, people often want absolute answers. For example, people often demand to know exactly what *will* happen, not what *might* happen.

A sixth factor is the reluctance of people to change strongly held beliefs. Strong beliefs about risks, once formed, change very slowly. They can be extraordinarily persistent in the face of contrary evidence.

Strategies for Overcoming Challenges to Effective Risk Communication

Strategies for Overcoming Selective and Biased Reporting by the Media about Risks

Risk communicators can use a variety of strategies to enhance the quality of media reporting. For example, if done in advance, the following strategies can result in better media stories:

- Appoint a skilled lead spokesperson with sufficient seniority, expertise, and experience to establish credibility with the media.
- Establish a positive, ongoing relationship with the media.
- Develop a comprehensive risk communication plan containing the elements found in Figure 18.2.

FIGURE 18.2

25 Elements of a Comprehensive Risk and Crisis Communication Plan

1. Identify all anticipated scenarios for which risk, crisis, and emergency communication plans are needed, including worst cases and low-probability, high-consequence events
2. Describe and designate staff roles and responsibilities for different risk, crisis, or emergency scenarios.
3. Designate who in the organization is responsible and accountable for leading the crisis or emergency response.
4. Designate who is responsible and accountable for implementing various crisis and emergency actions.
5. Designate who needs to be consulted during the process.
6. Designate who needs to be informed about what is taking place.
7. Designate who will be the lead communication spokesperson and backup for various scenarios.
8. Identify procedures for information verification, clearance, and approval.
9. Identify procedures for coordinating with important stakeholders and partners (e.g., with other organizations, emergency responders, law enforcement, elected officials, and state and federal government agencies).
10. Identify procedures to secure the required human, financial, logistical, and physical support and resources (such as people, space, equipment, and food) for communication operations during a short, medium, and prolonged event (24 hours a day, 7 days a week if needed).
11. Identify agreements on releasing information and on who releases what, when, and how; policies and procedures regarding employee contacts from the media.
12. Include regularly checked and updated media contact lists (including after-hours news desks).
13. Include regularly checked and updated partner contact lists (day and night).
14. Identify a schedule for exercises and drills for testing the communication plan as part of larger preparedness and response training.
15. Identify subject-matter experts (for example, university professors) willing to collaborate during an emergency, and develop and test contact lists (day and night); know their perspectives in advance.
16. Identify target audiences.
17. Identify preferred communication channels (for example, telephone hotlines, radio announcements, news conferences, Web site updates, and faxes) to communicate with the public, key stakeholders, and partners.
18. Include messages for core, informational, and challenge questions.
19. Include messages with answers to frequently asked and anticipated questions from key stakeholders, including key internal and external audiences.
20. Include holding statements for different anticipated stages of the crisis.
21. Include fact sheets, question-and-answer sheets, talking points, maps, charts, graphics, and other supplementary communication materials.
22. Include a signed endorsement of the communication plan from the organization's director.

(continued)

FIGURE 18.2
(*Continued*)

23. Include procedures for posting and updating information on the organization's Web site.
24. Include communication task checklists for the first 2, 4, 8, 12, 16, 24, 48, and 72 hours.
25. Include procedures for evaluating, revising, and updating the risk and crisis communication plan on a regular basis.

Strategies for Overcoming the Psychological, Sociological, and Cultural Factors That Create Public Misperceptions and Misunderstandings about Risks

A broad range of strategies can be used to help overcome distortions in risk information caused by psychological, sociological, and cultural factors.

Several of the most important strategies derive from the risk perception model. For example, because risk perception factors such as fairness, familiarity, and voluntariness are as relevant as measures of hazard probability and magnitude in judging the acceptability of a risk, efforts to reduce outrage by making a risk fairer, more familiar, and more voluntary are as significant as efforts to reduce the hazard itself. Similarly, efforts to share power, such as establishing and assisting community advisory committees, or supporting third-party research, audits, inspections, and monitoring, can be powerful means for making a risk more acceptable.

Additional strategies include

- developing only a limited number of key messages (ideally three or one key message with three parts) that address the concerns of key stakeholders;
- developing messages that are clearly understandable by the target audience, typically at or below their average reading grade level (see, for example, CDC's *Plain English Thesaurus for Health Communications* at www.nphic.org/files/editor/file/thesaurus_1007.pdf);
- adhering to the "primacy/recency" or "first/last" principle by putting the most important messages in the first and last positions in lists;

Risk Communication and Environmental Health 381

- citing credible third parties that support or can corroborate key messages;
- providing information that indicates genuine empathy, listening, caring, and compassion;
- using graphics, visual aids, analogies, and narratives (such as personal stories); and
- balancing negative information with positive, constructive, or solution-oriented key messages.

Because of institutional and other barriers, strong leadership is often required to implement these strategies. An excellent example of such leadership occurred on September 11, 2001. Mayor Rudolf Giuliani shared the outrage that Americans felt about the terrorist attack on the World Trade Center. He delivered his messages with the perfect mixture of compassion, anger, and reassurance. He displayed the risk communication skills needed to be effective as a leader in a crisis. These include

- listening to, acknowledging, and respecting the fears, anxieties, and uncertainties of the many public and key stakeholders;
- remaining calm and in control, even in the face of public fear, anxiety, and uncertainty;
- providing people with ways to participate, protect themselves, and gain or regain a sense of personal control;
- focusing on what is known and not known;
- telling people what follow-up actions will be taken if a question cannot be answered immediately, or telling people where to get additional information;
- offering authentic statements and actions that communicate compassion, conviction, and optimism;
- being honest, candid, transparent, ethical, frank, and open;
- taking ownership of the issue or problem;
- remembering that first impressions are lasting impressions—they matter;
- avoiding humor, because it can be interpreted as uncaring or trivializing the issue;

- being extremely careful about saying anything that could be interpreted as an unqualified absolute ("never" or "always")—it only takes one exception to disprove an absolute;
- being the first to share bad or good news;
- balancing bad news with three or more positive, constructive, or solution-oriented messages;
- avoiding mixed or inconsistent verbal and nonverbal messages;
- being visible or readily available;
- demonstrating media skills (verbal and nonverbal), including avoidance of major traps and pitfalls—e.g., speculating about extreme worst-case scenarios, saying "there are no guarantees," repeating allegations or accusations, or saying "no comment";
- developing and offering three concise key messages in response to each major concern;
- continually looking for opportunities to repeat the prepared key messages;
- using clear, nontechnical language free of jargon and acronyms;
- making extensive but appropriate use of visual material, personal and human-interest stories, quotes, analogies, and anecdotes;
- finding out who else is being interviewed and making appropriate adjustments;
- monitoring what is being said on the Internet as much as other media;
- taking the first day of an emergency very seriously—dropping other obligations;
- avoiding guessing—checking and double-checking the accuracy of facts;
- ensuring facts offered have gone through a clearance process;
- planning risk and crisis communications programs well in advance using the APP model (anticipate/prepare/practice)—conducting scenario planning, identifying important stakeholders, anticipating questions and concerns, training spokespersons, preparing messages, testing messages, anticipating follow-up questions, and rehearsing responses;
- providing information on a continuous and frequent basis;

Risk Communication and Environmental Health

- ensuring that partners (internal and external) speak with one voice;
- having a contingency plan for when partners (internal and external) disagree;
- when possible, using research to help determine responses to messages;
- planning public meetings carefully—unless they are carefully controlled and skillfully implemented, they can backfire and result in increased public outrage and frustration;
- encouraging the use of face-to-face communication methods, including expert availability sessions, workshops, and poster-based information exchanges;
- being able to cite other credible sources of information;
- admitting when mistakes have been made—being accountable and responsible;
- avoiding attacking the credibility of those with higher perceived credibility;
- acknowledging uncertainty; and
- seeking, engaging in, and making extensive use of support from credible third parties.

Another important risk communication skill demonstrated by Mayor Giuliani was his ability to communicate uncertainty. He recognized the challenge to effective risk communication caused by the complexity, incompleteness, and uncertainty of risk data. In addressing this challenge, Mayor Giuliani acknowledged uncertainty, explained that risks are often difficult to assess and estimate, and announced problems early.

Finally, Mayor Giuliani recognized the importance of risk communication planning, preparation, and practice. To communicate risks effectively, he believed organizations need to have a risk and crisis communication plan written and practiced in advance. At a minimum, the risk communication plan should address questions such as: What needs to be done? Who needs to know? Who is the spokesperson? And who needs to act? Ideally, the plan should include responses, reviewed and cleared in advance, to all anticipated questions about the risk in question. For example, if the risk in question relates to a disaster or to an environmental cleanup at a hazardous waste site, the plan should include answers to the questions listed in Figures 18.3 and 18.4.

FIGURE 18.3
77 Questions Commonly Asked by Journalists during an Environmental Emergency or Crisis

Journalists are likely to ask six types of questions in a crisis (who, what, where, when, why, how) that relate to three broad topics: 1) what happened, 2) what caused it to happen, and 3) what it means. Specific questions include the following:

1. What is your name and title?
2. What are your job responsibilities?
3. What are your qualifications?
4. Can you tell us what happened?
5. When did it happen?
6. Where did it happen?
7. Who was harmed?
8. How many people were harmed or killed?
9. Are those that were harmed getting help?
10. How are those who were harmed getting help?
11. What can others do to help?
12. Is the situation under control?
13. Is there anything good that you can tell us?
14. Is there any immediate danger?
15. What is being done in response to what happened?
16. Who is in charge?
17. What can we expect next?
18. What are you advising people to do?
19. How long will it be before the situation returns to normal?
20. What help has been requested or offered from others?
21. What responses have you received?
22. Can you be specific about the types of harm that occurred?
23. What are the names of those who were harmed?
24. May we talk to them?
25. How much damage occurred?
26. What other damage may have occurred?
27. How certain are you about damage?
28. How much damage do you expect?
29. What are you doing now?
30. Who else is involved in the response?
31. Why did this happen?
32. What was the cause?
33. Did you have any forewarning that this might happen?
34. Why wasn't this prevented from happening?
35. What else can go wrong?
36. If you are not sure of the cause, what is your best guess?

(continued)

FIGURE 18.3
(Continued)

37. Who caused this to happen?
38. Who is to blame?
39. Could this have been avoided?
40. Do you think those involved handled the situation well enough?
41. When did your response to this begin?
42. When were you notified that something had happened?
43. Who is conducting the investigation?
44. What are you going to do after the investigation?
45. What have you found out so far?
46. Why was more not done to prevent this from happening?
47. What is your personal opinion?
48. What are you telling your own family?
49. Are all those involved in agreement?
50. Are people overreacting?
51. Which laws are applicable?
52. Has anyone broken the law?
53. What challenges are you facing?
54. Has anyone made mistakes?
55. What mistakes have been made?
56. Have you told us everything you know?
57. What are you not telling us?
58. What effects will this have on the people involved?
59. What precautionary measures were taken?
60. Do you accept responsibility for what happened?
61. Has this ever happened before?
62. Can this happen elsewhere?
63. What is the worst case scenario?
64. What lessons were learned?
65. Were those lessons implemented?
66. What can be done to prevent this from happening again?
67. What would you like to say to those who have been harmed and to their families?
68. Is there any continuing danger?
69. Are people out of danger? Are people safe?
70. Will there be inconvenience to employees or to the public?
71. How much will all this cost?
72. Are you able and willing to pay the costs?
73. Who else will pay the costs?
74. When will we find out more?
75. What steps need to be taken to avoid a similar event?
76. Have these steps already been taken? If not, why not?
77. What does this all mean? Is there anything else you want to tell us?

FIGURE 18.4
The 103 Most Frequently Asked Questions at Environmental Cleanup and Hazardous Waste Sites*

Health Risk Concerns
1. Am I at risk from the contamination?
2. What are the risks to my children?
3. What are the risks to my pets?
4. What are the impacts to natural habitat (i.e., fish and other species)?
5. Can my children and pets play in the soil?
6. What health effects can I expect to see if I've been exposed to site contaminants?
7. What are the short-term effects?
8. What are the long-term effects?
9. I have a recent health problem (i.e., headaches, rashes, etc.) that I never had before; could the site contamination have caused this problem?
10. Have any health problems been reported so far?
11. How many people have become ill as a result of the site?
12. Are you going to test residents for exposure?
13. Can you set up a temporary, local health center or clinic where we can be tested?
14. I'm pregnant (or planning to be). Will the contaminants affect my unborn child?
15. Is it safe to garden in my yard?
16. Is it safe to eat vegetables grown in my garden?
17. Is it safe to drink the water?
18. Will you provide us with bottled water?
19. Is it safe to bathe or shower in the water?
20. Is it safe to water our lawns with the potentially contaminated water?
21. Is it safe to mow our lawns if the soil underneath is potentially contaminated?
22. Is it safe to use the river for fishing and other recreational purposes?
23. Is it safe to eat the fish?
24. What's being done right now to protect mine and my family's health?
25. Will capping the site protect my health?
26. How serious is the contamination?
27. What happens if my ventilation system shuts down?
28. Can I get sick from breathing the air?

Investigation/Data Concerns
1. Where did the contamination come from?
2. How bad is the problem?
3. How much contamination is there?
4. Is the contamination moving, and, if so, in what direction?
5. Are there any other contaminants besides the ones we were told about?
6. How can you be sure there are no other contaminants?
7. Will you conduct testing/sampling to make sure the soil in my yard is free of contaminants?
8. How will you decide where to sample and where not to sample?

(continued)

FIGURE 18.4
(*Continued*)

9. Who determines what levels of contamination are considered "safe"?
10. Why don't you clean up all of the contamination, instead of allowing some to remain?
11. How do you know whether my drinking water is contaminated?
12. How do you know whether my yard has contaminated soil?
13. How do you know that it's safe to breathe the air?
14. How do you know whether it's safe to go fishing?
15. Why hasn't my well been sampled?
16. Why have some people received bottled water and not others?
17. Can I see the results of the testing you've done on my property?
18. Can I see the results of testing you've done on other properties in the neighborhood?
19. Do I have to give you access to sample my property?
20. What if I refuse access to my property?
21. Do I need to be home and take time off work while you're sampling my property?
22. I'm moving into the area; can I see the results of sampling that's been done?
23. Who will be doing the sampling?
24. How can we be sure the sampling data are accurate?
25. How can we be sure that future sampling won't find things that you didn't find now?
26. Can you guarantee the accuracy of the sampling results?

Cleanup Concerns
1. How exactly are you going to clean up the site?
2. Why was this particular cleanup method chosen over other options?
3. How long will the cleanup take?
4. When are you going to start the cleanup?
5. Who is going to perform the cleanup?
6. What process was used (or will be used) to select contractors to perform the cleanup?
7. How will cleanup performance be monitored or evaluated?
8. How much will the cleanup cost?
9. Who will pay for the cleanup?
10. Will my tax dollars have to pay to address this problem that someone else caused?
11. Can taxpayers be reimbursed?
12. How will you know when everything is "clean"?
13. Why are you going to just "cap" everything and leave the contamination there?
14. Why not dig up the contamination?
15. Is dredging safe?
16. Won't dredging just "stir up" things and contaminate the water even more?
17. What if the cleanup doesn't work?

(*continued*)

FIGURE 18.4
(Continued)

---

18. Can you guarantee that all of the contamination will be removed?
19. How will my quality of life be affected during the cleanup (i.e., noise, traffic, odors, etc.)?
20. After you finish the cleanup, then what? (What happens next?)
21. After the cleanup, will you continue to test to make sure it's still working?
22. What happens if my water (or soil, etc.) is still contaminated after the cleanup?

Communication Concerns
 1. Why did it take you so long to tell us about the contamination?
 2. How can I trust what you're telling me about the site?
 3. How can I trust what you're telling me about my safety?
 4. What happens if you find high concentrations of contaminants near my home; how will I know?
 5. How will I be informed of what's going on?
 6. Will you share the testing data with residents?
 7. Will you let us know if something unexpected happens during the cleanup and things get worse?
 8. Is there someone local residents can talk to if we have questions or concerns?
 9. Where can I get more information about this site?
10. Where can I get more information about similar sites that have already been cleaned up?
11. If a cleanup plan is selected that residents disagree with, is there an appeal process?
12. How will you address public comments?
13. Will you address ALL of the public comments?
14. How do you decide which comments NOT to address?
15. If the majority of residents disagree with how the EPA [or other agency] is planning to cleanup the site, will the EPA [or other agency] change its mind?
16. There's another site down the road; can you tell me what's going on there?
17. When you first discovered there MIGHT be a problem, why didn't you tell us then?

Economic Concerns
 1. If soil is excavated from my yard, will I receive financial assistance to replace plants and shrubbery?
 2. My property value has decreased because of the site contamination problem. Will I be compensated for this?
 3. I'm concerned that cost will be the driving force behind the agency's selected cleanup option; does community opinion really matter?
 4. I was told residents might have to relocate during the site cleanup. Who will pay for my moving costs? What about other expenses I may be forced to incur (i.e., costs of transporting my children to school because they won't be able to take the bus, or daily food costs because I won't have access to my stove and refrigerator, etc.)?

(continued)

FIGURE 18.4
(*Continued*)

---

5. The site has placed a "negative stigma" on our community that may affect potential investors, developers, or homeowners; what will be done about this?
6. Will this keep our community from developing?
7. Can we get jobs helping with the cleanup?
8. If we can't eat the fish anymore because of health risks, can you give us a food subsidy?
9. Do you have enough money to cover the cleanup costs?
10. What if you discover the cleanup is going to cost more than estimated? What happens then?

---

*This list is based on an analysis of public documents and media reports and was presented by J. Ross and V. Covello at the 2007 EPA Remedial Project Managers Workshop.

Conclusion

Risk communication is central to informed decision making. It is a core practice and competency for all those involved in environmental health risk management. Risk communication guidelines help organizations engage in constructive two-way exchanges of information.

Notes

1. National Academy of Sciences, *Improving risk communication* (Washington, DC: National Academy Press, 1989).
2. V. T. Covello and F. Allen, *Seven cardinal rules of risk communication* (Washington, DC: Environmental Protection Agency, 1988).
3. J. Lehrer, *How we decide* (New York: Houghton Mifflin, 2009), 76–77.

19

The Role of Health Professionals in Protecting Environmental Health

Robert M. Gould

Introduction

The publication of Rachel Carson's *Silent Spring* in 1962 brought the link between environmental health and human health to the fore and is widely credited with inspiring the environmental movement that rose throughout the world in the wake of the tumultuous events and cultural shifts stemming from the civil rights, anti–Vietnam War, and feminist movements.[1] The April 22, 1970, Earth Day, an environmental event modeled on the teach-in protests against the Vietnam War, drew 20 million people in cities, towns, and college campuses throughout the nation and helped provide the political pressure to convince President Richard Nixon to create the U.S. Environmental Protection Agency (EPA) in 1970.[2] Together with the passage of the Clean Air Act in 1970, the Clean Water Act in 1972, and the Safe Drinking Water Act in 1974, the fundamental regulatory basis for protection of environmental health was created,[3] providing a prime target for the activities of grassroots movements and environmental advocacy organizations over the following decades.

By the 1980s the concept of *precaution* began appearing in international environmental agreements in the context of mounting evidence of unprecedented environmental changes surrounded by vast uncertainties. The idea at the core of the *precautionary principle* is that action should be taken to prevent harm to the environment and human health, even if scientific evidence is inconclusive.

In 1998 the precautionary principle was introduced into public discourse in the United States after participants at the Wingspread Conference in Racine, Wisconsin, issued this statement: "When an activity raises threats of harm to human health or the environment, precautionary measures should be taken even if some cause and effect relationships are not fully established scientifically."[4] In the years since the Wingspread statement, the precautionary principle and related strategies have had major influence in galvanizing work on the environmental health agenda and policies in many regions. The principle is now embedded in a wide range of international treaties and national and local laws, regulations, and policies.

Physicians and allied health professionals have served as credible spokespersons for the environmental health movement from its beginnings, and their ethical imperative to "at first, do no harm," is steeped in such a precautionary approach.

Notable examples of the critical role of harnessing the precautionary health voice in support of the broader environmental health movement include the work of Physicians for Social Responsibility (PSR) and its global partners within the International Physicians for the Prevention of Nuclear War (IPPNW), which has connected continued efforts against the global threats of nuclear war and radioactive pollution to the dangers posed by global climate change and toxic degradation from the environment.

Physicians and other health professionals have also organized within their professional associations to pass policies in support of numerous environmental health advocacy efforts. More recently, particular emphasis has been given to transforming the healthcare industry from being a major source of environmental pollution to becoming a major example and vehicle for sustainable environmental practices, epitomized by the campaigns of Health Care Without Harm (HCWH) and its international network. This chapter focuses on the role of all of these efforts as illustrations of how health professionals can make a unique and essential contribution to the broader societal movements for the protection of environmental health.

Physicians for Social Responsibility

Health Impacts of the Nuclear Age

Since the dawn of the nuclear age, physicians have been active in local and global efforts to bring the health threats posed by nuclear weapons to the attention of the public and policy makers. In 1961 Physicians for Social Responsibility (PSR) was founded to advance a health-centered, primary prevention approach toward combating the potentially globally annihilating impacts of nuclear weapons and the environmental health impacts of nuclear weapons research, testing, and production. The original leaders of PSR recognized that there was no adequate medical response to a thermonuclear war between major nuclear powers such as the United States and the former Soviet Union, and in 1962 they documented their concerns in a series of groundbreaking articles in the *New England Journal of Medicine (NEJM)*, entitled "The Medical Consequences of Thermonuclear War."[5–7]

The *NEJM* articles supported the position that given the impossibility of an adequate medical response to nuclear warfare, it was incumbent upon physicians to prevent such a catastrophe from occurring. PSR physicians were especially prominent in countering omnipresent government assurances that open-air nuclear weapons tests caused no harm to the population. To substantiate the pervasive and dangerous levels of radioactive fallout resulting from such testing (decades later estimated to have produced at least 430,000 fatal cancers by the year 2000 around the world),[8] PSR members around the country participated in larger efforts to gather the baby teeth of children. Tests on these teeth showed the presence of strontium 90, an exclusive by-product of nuclear testing. This finding helped build public support for a halt to U.S. atmospheric tests and for the Partial Test Ban Treaty, which ended above-ground nuclear tests by the United States, the Union of Soviet Socialist Republics (USSR), and Britain in 1963.[9]

In 1979 the near-meltdown of the nuclear reactor at Three Mile Island in Harrisburg, Pennsylvania, provoked a wave of popular revulsion to nuclear power, but shortly thereafter, with the election of President Ronald Reagan, the U.S. government embraced ever more aggressive nuclear weapons policies. These events led to mass movements in the United States and around the world that coalesced around the heightened dangers of the nuclear age. The rise of the Nuclear Freeze movement, which grew out of the 1970s "Mobilization for Survival," galvanized the revitalization of PSR, which expanded to tens of thousands of health professionals across the country.[9, 10]

PSR organized medical symposia featuring the showing of "The Final Epidemic: Physicians and Scientists on Nuclear War" in more than thirty cities throughout the country. Each event demonstrated how the cataclysmic effects of a nuclear attack on the United States would leave the medical community helplessly short of personnel, medical supplies, and hospital beds needed to treat victims and alleviate human suffering. These symposia built activist networks across the nation among healthcare workers and other concerned citizens and helped to foster widespread public support for nuclear disarmament, exemplified by the one million people participating in the 1982 antinuclear demonstration in New York City, organized around the theme "Freeze the Arms Race—Fund Human Needs."[9, 10]

The global movement of health professionals working to protect human health from the dangers of the nuclear age was greatly aided by the 1981 creation of International Physicians for the Prevention of Nuclear War (IPPNW). IPPNW affiliate PSR and its Soviet counterpart brought dozens of Soviet and U.S. physicians together in local communities throughout both countries, whereby participants shared medical and cultural information, discussed arms control strategies, and met capacity crowds at press conferences and public events. These efforts led to a marked cooling of tensions between the United States and USSR, which served to greatly reduce the dangers of a catastrophic nuclear exchange, and in 1985 IPPNW was awarded the Nobel Peace Prize for its central role in helping to open arms control discussions between the two countries.[9]

In the mid-1980s environmental threats to public health stemming from the nuclear weapons complex began to gain more widespread recognition, when the safety of U.S. nuclear weapons–related reactors was scrutinized after the explosion and meltdown of the Chernobyl nuclear reactor in Ukraine in April 1986. Over time, it was revealed that routine and sporadic operational releases of hazardous radioactive and toxic materials from weapons research, production, and testing caused widespread contamination of communities and ecosystems surrounding nuclear weapons and related facilities.[11]

In 1988 PSR called for a comprehensive and independent evaluation of health and safety problems in the DOE's nuclear weapons production complex, where radioactive and toxic wastes threatened workers and nearby residents. These efforts culminated in the 1992 publication of *Dead Reckoning*, PSR's critical review of the DOE's epidemiologic research on the health risks of nuclear weapons production, which helped prompt improvements in the oversight of research on the health hazards of making and testing nuclear weapons.[12]

Physicians in the United States and around the world continue to connect the health perspective to nuclear "security" issues. For example,

IPNNW's International Campaign to Abolish Nuclear Weapons utilizes physician experts to present scientific testimony to decision makers throughout the world about the evolving understanding of the environmental health impacts of nuclear weapons. To this end, IPPNW physicians presented evidence at the 2010 Nuclear Non-Proliferation Treaty Review Conference, held at the United Nations, about the potential global climate-disrupting impacts of even a regional nuclear conflict, such as between India and Pakistan, which could lead to widespread malnutrition throughout the planet.[13] This testimony underscored IPPNW's urgent call for a Nuclear Weapons Convention with the primary prevention goal of abolishing nuclear weapons, in line with similar conventions against chemical and biological weapons.[14]

The approximately 22,000 nuclear warheads currently possessed worldwide continue to provide a large and looming threat to environmental and human health.[15] Regarding the radioactive and toxic legacy of the nuclear age, a 2000 report published by the National Academy of Sciences stated: "At many [U.S. nuclear weapons complex] sites, radiological and non-radiological hazardous wastes will remain, posing risks to humans and the environment for tens or even hundreds of thousands of years" (p. 1).[16] The document also indicated that "Complete elimination of unacceptable risks to humans and the environment will not be achieved, now or in the foreseeable future" (p. x).[16]

Other Environmental Threats to Human Health

In the early 1990s PSR expanded its focus beyond nuclear weapons and related technologies to the health threat of ubiquitous environmental chemical pollution and unfolding global climate change. In so doing, it joined a robust movement of advocacy groups that had worked for decades providing the mass popular pressure for enforcing and expanding the reach of the health protective legislative and regulatory accomplishments of the twentieth century.

PSR's early work to mobilize the medical community on broader environmental health issues included collaboration with the Massachusetts Institute of Technology, the Harvard School of Public Health, Brown University, and PSR's Greater Boston chapter to convene more than 700 physicians and environmentalists to assess environmental health issues.

This process made a significant contribution to the development of *Critical Condition*, a pioneering and definitive volume on human health and the environment.[9, 17]

In recognition of the fact that physicians receive limited, if any, environmental health education in their training or subsequent careers, PSR launched a series of publications to provide clinicians with information on the impact of environmental pollution on human health. Leading PSR physicians produced a series of groundbreaking works that made the evidence-based case for how the environment impacts human health and articulated the role of clinicians in addressing the environmental contributors to health. The works included *Generations at Risk,* which explained the scientific evidence on the deleterious reproductive and developmental effects of solvents, heavy metals, pesticides, and other widespread toxicants;[18] *In Harm's Way*, which documented the known impacts of a wide variety of environmental exposures on traits associated with attention deficit hyperactivity disorder and other neurodevelopmental problems;[19] and *Pesticides and Human Health: A Resource for Health Professionals*, aimed at raising the awareness and diagnostic abilities of health providers regarding pesticide health issues.[20]

To make the science connecting the environment and human health readily useful to the busy practicing physician, the Greater Boston and San Francisco Bay Area chapters of PSR, in partnership with the Pediatric Environmental Specialty Unit at the University of California, San Francisco, and a team of pediatricians from around the United States, developed the *Pediatric Environmental Health Toolkit.*© The *Toolkit*, endorsed by the American Academy of Pediatrics, was designed as a combination of easy-to-use reference guides and user-friendly health education materials on preventing exposure to toxic chemicals and other substances that affect infant and child health.[21]

The examples above serve to illustrate the many and varied collaborative efforts to harness the clinical voice to environmental health advocacy efforts to prevent toxic exposures. These same types of efforts have also served to leverage support measures to prevent and mitigate the health and environmental impacts of global warming. For example, medical community organizing efforts have included letter sign-on campaigns enlisting the names of Nobel Laureates in medicine, deans of medical schools, and hundreds of prominent physicians for a strong climate-protective treaty.

PSR developed various *Death by Degrees* state-specific reports that publicized the regional health consequences of climate change predicted in sequential reports from the Intergovernmental Panel on Climate Change, focusing outreach on medical and university audiences to support strong legislative and regulatory measures to address such threats.[9, 22]

PSR's advocacy related to the health impacts of climate change naturally dovetailed with work against air pollution regarding the combustion of various fossil fuels associated with planetary warming. Recent efforts have included participation in California's Health Network for Clean Air to support strong efforts established under California law to concretely plan to drastically reduce statewide greenhouse gases and institute stringent regulations to improve air quality. In addition, PSR chapters in Iowa and Wisconsin have partnered with the Sierra Club to shut down coal-fired power plants or stop the construction of new ones, because of the combined air pollution and global warming concerns. The disastrous 2005 Hurricane Katrina has served as a warning of how unprepared the United States is for the predicted increase in extreme weather events, and this lesson has been integrated into the outreach of PSR and many environmental organizations that promote the development of prevention, mitigation, and adaptation strategies for population protection. Addressing the life-cycle toxicity of the coal industry, in 2010 PSR and Earthjustice issued the report *Coal Ash: Toxic Threat to Our Health and Environment*, about the health impacts of the 2008 Tennessee coal ash disposal pond spill, which released more than one billion gallons of coal ash slurry.[23]

Finally, PSR health professionals and their colleagues have worked actively in their respective medical and professional organizations to advance environmental health and primary prevention. This type of health professional advocacy is described in the next section.

Health Professional Associations

Health professional organizations serve as essential avenues for clinician advocacy in environmental health. Health professionals working within their own associations have spurred preventive action on a wide range of environmental health issues, working to develop guidelines for clinical practice and intervening more broadly in policy areas in local, national, and international arenas.

The health impacts of the nuclear age were an early focus of environmental health advocacy among leading health professional organizations. For example, the American Academy of Pediatrics (AAP) established an environmental health committee over a half century ago in recognition of the threat of nuclear fallout to children's health.[24] The nuclear freeze movement in the 1980s included the support of the American Nurses Association, American Pediatric Society, and American Public Health Association (APHA).[10] Illustrative of advocacy on the part of health professional organizations, on September 30, 1986, more than 500 APHA members and annual meeting attendees protested at the Nevada Test Site (NTS) in support of a Comprehensive Test Ban Treaty as a step toward the abolition of nuclear weapons. Approximately 130 protesters participated in nonviolent civil disobedience, including many APHA leaders and the renowned astrophysicist Carl Sagan. This demonstration provided impetus to years of continued demonstrations at the NTS that helped influence the United States to finally accede to a moratorium on underground explosive nuclear testing in the early 1990s. APHA has subsequently adopted numerous policies supporting nuclear disarmament and the abolition of all weapons of mass destruction.[25]

Health professional organizations have been active on many other environmental health issues. Since 1999 the AAP has published the *Green Book*, a clinician handbook for the prevention of childhood diseases linked to environmental exposures.[26] The third edition of the *Green Book* is scheduled for release in 2011 and will be available electronically. In 2009 the Endocrine Society published a scientific statement on endocrine disrupting chemicals (EDCs) that made a number of recommendations, including increasing basic and clinical research, invoking the precautionary principle, and advocating involvement of individual and scientific society stakeholders in communicating and implementing health-protective changes in public policy related to EDCs.[27]

Addressing the environmental contributors to health is deeply embedded within the purview of public health nurses, and organizations of nursing professionals have long been in the forefront of environmental health advocacy. In 1995 the Institute of Medicine Committee on Enhancing Environmental Health Content in Nursing Practice published a landmark report, *Nursing, Health and Environment*, which identified major themes; made recommendations for practice, education, and research; and provided a beginning set of resources for practitioners and educators.[28] In December

2008 a group of national nursing leaders declared the need to formalize their relationship in order to work on environmental health and nursing issues. They created the Alliance of Nurses for Healthy Environments (ANHE) as a way to convene individual nurses and nursing organizations around environmental health issues. Using the blueprint articulated in the Institute of Medicine report, the Alliance developed a strategic plan for the integration of environmental health into nursing education, practice, research, and policy/advocacy work.[29]

The California Medical Association (CMA) has a long history of bringing the clinical voice to bear on environmental health issues that exemplifies the reach of state-based organizations. In 1968, two years before the 1970 Clean Air Act, the CMA established a policy to "vigorously support all rational efforts for the control of air pollution," as well as to "urge the support of studies and the enactment of laws that will assure a healthful air supply in the future."[30] The CMA has since enacted additional policies calling for increasingly comprehensive steps to protect the public from the health effects of air pollution.

Over the last two decades physician activists were successful in getting the CMA to adopt policies calling for the reduction and eventual elimination of mercury and certain polyvinyl chloride (PVC) plastics in hospital practice to prevent toxic impacts stemming from their use and disposal.[31] More recent policies in the CMA and other state medical associations have called for the adoption of national public-health protective chemical policy, exemplified by ongoing efforts for reforming the 1976 Toxic Substances Control Act (TSCA). The limitations of TSCA have resulted in a situation wherein the vast majority of the over 87,000 chemicals that are registered for use in U.S. commerce have entered the marketplace without comprehensive and standardized information on their reproductive or other chronic toxicities.[32, 33] Other CMA policies have aimed to protect schoolchildren, farmworkers, and agricultural communities from the dangers of pesticides.[30]

The CMA has also adopted comprehensive policies encouraging hospitals to take the lead in improving the health of patients and the general population by implementing food purchasing and menus that promote health and prevent disease. For example, such policies call for sourcing meat from free-range animals instead of concentrated animal feeding operations and purchasing food grown on small and medium-sized local farms, according to organic or other methods that emphasize renewable resources, ecological diversity, and fair labor practices. In 2009 the CMA responded to the rapidly accumulating evidence linking the overuse of antibiotics to serious outbreaks of drug-resistant infections by opposing the use of nontherapeutic antibiotics in livestock.[30]

The CMA has called for hospitals to use the cleanest and most sustainable forms of energy and encouraged physician support for binding reductions in national and global greenhouse emissions. In 2009 CMA adopted a policy in support of "smart growth" strategies and endorsed the education of health professionals with resources aimed at mitigating the impacts of healthcare system contributions to climate change and toxic pollution.[30]

These pioneering policies on the state level in California have been complemented by similar efforts of physician activists working in state medical associations in Oregon, Washington, and Illinois, and in many state public health associations. Together, these activities have helped to influence the historically conservative American Medical Association (AMA) to support major national environmental health initiatives with the potential to transform both the practices of mainstream medical practice and the direction of U.S. environmental policies.

For example, the AMA has adopted policies promoting the incorporation of environmental health into medical education, supporting reforms in chemical policy, and addressing mercury exposure and other key environmental health issues.[30, 34] In November 2009 the AMA underscored its major commitment to supporting expedient action to address climate change by sending a letter to President Obama citing the "significant public health implications" of climate change.[35] This was immediately followed by the AMA president-elect, in a Capitol press briefing sponsored by Senator Tom Udall, stating: "The health effects from these [climate change] events should be a major concern to the medical community . . . The AMA supports educating the medical community on the potential adverse public health effects of climate change and incorporating the health implications into the spectrum of medical education."[36]

The AMA has also adopted a comprehensive policy to promote the engagement of clinicians and policy makers in creating a healthy and sustainable food system that simultaneously would mitigate the impacts of climate change and protect human health. As stated in its 2009 "Sustainable Food" report of the AMA Council on Science and Public Health,

"Locally produced and organic foods are considered part of a healthy sustainable food system for many reasons. They reduce the use of fuel, decrease the need for packaging and resultant waste disposal, preserve farmland, and/or support a greater diversity of crops. The related reduced fuel emissions contribute to cleaner air and in turn lower the incidence of asthma attacks and other respiratory problems." In this report, the AMA also supports organic meat production to help "reduce the development of antibiotic resistance, as well as air and water pollution," and promotes organic and local foods, as they "can have improved nutrient profiles," and "may encourage increased consumption of fruits, vegetables, and lean meat, while also decreasing exposure to pesticides and hormones."[37]

This successful policy work of committed health professionals has created space for environmental health issues to be considered appropriate subjects for "Grand Rounds" in medical centers and curriculum for medical, nursing, and public health students. It has also contributed to efforts by the healthcare industry to incorporate sustainable and environmentally sound practices, epitomized by the groundbreaking work *Health Care Without Harm*, discussed in the next section.

Healthcare Without Harm: Promoting Environmental Stewardship in Healthcare

Historically, the healthcare industry has been a main contributor to environmental pollution through its operations and waste disposal. Medical waste incineration caused highly toxic combustion products, particularly heavy metals and dioxins. In the 1990s a draft EPA report identified medical waste incinerators as the largest U.S. source of dioxins.[38]

In recognition of the environmental impacts of the healthcare industry in general, and medical waste incineration in particular, in 1996 a broad environmental movement was formed to align industry policies and practices with the goal of preserving human health. The coalition, named Health Care Without Harm (HCWH), initially focused its efforts on stopping medical waste incineration. HCWH members spanned "downstream" communities near medical waste incinerators through health professionals and environmental advocacy organizations.[38, 39]

Over the past fifteen years HCWH has developed into a coalition of approximately 500 hospitals, universities, health professional organizations, and environmental groups working in fifty-two countries. HCWH "shares a vision of a health care sector that does no harm, and instead promotes the health of people and the environment," and "works to implement sound and healthy alternatives to health care practices that pollute the environment and contribute to disease." Understanding the health sector's massive buying power and institutional interest in preventing disease, HCWH has sought to utilize the healthcare industry as a key lever to transform the entire economy toward safer and sustainable products and practices, without compromising patient safety and care.[38, 39]

To reduce the volume and toxicity of the medical waste stream, HCWH has worked with healthcare systems to reduce their use of PVCs by substituting nonchlorinated plastics in intravenous materials, reducing the amount of packaging materials, instituting methods of recycling, and autoclaving medical waste instead of incinerating it. As a result, the number of U.S. medical waste incinerators declined from approximately 5,000 in the 1990s to 83 in 2006.[38]

HCWH campaigns have also resulted in the virtual elimination of mercury thermometers in the United States, and HCWH is now partnering with the World Health Organization (WHO) to eradicate mercury from healthcare worldwide. The enormous purchasing power of the hospital sector has been utilized to bring safer products into hospital settings, such as "green" cleaning supplies and PVC-free rugs that provide the basis for making such products available for general consumer use.[38]

HCWH has enlisted the healthcare industry into embracing environmental stewardship as a guiding principle of operations focused on protecting the fundamental health needs of patients, workers, and communities, including addressing the challenges posed by global warming. The Healthy Hospital Initiative, which includes HCWH and major healthcare systems, works to transform the way the sector designs, builds, and operates its facilities. This includes reducing energy use, utilizing "green" design, materials, and products, while actively promoting health-protective public policy, exemplified by 2010 congressional testimony from Kaiser Permanente in support of Toxic Substances Control Act reform legislation.[38]

HCWH has recently distilled the vast knowledge of its associated scientists, doctors, and nurses about various health aspects of climate warming into the creation of standardized presentations aimed at activating health professionals and hospital administrators to support necessary institutional changes. Nurses, who have such an intensive hands-on role in patient care and operational support, have been the clinical backbone of HCWH's work, with critical support from ANA and the Alliance of Nurses for Healthy Environments.

HCWH's Healthy Food in Health Care campaign aims to leverage the $12 billion-per-year hospital food industry to become a leading force for transforming the nation's food supply into one that is nutritious, sustainable, and free of toxic environmental contaminants, and which addresses the rising health toll and costs of obesity and diabetes. The Balanced Menus project of HCWH and the San Francisco-Bay Area chapter of PSR includes a partnership with hospitals and health networks around Northern California. Recognizing the significant contribution of livestock production to antibiotic resistance and global climate change, hospitals aim to reduce the overall amount of meat served by 20 percent over twelve months, while increasing the proportion of sustainable meat and produce. By developing networks with community farmers for local sourcing of food, the global warming contribution of "food miles" has been reduced. An evaluation of these efforts at four institutions has shown that implementation of Balanced Menus can yield substantial savings in costs and greenhouse gas emissions.[40]

Healthy Food in Health Care has partnered in efforts aimed at heightening health provider sensitivity to the dangers of pesticides and other toxic components of the food supply, including webinars and reports organized by the APHA, AMA, and University of California, San Francisco, Program on Reproductive Health and the Environment.

HCWH's clinical outreach has been bolstered by the Collaborative on Health and the Environment (CHE), which represents an alliance of environmentally impacted communities, environmental health activists, and health professionals. For example, CHE has pioneered a series of webinars covering breaking environmental health issues, connecting knowledge of gene–environment interactive effects and related endocrine disruption that increasingly is understood as a basis for much human disease.[41]

Conclusion

Further advances in the protection of environmental health, including meeting the challenges posed by unfolding climate change, face considerable hurdles from global budgetary constraints and job losses occurring in the wake of an unprecedented worldwide economic crisis. The continued global waste of resources and associated environmental devastation represented by unabated preparations for, and conduct of, warfare, including unchecked proliferation of nuclear weapons, greatly exacerbate the dangers faced by all.

In these times, the role of health professionals is increasingly critical for providing the key scientific and ethical arguments for broad movements necessary to prevent the otherwise inevitable intertwining of human and environmental devastation. Such heightened responsibility in many ways profoundly reconnects with the core messages of the earlier environmental movement.

Indeed, in October 1963 Rachel Carson implored the clinical community to take action in what would become her last speech before succumbing to breast cancer. Speaking to 1,500 physicians and allied health professionals in San Francisco on "The Pollution of Our Environment," Carson articulated the inextricable link between human health and the health of the environment, saying,

> Underlying all of these problems of introducing contamination into our world is the question of moral responsibility—responsibility not only to our generation, but to those of the future. We are properly concerned about somatic damage to generations now alive; but the threat is infinitely greater to the generations unborn; to those who have no voice in the decisions of today, and that fact alone makes our responsibility a heavy one.[42]

Notes

1. R. Carson, *Silent Spring* (New York: Houghton Mifflin, 1962).
2. J. Picard, EPA is 40 and no less controversial, *International Business Times*, December 5, 2010, http://www.ibtimes.com/articles/88833/20101205/environment-carbon-dioxide.htm# (accessed February 4, 2011).
3. E. Yehle, Drastic enforcement cases heralded arrival of EPA 40 years ago, *New York Times*, December 2, 2010, http://www.nytimes.com/gwire/2010/12/02/02greenwire-drastic-enforcement-cases-heralded-

arrival-of-50613.html?scp=1&sq='Drastic%20Enforcement%20 Cases'%20Heralded%20Arrival%20of%20EPA%2040%20Years%20 Ago&st=cse (accessed February 4, 2011).
4. Wingspread statement on the precautionary principle, January 1998, http://www.gdrc.org/u-gov/precaution-3.html (accessed February 4, 2011).
5. D. G. Nathan, H. J. Geiger, V. W. Sidel, et al., The medical consequences of thermonuclear war: Introduction, *New England Journal of Medicine* 266 (1962): 1127.
6. F. R. Ervin, J. B. Glazier, S. Aronow, et al., The medical consequences of thermonuclear war: I. Human and ecological effects in Massachusetts of an assumed nuclear attack on the United States, *New England Journal of Medicine* 266 (1962): 1127–1136.
7. V. W. Sidel, H. J. Geiger, and B. Lown, The medical consequences of thermonuclear war: II. The physician's role in the postattack period, *New England Journal of Medicine* 266 (1962): 1137–1145.
8. International Physicians for the Prevention of Nuclear War, Institute for Energy and Environmental Research, *Radioactive heaven and earth: The health and environmental effects of nuclear weapons testing in, on and above the earth* (New York: Apex Press, 1991), 163.
9. Physicians for Social Responsibility, *A history of accomplishments* (Washington, DC: Physicians for Social Responsibility, May 2000), https://secure2.convio.net/psr/documents/history.pdf (accessed February 4, 2011).
10. L. S. Wittner, *The nuclear freeze and its impact* (Washington, DC: Arms Control Association, December 2010), http://www.armscontrol .org/act/2010_12/LookingBack (accessed February 4, 2011).
11. P. M. Sutton and R. M. Gould, Nuclear weapons, in *War and public health*, 2nd ed., edited by B. S. Levy and V. W. Sidel (New York: Oxford University Press, 2008), 161–164.
12. H. J. Geiger, D. Rush, and D. Michaels, *Dead reckoning: A critical review of the Department of Energy's epidemiologic research; A report by the physicians task force on the health risks of nuclear weapons production* (Washington, DC: Physicians for Social Responsibility, 1992).
13. A. Robock and O. B. Toon, Local nuclear war, global suffering, *Scientific American* (January 2010): 74–81.

14. Non-Proliferation Treaty Review Conference 2010, *Towards nuclear abolition: A report by the international campaign to abolish nuclear weapons* (Cambridge, MA: International Physicians for the Prevention of Nuclear War, June 2010), http://www.icanw.org/files/RevCon 2010_1.pdf (accessed February 4, 2011).
15. R. S. Norris and H. M. Kristensen, Global nuclear weapons inventories, 1945–2010, *Bulletin of the Atomic Scientists* (July/August 2010).
16. Commission on Geosciences, Environment, and Resources, *Long-term institutional management of U.S. Department of Energy waste sites* (Washington, DC: National Academy Press, August 2000).
17. E. Chivian, M. McCally, H. Hu, et al. (eds.), *Critical condition: Human health and the environment* (Cambridge, MA: The MIT Press, 1993).
18. T. Schettler, G. Solomon, M. Valenti, et al., *Generations at risk: Reproductive health and the environment* (Cambridge, MA: The MIT Press, 1999).
19. T. Schettler, J. Stein, F. Reich, et al., *In harm's way: Toxic threats to child development; A report by Greater Boston Physicians for Social Responsibility* (Cambridge MA: Greater Boston Physicians for Social Responsibility, May 2000), http://action.psr.org/site/DocServer/ihw complete.pdf?docID=5131 (accessed February 4, 2011).
20. G. Solomon, *Pesticides and human health: A resource guide for health care professionals* (Physicians for Social Responsibility and Californians for Pesticide Reform, 2000), http://www.sfbaypsr.org/pdfs/pahh.pdf (accessed February 4, 2011).
21. Physicians for Social Responsibility, *Pediatric environmental health toolkit* (Washington, DC: Physicians for Social Responsibility, 2009), http://www.psr.org/resources/pediatric-toolkit.html (accessed February 4, 2011).
22. Physicians for Social Responsibility, *Death by degrees* (Washington, DC: Physicians for Social Responsibility, n.d.), http://www.psr.org/resources/death-by-degrees.html (accessed February 4, 2011).
23. B. Gottleib, S. G. Gilbert, and L. G. Evans, *Coal ash: Toxic threat to our health and environment; a report from Physicians for Social Responsibility and Earth Justice* (Washington, DC: Physicians for Social Responsibility, September 2010), http://www.psr.org/assets/pdfs/coal-ash.pdf (accessed February 4, 2011).

24. American Academy of Pediatrics, Section on Emergency Medicine, *Section and committee history*, http://www.aap.org/sections/pem/PEM-history.htm (accessed February 4, 2011).
25. American Public Health Association, *APHA policy statement 20095: The role of public health practitioners, academics, and advocates in relation to armed conflict and war* (Washington, DC: American Public Health Association, November 10, 2009).
26. R. A. Etzel and S. J. Balk (eds.), *Pediatric environmental health*, 2nd ed. (American Academy of Pediatrics Committee on Environmental Health, 2003).
27. E. Diamanti-Kandarakis, J. P. Bourguignon, L. C. Giudice, et al., Endocrine-disrupting chemicals: An endocrine society scientific statement, *Endocrine Reviews* 30 (4) (2009): 293–342.
28. A. M. Pope, M. A. Snyder, and L. H. Mood (eds.), *Nursing, health and the environment: Strengthening the relationship to improve the public's health*, Institute of Medicine (Washington, DC: National Academy Press, 1995).
29. Alliance of Nurses for Healthy Environments, http://e-commons.org/anhe/about-the-e-commons/ (accessed February 4, 2011).
30. California Medical Association, House of Delegates resolution 18–68, air pollution (1968), http://www.cmanet.org/member/memberdoc.cfm?templateinc=POLICYNEW&docid=28&parent=26&policyid=1225 (accessed February 5, 2011).
31. R. Gould and C. Russell, Taking action to prevent harm: County medical associations and environmental health, *San Francisco Medicine* 83 (3) (2010): 27–29.
32. U.S. Environmental Protection Agency, *What is the TSCA Chemical Substance Inventory?* (2006), http://www.epa.gov/opptintr/newchems/pubs/invntory.htm (accessed February 5, 2011).
33. M. P. Wilson, D. A. Chia, and B. C. Ehlers, Green chemistry in California: A framework for leadership in chemicals policy and innovation, *New Solutions* 16 (4) (2006): 365–372.
34. Council on Science and the Public Health, *Report 3-I-08: Green initiatives and the health care community* (American Medical Association; 2008).
35. M. D. Maves, Letter to U.S. President Barack Obama (American Medical Association, November 19, 2009).

36. C. Wilson, The health community speaks out about climate change: Implications for U.S. states and for the nation (press release, Harvard Medical School, Center for Health and the Global Environment, November 20, 2009).
37. Council on Science and the Public Health, *Report 8-A-09: Sustainable food* (American Medical Association, 2009).
38. C. Elton, Cohen's nonprofit lets hospitals go green, *Miller-McCune*, December 27, 2010, http://www.miller-mccune.com/health/cohens-nonprofit-helps-hospitals-go-green-26380/ (accessed February 4, 2011).
39. Health Care Without Harm, http://www.noharm.org/ (accessed February 5, 2011).
40. C. Danzig, Healthier hospital food—for us and the earth, *Miller-McCune*, December 27, 2010, http://www.miller-mccune.com/health/healthier-hospital-food-for-us-and-the-earth-26384/ (accessed February 5, 2011).
41. The Collaborative on Health and the Environment, http://www.healthandenvironment.org/ (accessed February 5, 2011).
42. R. Carson and L. Lear (eds.), *Lost woods: The discovered writing of Rachel Carson* (Boston: Beacon Press, 1998), 227–228, 242.

20

Renewable Energy

Brian C. Black and Richard E. Flarend

Introduction

It can be unnerving to see the ridges of Pennsylvania's Allegheny Mountains lined with some of the largest propellers humans have ever constructed. We are used to spinning propellers lifting the cargo to which they are attached, whether it is the chassis of a helicopter or airplane. Therefore, one might feel compelled to ask: Will they be strong enough to lift the long, slight, tree-covered ridges? Is that the intention? In fact, after years of being mined to provide coal for energy production, these mountains may now have a respite—while still being a major component of America's energy future.

During the industrial era, Pennsylvania has produced 20 billion short tons of coal. It has been removed through a variety of means, including underground mining and, more recently, strip-mining and mountaintop removal. No matter how it was removed, though, the coal from these mountains was most often fed into long lines of railroad hopper cars and carried to sites all over the nation to be burned, its released power then used by industry or by utilities to make electricity for consumers. Today, Pennsylvania remains one of the nation's largest producers of anthracite, or hard coal. Now, however, this new energy crop has also crept into the state: hundreds of turbines have been built throughout the state, and a number of international manufacturers of wind turbines have made Pennsylvania their U.S. headquarters.

The turbine-decked mountains are one of many pieces of evidence that help to prove a fact about current times: Americans are fully engaged in a significant and potentially seminal energy transition. The energy transition initiated in the 1970s did not bring immediate changes to American attitudes toward renewable energy. For many observers, this reality marks a failure of our species to pursue energy paths with less impact on planet Earth. Although this perspective is, on the whole, correct, it fails to appreciate the cultural and technical nuances of energy transitions. To complete an energy transition, technical advances are needed that require many years of research. This research began in earnest in the 1970s, so that by 2010 the mass production of thoroughly developed wind turbines was possible. From a historical standpoint—in terms of centuries and even millennia—a shift in the basic habits of the human species can drag on for decades and even centuries. Clearly we are in the midst of an energy transition that very likely began thirty years ago, and it may continue for decades longer.

In hindsight, the temporary scarcity of the 1970s provided a catalyst for disrupting the paradigm that had classified most renewable sources as energy alternatives. Many experts believe these methods for harvesting energy promise the most likely paradigm for future development. Or at least they represent a segment of our energy use that must slowly—and strategically—expand and incrementally replace our use of fossil fuels. As we begin to forecast the future of alternative fuels, we can't help but also cast a look backward to the history of such energy sources. With even a passing glance backward, it becomes obvious that there is a long and complex history related to the emergence of alternative energy. Often these are very old technologies used in a new fashion. Most important, contained within the very terminology of the name of these sources of power is their basic reality: To what are these energy sources an alternative?

Defining Alternatives

Fossil fuels, which currently supply about 85 percent of our energy, are all concentrated in locations that are out of view of the general public. The environmental impact of extracting these energy sources goes unseen, not to mention that most waste emissions are dumped into the atmosphere and rivers in invisible amounts—but not with invisible harm. The economic and environmental costs of these emissions and the societal impact of this extraction of energy are not factored into the consumer cost of electricity or gasoline. Therefore, the chain of connectivity between the extraction and supply of energy to society and the resulting negative impact on society is broken. Many scholars have sought to create a more complete accounting of energy production. In short, to prove the point, a single place must currently absorb or suffer from much of the cumulative environmental or social effects caused by the extraction of energy. This out-of-sight, out-of-mind pattern misleads the public, particularly when that impact is exerted mostly on regions distant from the populations that reap the benefits of the energy created.

New ideas in energy accounting take these impacts into consideration and thereby provide even more substantiation to alternative methods of creating power. Finally, the pollution or end product of each method has also begun to be used as a quantifiable entity. Many experts believe that using carbon accounting to add these costs to fossil fuels will make alternative sources even more competitive in the energy marketplace. Many are now promoting the idea of knowing what we eat by being familiar with where and how the food is grown or produced. Similarly, the public needs to be aware of where their energy comes from and what impact the entire fuel chain has on land use, the environment, and the unintended impacts on human health.

This chapter demonstrates how such an accounting process can make alternative energy much more viable and cost-effective than those sources on which we now rely. However, we also must factor in that most renewable sources would require the construction of infrastructure in addition to that which has been built and is currently maintained, to take advantage of the flexibility found in hydrocarbon-based power.

By contrast, most renewable resources are spatially concentrated and immobile. This lack of flexibility has contributed to the public resistance that such development has encountered at various sites. Renewable resources are presently confronted with systemic limitations that are very similar to those that befell other resources earlier in their usage. In the case of fossil fuels, we found technical solutions to their problems with location and use; now we must do so with alternatives as well. In this fashion, the paradigm of cheap energy—created over the last few centuries—that forced the title "alternative" on renewables decades ago, has been shaken to the point of fracture. Our future energy paradigm, most experts agree, is in play, and possesses the potential to shift considerably.

Rising Costs of Fossil Fuels Promote Alternatives

Energy costs in the early twenty-first century have risen at staggering rates. Since 2000, natural gas rates have risen by 80 percent, and gasoline has more than doubled in price. Even electricity rates have risen by nearly 40 percent after actually declining during most of the 1990s. For a typical household, direct expenditures on energy (gasoline, electricity, and home heating/cooling) increased a whopping $300 per month from 2000 to 2008, just prior to the economic collapse. It is no wonder that many households had financial trouble, especially when the accompanying increase in food prices is considered. This increased spending on energy was at least one of the reasons for the economic collapse that occurred just a few months later. Some argue that energy prices were the primary cause of this collapse.

Americans have found evidence of these increased costs in all types of related goods as well, particularly agriculture. Although agriculture begins with photosynthesis, most American food products are now dependent on petroleum and natural gas, which means that we rely on fossil fuels to eat. In the United States, in 2004, commercial farming accounted for 12 percent of our annual energy use, most of that being natural gas used to make artificial fertilizer. After natural gas shortages develop, the artificial fertilizer plants shut down or price their product so high that most farmers can't afford it. Food prices increase because of higher transportation costs as well as increased costs for herbicides and pesticides. Energy-related price increases are not limited to fruits, vegetables, and grains; they impact dairy and meat products as well. In fact, the energy required to produce meat is 100 times as much as the energy required to produce a similar diet of grains. Across the board, increased fossil fuel prices have trickled into the lives of American consumers and reminded them that we live an energy-intensive lifestyle. Based on cheap fuels, this lifestyle has defined American life for a century. As the prices now rise, the basic cost of living stresses many in the middle class to the breaking point.

Unfortunately, though, the rising costs of energy sources that are destined to expire is only one aspect of a high-energy life. Scientists have now demonstrated to us that burning fossil fuels has created emissions and pollution that imperil or at least corrupt Earth's natural systems. Many health and environmental problems that our country faces today, we have learned, are a result of our fossil-fuel dependence. The coal industry's most troublesome problem today is removing organic sulfur, a substance that is chemically bound to coal. All fossil fuels, such as coal, petroleum, and natural gas, contain sulfur. When these fuels are burned, the organic sulfur is released into the air, where it combines with oxygen to form sulfur dioxide. Sulfur dioxide is an invisible gas that has been shown to have adverse effects on the quality of air we breathe and leads to the premature deaths of tens of thousands. It also contributes to acid rain, an environmental problem that adversely affects fish, wildlife, and forests.

In an effort to solve the problem, some coal-burning power plants are installing scrubbers to remove the sulfur in coal smoke. Scrubbers are installed at coal-fired electric and industrial plants, where a water and limestone mixture reacts with sulfur dioxide to form a sludge. Scrubbers eliminate up to 98 percent of the sulfur dioxide, and though they are expensive to build, they are far cheaper—in monetary value, not to mention in intrinsic value—than the healthcare problems caused by the release of the sulfur dioxide (p. 45).[1] Efforts to create regulations to require the addition of scrubbers were resisted by the administration of President George W. Bush, particularly due to the increased cost to energy producers, and without regard to the increasing cost of dealing with the health problems caused by their absence. In subsequent years, the administration of President Barack Obama has clearly sought to implement new accounting measures into America's energy future; however, it remains unclear who should pay for the improved technologies.

The environmental implications of sulfur dioxide and nitrous oxides are particularly problematic because they are transboundary issues: the air pollution from one area may create acid rain problems in other geographical areas. In addition, these pollution problems are difficult to trace to their exact source and even more problematic to quantify. The new field of environmental accounting has attempted to create a rubric for such patterns, and this is discussed in the section "Carbon Counting," below. In more-developed countries, modern emission control technologies and the greater use of low-sulfur coal have greatly reduced acid rain. The best available technology for removing sulfur from power plant emissions is to use wet scrubbers. Wet scrubbers are an expensive pollution control technology that costs about $200 million for a typical power plant plus another $15 million per year to operate.[2] To put this amount into perspective, a typical power plant generates about $600 million worth of electricity per year. Thus, while expensive by most standards, this cost is not out of balance with the economics of a large power plant.

Among the gases emitted when fossil fuels are burned, the most significant in the long term is carbon dioxide, a gas that traps heat in the earth's atmosphere. Over the last 150 years, burning fossil fuels has resulted in more than a 25 percent increase in the amount of carbon dioxide in our atmosphere. Fossil fuels are also implicated in the increased levels of atmospheric methane and nitrous oxide, although they have less importance as greenhouse gases than carbon dioxide.

Finally, researchers have connected these changes in atmosphere to a global rise in temperature and ocean levels. Since reliable records began in the late 1800s, the global average surface temperature has risen 0.5–1.1°F (0.3–0.6° C). Scientists with the Intergovernmental Panel on Climate Change (IPCC) concluded in a 2007 report what is now considered unequivocally true, that the earth's climate is indeed warming, and that "most of the observed increase in global average temperatures since the mid-20th century is very likely due to the observed increase in anthropogenic [human] greenhouse gas concentrations."[3] Scientists from around the world who make up the IPCC panel unanimously support the conclusion that it is virtually certain that the earth will continue to warm if carbon dioxide levels continue to rise. They also say that projected temperature increases will very likely result in an increased frequency of heat waves and severe rainfalls. These patterns will likely result in an increase in areas affected by drought, occurrences of intense tropical storms, and occurrences of extreme events on the high seas.

It is worth noting that the findings of the IPCC are inherently conservative, because all of the members who make up the committee must unanimously support its conclusions. Many of the members believe that human-induced climate change is actually worse than what is indicated by the official committee findings. The warming of the planet will cause a variety of impacts. The warmth itself continues melting glaciers, ice sheets, and permafrost, as well as warming oceans and lakes. This will lead to the inundation of wetlands, river deltas, and even populated areas. The warmth will cause increased evaporation of moisture from both land and sea, resulting in more droughts and more severe precipitation in the form of rain, freezing rain, and snow. Since parched, dry land is less absorbent, runoff from heavy rains will be more likely to cause flooding. Many agricultural lands will be faced with this cycle of alternating droughts and floods.

Although there are environmental impacts from mining for any mineral, it appears that the greatest impact of the fossil fuel era will be the pollution that burning these resources for energy placed in our ecological commons, including the air and ocean that all humans need to survive. When these additional costs are accounted for, fossil fuels are no longer cheap, and they certainly are not without detrimental effects. Estimates have been made that when energy producers prevent these harmful emissions or otherwise pay for their effects, the cost of fossil fuels doubles.

Methods for a Full Accounting of Energy Production

With the full accounting of fossil fuel, energy sources, and their impacts on human health, the environment, and climate change, alternative energy sources have become mainstream. This full accounting of the price of fossil fuels can be done in a variety of ways. Ideally, the producer of a certain type of energy should be required to pay for its production and all its detrimental effects to society and the environment. Were this done, the producer would then pass this cost along to the consumer. The consumer would then be able to reap the financial benefit of choosing a low-energy-existence life.

Without this production-side accounting, well-meaning consumers who choose to live off the grid in a solar-powered home with electric vehicles will not reap the benefits of their lifestyle. Although they would not be responsible for the daily emission of pollution, they would still be forced to breathe the same air as their neighbors living in an inefficient home with a 10,000-pound SUV. Thus, without production-side accounting of energy, there is no way for those with a low-energy life to reap the full benefits of their lifestyle. Of course, the owner of that 10,000-pound SUV will also unfairly reap the clean air rewards of all the other people who drive around in hybrids.

Even without a complete production-side accounting, the government plays an important role in energy accounting, using several different methods. The government can provide incentives to those who use renewable energy and purchase more efficient products. These incentives are nearly always financial in nature, so they don't technically provide for cleaner air or a cleaner environment. And though these incentives have not been valued highly enough in relation to the health and environmental impacts of the use of fossil energy, they have promoted alternative energy and conservation.

Another way for the government to promote a full accounting of energy production is to establish a carbon tax or carbon-trading scheme. The emission of carbon dioxide is the leading cause of global climate change and will have an impact of massive proportions on future generations. By enacting a carbon tax, the government doesn't stop the emission of carbon dioxide and the accompanying climate change, but it does make those emissions more expensive. The producer of energy that emits carbon dioxide must then pass this cost along to the consumer. This is similar to production-side accounting and encourages energy use from producers who do not emit carbon dioxide.

A third way for the government to be involved is to pass laws to prevent the emission or release of harmful pollutants. This is sometimes called a command and control structure by those opposed to it. With this legal requirement, an energy producer must take the necessary steps, at whatever cost, to prevent the harmful pollution. This cost is then passed on to the consumer. This type of accounting is production-side accounting. If this were done, it would not be necessary for renewable incentives or carbon taxes to be provided. However, this type of legal requirement to prevent harmful pollution has proven very difficult to enact and enforce.

In practice, the government employs a mix of these accounting schemes, and they have had the effect of making alternative energy production cost-competitive. As more of these schemes are employed to account for additional harmful pollution from the use of fossil fuels, alternative energy will continue to become more cost-effective, and perhaps fossil fuels will soon be cost-prohibitive. In this context, each of the existing methods for harvesting renewable energy enters the second decade of the twenty-first century at a most dynamic moment of opportunity.

Wind Energy

New electricity production in many states is now more likely to be from wind turbines than from any other source of energy. This surge in new wind turbine construction is only the result of technological advances begun in the 1970s having made wind energy the cheapest form of electricity in wind-favorable locations. Most of this expansion has been by utility-scale wind farms with a capacity of many megawatts. In most areas, this new energy source has been welcomed, but in some areas, mostly mountaintop locations, it has not been welcomed by all.

Although there are enough wind resources in the United States to provide all of the nation's electrical and transportation energy, there are practical considerations that limit the use of wind. Using the current electrical grid, it is estimated that wind can supply about 20 percent of the nation's electricity. Beyond this, transmission lines will become overloaded trying to get additional wind-generated electricity to more distant consumers, or bringing in backup sources of electricity on nonwindy days. However, a greatly expanded electrical grid with smart controls could be used to increase the amount of wind energy that can be utilized.

The bizarre nature of our energy transition reached a new level when an actor entered who had been a major player in the previous transition. Texas oil tycoon T. Boone Pickens unleashed a national series of television commercials during the summer of 2008 that scolded Americans for not having an energy plan. In the place of government leadership, Pickens offered his own plan on July 18, 2008, which called for huge investments in the development of alternatives, particularly wind, and the shifting of natural gas from electricity generation to powering vehicles. His plan, clearly, was about everything but the petroleum that had made Pickens wealthy. The centerpiece of the plan is wind development on the Texas plains, which is, in Texas style, gargantuan. The construction of a wind future on the plains of Texas, of course, possesses some of the irony of the wind turbines atop Pennsylvania's Appalachian Mountains, discussed above. The complaints leveled against wind development in other areas of the United States have little traction in West Texas, a sparsely populated region also pockmarked with oil drilling and exploration equipment.

Texas already generates about 5,000 megawatts of wind power, more than any other state. Most of Texas's wind-energy production is in petroleum-producing West Texas, where nearly 4,000 wind turbines tower over oil-pump jacks and capture the breeze that blows across the flat and largely barren landscape. The new plan would not only build a slew of new turbines, but would also add transmission lines capable of moving electricity all over the country. State funds have been directed at building transmission lines that would carry wind-developed power to other regions. The economic stimulus bill of 2009 provided $4.5 billion for improvements to the national electrical grid, thus manufacturing the opportunity for turbine developers to enter the picture.

Rebirth of Nuclear Power

Perhaps the clearest sign of a sea change in public acceptance of nuclear power was during the 2008 U.S. presidential campaign. During this campaign, the candidates of both major parties made clear their support for the use of nuclear power as part of the energy mix in America. Prior to this, such public support for nuclear power had often marked the end of a political career. Even before this, the company Areva began a national advertising campaign for nuclear power. This sea change was also evident in the actions of many different companies in the energy industry, many of which submitted proposals to the government for new and expanded nuclear power plant construction. Prior to 2008, there had been no such proposals made for nearly thirty years.

The driving force behind this desire for more nuclear power is varied. For some it just represents the lesser of two evils (coal and nuclear), while for others it represents a steady, clean, and relatively cheap source of energy; and of course there are those who remain opposed to nuclear power under any circumstances. Nuclear power does solve many of the problems associated with both fossil fuel and renewable energy sources, but it also presents a set of new problems. There are no emissions of any pollutants like there are with fossil fuels. The nuclear waste that is generated is completely contained and not released into the environment. It is the steadiest of all the sources of energy and is independent of weather (as well as, for the most part, of geography). New problems presented include the long-term storage of waste and the proliferation of nuclear weapons. Proponents of nuclear power say that these problems have been solved from a technological viewpoint, just not acted upon for political purposes.

The reemergence of nuclear power has been decades in the making. Plant designs are more advanced and fail-safe than those of decades ago. Also, designs allow for faster construction, which reduces costs. But perhaps most important, nuclear power plants are cost-competitive, if not cheaper, than the full accounting of fossil fuel power plants. Because nuclear power is now viewed as cost-competitive, industry is choosing to invest in this technology, and it appears that nuclear power will meet a larger portion of our electricity needs in the future.

In addition, nuclear power is viewed as a replacement for coal and natural gas electricity generation. Nuclear power plants operating now are over 90 percent reliable, which is much greater than any other type of power plant. They also run independent of weather conditions, making them ideal for base-load power. These factors have led to the rebirth of nuclear power, and time will tell how completely nuclear power will be embraced by society. Throughout the history of nuclear power, accidents have demonstrated the energy's volatile dimensions just as it seemed most viable. Following Three Mile Island, Chernobyl, and other incidents, the 2010 tsunami in Japan created a true nuclear disaster at the Fukushima Daiichi reactor. The accident threw Japan's nuclear program into disarray and led nations including Germany to abort plans for future expansion.

Even if it gets back on track, nuclear power cannot be the single solution to our energy crisis, at least not in the form currently used in the United States. There is not enough uranium in the world to supply a vastly expanded use of nuclear power for a time period of a century or so. In order for uranium to be a lasting part of our energy mix, it will become necessary for the science of breeder reactors and reprocessing nuclear waste into new plutonium and thorium fuel to take precedence over the politics of not wanting to reprocess nuclear waste. This reprocessing of nuclear waste is sometimes called a "closed fuel cycle," to indicate that fuel is used to make more fuel.

Although other nations such as Japan, France, and Russia currently reprocess their nuclear waste, the United States has had a policy for over thirty years of not reprocessing nuclear waste. This political policy was adopted in the hope of stopping the spread of nuclear weapons around the world. However, as is evidenced by developments in North Korea, Pakistan, India, Israel, and South Africa, this policy has failed. Many of the latest reactor designs being pursued internationally assume the future use of reprocessed nuclear waste, so that nuclear power can provide energy for centuries more while reducing the amount of high-level waste that must be stored long-term.

Increasing Use of Biofuels

Possibly the most significant change in the energy transition of 2010 was the broadening of production and use of biofuels. In 2006, when President George W. Bush castigated Americans for their addiction to oil, he called for the use of alternatives to produce biofuels, including switchgrass. Most experts expect that in the first decades of the twenty-first century there will be a mad rush to biofuels, homegrown gasoline, and diesel substitutes made from crops like corn, soybeans, and sugarcane. These technologies have been around for a century, but now are being thrust forward as the most effective transitional energy source as humans consider other ways to power transportation. Although most were never intended for use on a massive scale, biofuels have become major players in the energy sector because of high gas prices.

The image is enticing to many Americans: not only liberating Americans from Middle East oil but also pumping that revenue into the declining rural economy of the country. The entire industry, though, remains based in speculation and uncertainty. Biofuels as currently rendered in the United States are doing great things for some farmers and for agricultural corporations, including Archer Daniels Midland and Cargill. Most Americans see ethanol as a green alternative; however, ethanol plants burn natural gas or, increasingly, coal, to create the steam that drives the distillation. In addition, diesel farm machinery is used to tend the fields, and natural gas–based fertilizers and herbicides are used to maximize the crop yield, leading to substantial use of fossil fuels to make ethanol. "Biofuels are a total waste and are misleading us from getting at what we really need to do: conservation," says Cornell University's David Pimentel, who is one of ethanol's harshest critics. "This is a threat, not a service. Many people are seeing this as a boondoggle."[4]

Fortunately, with improvements in technology the ethanol yield has improved and is now approaching 500 gallons per acre for corn, and the energy content of that yield is approaching a 50 percent increase over the total fossil energy required to produce the ethanol. Perhaps even more important is that most of the fossil energy put into ethanol production is in the form of natural gas and coal. Thus, ethanol effectively serves as a method of converting domestic natural gas and coal into a somewhat larger amount of liquid fuel for transportation, which displaces imported petroleum.

Three factors came together in the early 2000s to make ethanol less an alternative fuel and to move it into the mainstream: record high prices for petroleum, the phaseout of the MTBE gasoline additive, and society's desire to become more energy independent. Ethanol production has responded to these factors, increasing from 50 million barrels in 2002 to over 200 million barrels in 2008. Continued increases in production will be limited by the ability to grow suitable feedstock for biofuels. For instance, experts estimate that even if we turned our entire corn and soybean crop into biofuels, together they would replace only 12 percent of our gasoline and 6 percent of our diesel. And getting just to this point would lead to severe food and meat shortages, as the crops would no longer be used for animal feed or other food stocks (including pork, beef, and poultry).

The push to produce more ethanol has quickly revealed some of the problems associated with biofuels. The growth in ethanol production has pushed corn demand to heights not seen in years, affecting food prices and spurring U.S. growers to plant the largest crops since World War II. Around a fifth of the harvest will be brewed into ethanol—more than double the amount only five years ago. Corn is not the only crop that is problematically being made into fuel. From an environmental perspective, biodiesel from soybeans fares only slightly better. Rising prices for both crops pushed farmers to plow up more land than in previous years—approximately 35 million acres of marginal farmland now set aside for soil and wildlife conservation and in areas too arid for farming without depleting subsurface aquifers. But most disturbing of all is the impact on global trade as the United States exports less corn and soybeans. This lack of U.S. food exports has led to increased crop production elsewhere in the world, namely Brazil and Indonesia, where rainforests are clear-cut and plowed into new farmland. The carbon footprint of an acre of rainforest being turned into cropland, effectively for biofuel production, is much worse than if fossil fuels had been used in the first place.

These considerations have led pilot projects in the United States to experiment with making ethanol from cellulose acquired from noncrop biomass (switchgrass, wood). One ton can be converted into seventy gallons of ethanol in about a week. Overall, the current process is about half as efficient as that of deriving the energy from crude oil. If the technology is improved, noncrop biomass feedstock can be grown on land without displacing current crops. Furthermore, switchgrass and fast-growing trees can also be grown with a much lower environmental impact when the actual planting, tending, and harvesting procedures are taken into account. Another potential plant that scientists are experimenting with as a biofuel feedstock is much simpler: algae—single-celled pond scum. Because the plant does not require farmable land resources and can instead be grown even in wastewater, many experts believe algae-based fuels are the only feedstock with the potential to reach the supply levels required to make a significant impact on our energy use.

Giving Solar a Chance

Another slice of the new energy supply pie will likely derive from the oldest source of power. New, large-scale efforts to put solar power to work have recently taken shape in California. Two companies are constructing solar plants that will be ten times bigger than those now in use. Spurred by state mandates to derive 20 percent of its electricity from renewable sources by 2010, Pacific Gas and Electric will purchase the plants' electricity.

Each plant uses photovoltaic technology, which turns sunlight directly into electricity instead of using it to heat water. OptiSolar, a company that has just begun to make thin-film solar panels—with a layer of semiconductor material thinner than a human hair on the back of a glass panel—will install 550 megawatts in San Luis Obispo County, in central California. And the SunPower Corporation, which uses crystalline cells, will build 250 megawatts in the same county. The OptiSolar plant will cover about nine square miles, and the SunPower plant about 3.5, although the actual cell area will be smaller. Together, these plants will generate a total of 800 megawatts.

A megawatt is enough power to run a large Wal-Mart. At peak hours, together the plants will produce as much power as a large coal or nuclear power plant. But they will run far fewer hours of the year, so output will be at least a third less than that of a coal plant of the same size. SunPower's panels are mounted at a 20-degree angle, facing south, and pivot over the course of the day so they continuously face the sun. OptiSolar's panels are installed at a fixed angle. They are larger and less efficient, but much less costly, so that the cost per watt of energy is similar.

Solar energy, both photovoltaic and thermal, which uses the sun's heat to make steam, is bounding ahead, driven mostly by state quotas and government incentives. California required that 20 percent of the kilowatt-hours sold by investor-owned utilities come from renewable sources by 2010, a goal that some companies are still struggling to meet. Pacific Gas and Electric expects that when these two solar plants are completed, their total will rise to 24 percent, but that will not be until 2013.

Conclusion

It would appear that the life cycle of alternative fuels has arrived at a new juncture in human history. Just as wind turbines, a symbol of alternative approaches to power production, are appearing along the ridgelines of Central Pennsylvania, they can now be found revitalizing one of their primary points of origin: The Netherlands. In the early history of humans' use of energy, almost all the power available derived from renewable sources. The windmills of early industry in places such as The Netherlands were private or community enterprises. Today's efforts are most often developed by private companies, but as part of, or with the help of, large government initiatives.

In The Netherlands, for instance, the government has invested more than $80 million to restore some of the 1,040 older mills already in existence. Many of them have been retrofitted to generate electricity instead of to grind grain. In addition, the government has constructed one large-scale wind farm off the coast and has plans for others. Making The Netherlands' adoption of alternative power easier, of course, is the nation's small population, size, and, commensurately, small carbon footprint. Such changes are more complicated in nations that have allowed themselves to grow more dependent on fossil fuels.

The United States, ground zero for humans' high-energy lifestyle in the twentieth century, has been slower than The Netherlands and many other European nations in creating effective government stimuli for the development of wind power and other alternative energy. One of the most recent developments in our energy transition, though, has been a clear sea change in Americans' interest in and openness toward deriving their energy from sources other than fossil fuels. Linked to the ethic of modern environmentalism, green power options have moved to the mainstream in the twenty-first century, including incorporation into the economic stimulus initiatives of 2009 that grew from the business potential of these new opportunities. More broadly, however, the presidency of Barack Obama has returned Americans to a central question: How does one lead an energy transition forward?

President Jimmy Carter, of course, demonstrated the difficulty of the Oval Office attempting to lead technological innovation. It appears that the Obama administration has adopted a more integrated approach than that of Carter or any other U.S. president. Such initiatives, though, succeed or fail based on the public reaction to them. In order to further this transition, we must return to some of the basic roots of Americans' twentieth-century high-energy binge: the culture of consumption.

With informed consumption, consumers might play the most critical role in America's energy future. Since Americans first considered energy conservation to be part of their lifestyle in the 1970s, modern environmentalism has bred an entirely new genre of consumption, referred to as "green consumerism." In fact, across the board, mass consumption contains a thread of greenness—conservation thought—that runs diametrically opposed to the ethic behind our expansion into the high-energy lifestyle of the mid-twentieth century. History has taught us that such revisionary shifts in lifestyle do not fare well when presented to Americans from the top down; instead, we now operate in an information era in which well-informed consumers might steer producers toward more sustainable and, often, economical uses of energy.

A one-size-fits-all energy strategy neither can nor should be mandated by the U.S. federal government. Neither can society wait for a perfect solution to present itself as the path to a new energy future. By waiting for a perfect solution, America will fail to move forward and will ultimately rely on technologies developed in nations that have more actively pursued alternative sources of energy. The successful freeway to America's energy future will have many lanes representing a variety of energy sources, including even the clean use of the remaining fossil fuels. Each energy source will have its own set of imperfections. Perhaps the only technology that must be pursued is an expanded and modernized smart grid, which benefits all sources of power by helping to more efficiently meet the demands of society. A smart grid will allow the many energy sources to both compete and coordinate with each other. Such a competitive energy economy, including the consideration of the full life cycles of each energy source, holds the most promise for American society. No more fear of dwindling supplies, high prices, and reliance on other nations. The United States would move forward on many fronts to a diversified energy future.

In such an energy market, alternative sources of energy can no longer remain in their current state. As our energy transition proceeds, the most likely outcome is a diverse energy mix built on the backbone of a modernized, smart electrical grid that draws power from a wide variety of sources, prioritizing those that are sustainable and even renewable, and sends that power along to the consumer. Government must play an even more significant role in regulating and enforcing a fuller accounting of all energy sources, because individuals are too far removed by both geography and generations from observing the negative impacts of using cheap energy. When the entire life cycle of energy sources is priced correctly and Americans are given a more honest choice of various energy sources, the alternatives with which humans lived a few centuries ago will rise to the top and demand innovation and mainstream use.

Notes

1. R. Gelbspan, *The heat is on: The climate crisis* (Reading, MA: Perseus Books, 1995).
2. U.S. Energy Information Administration, *Electric power annual, electric power industry 2009: Year in review*, http://www.eia.doe.gov/cneaf/electricity/epa/epa_sum.html (accessed March 10, 2011).
3. IPCC-Intergovernmental Panel on Climate Change, IPCC fourth assessment report (AR4), http://www.ipcc.ch/publications_and_data/publications_ipcc_fourth_assessment_report_synthesis_report.htm (accessed March 10, 2011).)
4. J. K. Bourne Jr., Green dreams, *National Geographic* (October 2007), http://ngm.nationalgeographic.com/2007/10/biofuels/biofuels-text (accessed December10, 2010).

For Further Reading

Adams, D. A. 1996. *Renewable resource policy: The legal-institutional foundation.* New York: Island Press.

Andrews, R. N. L. 1999. *Managing the environment, managing ourselves.* New Haven, CT: Yale University Press.

Athansiou, T. 1998. *Divided planet: The ecology of rich and poor.* Athens: University of Georgia Press.

Black, B .C., and R. Flarend. 2009. *Alternative energy.* Westport, CT: Greenwood Press.

Brower, M. 1992. *Cool energy: Renewable solutions to environmental problems.* Rev. ed. Cambridge, MA: MIT Press.

Hughes, T. 1983. *Networks of power: Electrification in Western society, 1880–1930.* Baltimore, MD: Johns Hopkins University Press.

Hughes, T. 1989. *American genesis.* New York: Penguin.

Landes, D. 1969. *The unbound Prometheus: Technological change and industrial development in Europe.* New York: Cambridge University Press.

Lovins, A. 1979. *Soft energy paths.* New York: HarperCollins.

McNeil, J. R. 2001. *Something new under the sun: An environmental history of the twentieth-century world.* New York: Norton.

Montrie, C. 2003. *To save the land and people: A history of opposition to surface coal mining in Appalachia.* Chapel Hill: University of North Carolina Press.

Motavalli, J. 2001. *Forward drive: The race to build "clean" cars for the future.* San Francisco: Sierra Club Books.

National Geographic. Based on the story "Green Dreams" by Joel K. Bourne Jr. This site contains information about biofuels: http://ngm.nationalgeographic.com/2007/10/biofuels/biofuels-interactive.

Nye, D. 1999. *Electrifying America.* Boston: MIT Press.

Nye, D. 1996. *Technological sublime.* Boston: MIT Press.

Rifkin, J. 2003. *The hydrogen economy.* New York: Penguin.

21

Eco-Friendly Transportation and the Built Environment

*David A. Sleet, Rebecca Naumann, Grant Baldwin,
T. Bella Dinh-Zarr, and Reid Ewing*

Introduction

Public health strategies that decrease traffic injury risk through environmental design are the focus of this chapter. It is increasingly recognized that the design, layout, and use of communities' structures and transportation systems affect behavior, and ultimately, health.[1] Although built environment modifications designed to prevent injuries, like pedestrian overpasses, may be expensive initially, once in place they have the potential to protect people for a long time. Well-planned designs with built-in safety features at the outset are often cheaper than retrofitting the environment to correct problems later. The built environment can be modified to help prevent injuries, such as young children falling from balconies or pedestrians and vehicles colliding. Such modifications help prevent predictable injury events.

Injury Prevention from a Public Health Perspective

Injuries are not accidents. Most injuries are predictable and preventable. Although an injury may result from an unexpected event, it can usually be traced to specific actions (e.g., alcohol-impaired driving), hazardous environments (e.g., lack of sidewalks), or faulty products (e.g., vehicles with unsafe tires). To this extent, injuries are within human control. Injuries result from interactions among persons (host factors), energy (agent and vehicle factors), and the environment.[2] Thus, injury prevention is most effectively achieved when programs are implemented with elements of behavior change, policy, product safety, and environmental design.

Traffic Injuries

Traffic injuries are the leading cause of death for ages one through thirty-four in the United States (averaging 43,000 deaths annually) and a leading cause of death for children and young adults around the world.[3,4] Globally, 1.3 million people die each year on the world's roads, or about 150 people every hour of every day.[3] Moreover, there are between 20 and 50 million nonfatal injuries resulting in disability and diminished quality of life.[3] All told, global road traffic crashes cost an estimated $518 billion U.S. annually and cost governments between 1 and 3 percent of their GNP.[3]

Although low- and middle-income countries have fewer than 50 percent of the world's registered vehicles, over 90 percent of fatalities happen in those countries.[3] Unlike the United States, where drivers and passengers are at greatest risk, the most vulnerable road users in low- and middle-income countries are pedestrians, cyclists, and users of motorized two-wheelers like scooters.[3] The human and economic burden underscores the need for broad-based, global solutions, including a special recognition of the importance of the built environment.

From the epidemiologic perspective in public health, physical damage to the host (the person harmed) is usually brought about through a rapid transfer of kinetic energy. This energy transfer can be modified by making the host more resistant to it (e.g., by increasing human injury tolerance), by separating the host from the kinetic energy exchange (e.g., by using a seat belt), or by eliminating the source of energy exchange by changing the environment (e.g., by incorporating road design that either decreases the likelihood of a crash or protects people by buffering the energy exchange in a crash). Multi-component approaches that promote behavioral, vehicle, and environmental change are likely to have the most success.

The Role of Environmental Planning

Changes in vehicles and roadways have been among the most successful strategies to reduce traffic-related injuries. Since 1925 the annual death rate per million vehicle miles traveled has decreased over 90 percent in the United States, largely because of modifications in driver behavior, vehicle crashworthiness, road design, and changes in the built environment.[5] Motor vehicle safety modifications have included lap and shoulder belts, air bags, center-mounted brake lights, electronic stability control, and daytime running lights. Roadway safety changes have included divided highways, breakaway sign and utility poles, improved lighting, barriers separating oncoming traffic, and guardrails.

However, the extensive network of well-built high speed roads may have indirectly contributed to increased motor vehicle injuries by fueling urban sprawl, thereby increasing commute time, vehicle miles traveled, and exposure to traffic crashes.[6] Studies have shown that communities with less sprawl and fewer vehicle miles traveled (less exposure) have lower traffic fatality rates per population.[7] In the United States, road conditions are a contributing factor in 53 percent of all road deaths and 38 percent of injuries, and it is the single deadliest contributing factor to crash severity, ahead of speeding, alcohol, or lack of seat belt use.[8] In Sweden, road conditions are a contributing factor in at least 59 percent of fatal crashes.[7] The situation is potentially even worse for developing countries, where poor roads are combined with an influx of motor vehicles, new drivers, and poor infrastucture for emergency response and care.

To achieve further reductions in traffic-related injury and death, a renewed focus on the built environment and community modifications that can improve safety for motorized and nonmotorized road users will be important.

Creating "Green" Environments That Enhance Pedestrian and Bicycle Safety

Built environment strategies for preventing pedestrian injury include separating pedestrians from motor vehicles, installation of traffic signals, in-pavement flashing lights, four-way stops, pedestrian overpasses, fences to inhibit street access, and sidewalks.[9] In addition, pedestrian safety zones are increasingly being used as an effective strategy for separating vehicles from pedestrians and bicyclists.[10] In countries like Germany and The Netherlands, where reduced vehicle speeds and the design and building of "safe systems" that separate motor vehicles from pedestrians and bicyclists are more prevalent, rates of pedestrian injury and death are much lower than in the United States.[11]

Engineering measures designed to increase visibility of pedestrians, such as increased roadway illumination and relocating bus stops to the far side of intersections, also decrease injury risk. Of the engineering measures to manage vehicle speed, small roundabouts on residential roads and four-way stops at intersections are effective.[9] For children, speed humps can reduce overall child pedestrian injuries in a neighborhood setting.[12] Modern roundabouts, which have two or fewer lanes, can effectively reduce traffic crash injuries by slowing speeds.[13] Additional environmental strategies to decrease both pedestrian and bicyclist injuries can be found in Ewing and Dumbaugh.[14]

It is important to note that strategies should be selected and carefully implemented based on evidence of their effectiveness, as some environmental modifications originally thought to decrease injury risk have been found to increase it. Crosswalks without traffic signals increase risk for elderly pedestrians,[15] and crosswalks without traffic signals located on busy streets and/or on streets with more than two lanes increase risk for all pedestrians.[16] Roundabouts with multiple lanes can increase crashes because they increase the difficulty of negotiating lane changes safely.[13] In addition, some promising designs need further research, including routing traffic away from residential settings, off-road trails for pedestrians and bicycles, and area-wide traffic calming.[17]

Improving Eco-Friendly Transportation Environments in the United States

Efforts to improve the built environment with a focus on the safety of pedestrians and bicyclists have increased in recent years. Approaches such as "New Urbanism," "Smart Growth," and "Active Living" have emerged, which encourage the building of walkable communities in which people can reach destinations without driving. Designing communities with the comfort, enjoyment, and safety of the pedestrian or bicyclist as a priority

Eco-Friendly Transportation and the Built Environment 431

is an important aspect of these new movements (www.newurbanism.org/pedestrian.html; www.smartgrowth.org; www.activelivingresearch.org).

In a major about-face for the U.S. Department of Transportation, in March 2010 Secretary of Transportation Ray LaHood announced that the needs of pedestrians (and cyclists) will be considered along with those of motorists, and that the automobile will no longer be the primary consideration in federal transportation planning. He emphasized that walking and biking are important components for livable communities and that "this is the end of favoring motorized transportation at the expense of non-motorized transportation" (http://fastlane.dot.gov/2010/03/my-view-from-atop-the-table-at-the-national-bike-summit.html). Similarly, efforts to encourage and promote alternative transportation can be seen in Europe, where programs like the City Bike System are available in many cities. As these programs and efforts grow, it will be important to ensure that they are accompanied by built environment features that keep pedestrians and cyclists safe so that the benefits of active transportation may be obtained with minimal injury risk.

Denver, Colorado, and many other cities around the United States have adopted a "rent a bicycle" program to facilitate the ease of using bicycles to travel around the city-center (see Figure 21.1). These programs have been modeled after successful City Bike Systems, such as those in Copenhagen, introduced in 1995, that allow anyone to borrow a bike from stands around the city for a small coin deposit. When finished, they leave the bike at any of the 110 bike stands, and their money is refunded. Vienna, Austria, has sixty bike stations, some giving the rider the first hour free. Nearly 100 other cities around the world have replicated the system.

Building Safe Environments for Automobile and Motorcycle Travel

Built environments that address the safety of those using motorized vehicles are an important public health priority for both developed and developing countries. Though many of the strategies discussed above (e.g., speed reduction, traffic signals, and roadway illumination) can improve safety not only for pedestrians and cyclists but for all road users, additional strategies, as well as methods for prioritizing the implementation of these strategies, are available.

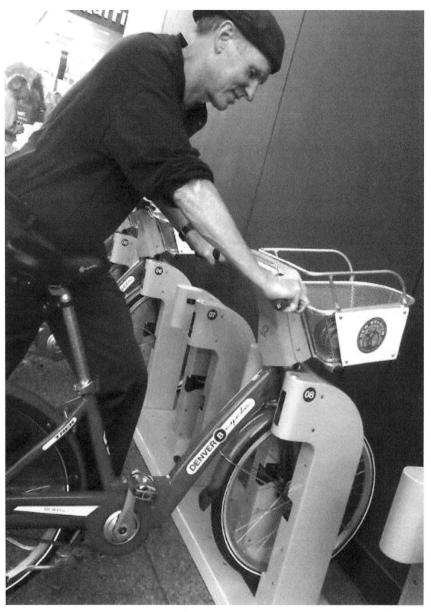

Figure 21.1 Bike rentals in Denver, Colorado, 2011. (photo credit: David E. Corbin, by permission)

Eco-Friendly Transportation and the Built Environment 433

In the developing world, automobile and motorcycle ownership and use has quickly grown, while road infrastructure in many countries has remained poor or very limited in capacity. In addition to the behavioral changes that are key to establishing a safety culture in developing countries, improving the built environment, through solutions such as safer pavement or separate ("segregated") motorcycle lanes, is important. A useful method for prioritizing and strategically planning these improvements is through road safety audits, such as those conducted by the International Road Assessment Program (iRAP).[18] iRAP promotes safe road design (for pedestrians, bicycles, motorcycles, and automobiles) by inspecting high-risk roads and recommending, in consultation with government officials and stakeholders, affordable engineering improvements. Motorcycles continue to be used in high volumes, so iRAP is recommending motorcycle lanes as a primary prevention strategy (separation of agents). Figure 21.2 is from Malaysia, where motorcycle volumes are high. These road safety audits have been conducted in both developed countries (AusRAP in Australia, EuroRAP in various European countries, KiwiRAP in New Zealand, and usRAP in the United States) and low- and middle-income countries in Africa, Asia,[19] and Latin America. In iRAP, teams of engineers and experts ride in a vehicle capable of capturing high-resolution digital images of the road at ten-meter intervals. The images, linked to GPS and road geometry data, are then assessed by a team of "raters" and used to generate road safety Star Ratings for vehicles, motorcycles, pedestrians, and bicyclists. These assessments provide the basis for decisions by road authorities and local stakeholders about where affordable and effective road improvements can be made. Completed iRAP projects have shown that small investments can result in lives and money saved. For example, in Kenya, a country where three out of four people killed on the road are economically productive young adults, iRAP recommendations are demonstrating a cost of less than U.S. $1,000 for each death and serious injury prevented.[18]

Another important area to focus built environment improvements on in the coming years will be road safety enhancements for older adult drivers. The "baby boom" generation (born between 1946 and 1964) begins turning sixty-five in 2011, and it is estimated that by 2050 the population of those eighty and older will have tripled in most Organization for Economic Cooperation and Development (OECD) countries.[20] Built environments will need to include "senior-friendly" road design, such as better

Figure 21.2 Following a successful campaign in 2007 which included distribution of low-cost helmets, education, and a law, helmet use rates in Vietnam rose from 3% to 99% and deaths decreased by 12%. Motorcycles continue to be used in high volumes so iRAP is recommending motorcycle lanes as a primary prevention strategy (separation of agents). The photo is from Malaysia, where motorcycle volumes are not significantly different from rural sections of road in Vietnam. (21) (Photo source: iRAP [by permission])

lighting, signage, and road markings; intersection improvements; and protected left-turn lanes.[21] Such improvements in the built environment, although designed to address declines in older adults' driving abilities, will also positively impact road users of all ages.

Implications for Designing More Eco-Friendly Traffic Environments

Conventional traffic engineering practices are oriented toward mobility, with safety identified as a complementary goal. Because safety and efficiency are treated as mutually supportive goals, most conventional transportation planning applications begin by identifying vehicle mobility needs and then proposing mobility-oriented solutions, such as road widening. Once a mobility need is identified, safety is addressed by designing these improvements for higher design speeds under the presumption that higher design speeds equate to enhanced safety performance. To the extent that the built environment is considered at all, it is solely for forecasting future levels of traffic demand to identify needed mobility improvements.[14]

Yet the empirical evidence on traffic safety suggests that safety and mobility may be conflicting goals, at least in urban areas. Contrary to accepted theory, the stop-and-go, high-volume traffic environments of dense urban areas appear to be safer than the lower-volume environments of the suburbs. The reason is that many fewer miles are driven on a per capita basis, and the driving that is done is at lower speeds that are less likely to produce fatal crashes. Also contrary to accepted theory, at least in dense urban areas, less "forgiving" road designs—such as narrow lanes, traffic calming measures, and street trees close to the roadway—appear to enhance a roadway's safety performance when compared to more conventional roadway designs.

One of the fundamental shortcomings of conventional traffic safety theory and planning is that it fails to account for the moderating role of human behavior. Decisions to reduce development densities and segregate land uses, or to widen specific roadways to make them more forgiving, are based on the assumption that in so doing, human behavior will remain unchanged. And it is precisely this assumption—that human behavior can be treated as a constant, regardless of design—that accounts for the failure of conventional safety practice.[22, 23] If safety is to be meaningfully addressed, it must consider how the built environment influences both the incidence of traffic crashes, injuries, and deaths, as well as the specific behaviors that cause them.

Future research on eco-friendly transportation and the built environment should address a number of factors. We need to better develop our understanding of how community design patterns influence travel and the mode of transportation used, and how these travel patterns impact traffic speed, traffic crashes, and injuries. More eco-friendly modes and lower-speed transportation users, including bicyclists and pedestrians, may be forced to share the road with higher-speed vehicle traffic, a user mix that is likely to increase crash incidence. It is likely that the safety gains achieved through eliminating neighborhood traffic may be offset by conflicts between bicyclists, pedestrians, and vehicles on arterial thoroughfares.

Finally, we must begin to develop our understanding of how design influences the behavior of specific roadway users and how these behaviors in turn influence crash incidence. Modifications in the built environment can profoundly influence vehicle speeds and traffic conflicts, which in turn have a profound effect on crash incidence. Yet there has been little research aimed at relating specific pre-crash behaviors to the environments in which they occur, and almost no attempt to understand how the characteristics of the built environment may encourage or discourage these behaviors from occurring in the first place.[14] Rather than superimposing rural design solutions on urban environments, as is often done, we must begin to develop an understanding of the moderating role of behavior on the relationship among environmental design, the built environment, and transportation safety.

Conclusion

Environmental modifications can effectively reduce traffic injuries, as the built environment and traffic safety are closely linked. Encouraging safe, active transport by foot and bicycle can promote physical and mental health[24] and improve air quality, but must be accompanied by features that are designed-in to protect pedestrians and cyclists from injury, such as separating them from traffic. Building and retrofitting communities to promote walking, biking, and greater use of public transit will not necessarily reduce injuries. Injuries could increase as more people use "active transportation" modes, like walking and biking, unless safety design features are "built-in" to environmental changes. Working toward "green" environments that promote "active transportation" is increasingly recognized as a key component of public health worldwide, but we must consider carefully the impact it could have on injuries. Traffic engineers and urban planners should consider the impact of neighborhood development patterns and the traffic environments of dense urban areas when they build environmental solutions for eco-friendly transportation alternatives. For cities whose populations want more nonmotorized transport, additional environmental infrastructure safety nets will be required as the need for mobility will have to be balanced with the need for safety.

Notes

Disclaimer: The views expressed in this chapter are those of the authors and do not necessarily represent the official policies of the Centers for Disease Control and Prevention, the FIA Foundation, or the University of Utah.

David A. Sleet is the corresponding author for this chapter

1. Prevention Institute, *The built environment and health: 11 profiles of neighborhood transformation* (2004), http://www.preventioninstitute.org/index.php?option=com_jlibrary&view=article&id=114&Itemid=127 (accessed August 10, 2010).
2. W. Haddon, On the escape of tigers: An ecologic note, A*merican Journal of Public Health* 60 (1970): 2229–2234.
3. World Health Organization, *Global status report on road safety: Time for action* (Geneva: World Health Organization, 2009), www.who.int/violence_injury_prevention/road_safety_status/2009. Accessed 10 August 2010.
4. Centers for Disease Control and Prevention, National Center for Injury Prevention and Control, Web-based injury statistics query and reporting system (WISQARS) (2010), www.cdc.gov/injury/wisqars/. Accessed 10 August 2010.
5. A. Dellinger and D. A. Sleet, Preventing traffic injuries, *American Journal of Lifestyle Medicine* 4 (2010): 82–89.
6. R. Ewing, R. A. Schieber, and C. V. Zegeer, Urban sprawl as a risk factor in motor vehicle occupant and pedestrian fatalities, *American Journal of Public Health* 93 (2003): 1541–1545.
7. H. Stigson, M. Krafft, and C. Tingvall, Use of fatal real-life crashes to analyze a Safe Road Transport System model, including the road user, the vehicle and the road, *Traffic Injury Prevention* 9 (5) (2008): 463–471.
8. T. Miller and E. Zaloshnja, *On a crash course: The dangers and health costs of deficient roadways,* Pacific Institute for Research & Evaluation (PIRE), 2009, http://www.pire.org/detail.asp?core=39633 (accessed August 10, 2010).
9. R. A. Retting, S. A. Ferguson, and A. T. McCartt, A review of evidence-based traffic engineering measures designed to reduce pedestrian-motor vehicle crashes, *American Journal of Public Health* 93 (2003): 1456–1463.

10. National Highway Traffic Safety Administration, *Countermeasures that work*, 3rd ed. (Washington, DC: U.S. Department of Transportation, 2008).
11. J. Pucher and L. Dijkstra, Promoting safe walking and cycling to improve public health: Lessons from The Netherlands and Germany, *American Journal of Public Health* 93 (2009): 1509–1516.
12. J. M. Tester, G. W. Rutherford, Z. Wald, and M. W. Rutherford, A matched case-control study evaluating the effectiveness of speed humps in reducing child pedestrian injuries, *American Journal of Public Health* 94 (2004): 646–650.
13. R. Ewing, personal communication, 2010.
14. R. Ewing and E. Dumbaugh, The built environment and traffic safety: A review of empirical evidence, *Journal of Planning Literature* 23 (2009): 347–367.
15. T. Koepsell, L. McCloskey, M. Wolf, et al., Crosswalk markings and the risk of pedestrian-motor vehicle collisions in older pedestrians, *JAMA* 288 (2002): 2136–2143.
16. C. V. Zegeer, J. R. Stewart, H. Huang, and P. Lagerwey, Safety effects of marked versus unmarked crosswalks at uncontrolled locations, *Transportation Research Record* 1723 (2001): 56–68.
17. F. Bunn, T. Collier, C. Frost, et al., Area-wide traffic calming for preventing traffic related injuries, *Cochrane Database Systems Review* 1 (2003): 1–21.
18. International Road Assessment Programme (iRAP), Countries, fact sheet, www.irap.net (accessed August 10, 2010).
19. G. Smith, Le Minh Khoa, L. V. D. Chai, et al., iRAP Vietnam: A life saving partnership, in *[Proceedings of] 24th ARRB Conference: Building on 50 years of road and transport research*, Melbourne, Australia, 2010.
20. Organisation for Economic Co-operation and Development, *Ageing and transport: Mobility needs and safety issues* (An Organization for Economic Cooperation and Development publication, 2001), http://www.oecd.org/dataoecd/40/63/2675189.pdf (accessed August 10, 2010).
21. L. Staplin, K. Lococo, S. Byington, and D. Harkey, *Highway design handbook for older drivers and pedestrians*, FHWA-RD-01-103 (Federal Highway Administration, U.S. Department of Transportation

Publication, 2001), http://www.fhwa.dot.gov/publications/research/safety/humanfac/01103/index.cfm (accessed August 10, 2010).
22. E. Dumbaugh, Safe streets, livable streets: A positive approach to urban roadside design (PhD thesis, Georgia Institute of Technology, Department of Civil and Environmental Engineering, Atlanta, 2005).
23. E. Dumbaugh, Enhancing the safety and operational performance of arterial roadways in the Atlanta Metropolitan Region (report submitted to the Atlanta Regional Commission, 2006).
24. U.S. Department of Health and Human Services, *Physical activity and health: A report of the surgeon general* (Atlanta, GA: U.S. Department of Health and Human Services, Public Health Service, CDC, National Center for Chronic Disease Prevention and Health Promotion, 1996).

22

Lighting and Astronomy

Christian B. Luginbuhl, Constance E. Walker, and Richard J. Wainscoat

Introduction

The sky is fading. Prime sites for astronomical observatories are rare. A stable, clear, dry atmosphere is crucial. Yet many of the best sites worldwide are slowly losing their view of the most distant astronomical objects as more and more stray light appears in the last microseconds of what may have been a ten-billion-year journey. That intruding light comes from outdoor lighting used for roadways, parking lots, advertising, and decoration and from automobile headlights. It gets into the sky, and ultimately into telescopes, either directly from light fixtures or through reflection off the ground or other surfaces, followed by scattering from molecules and aerosols in the atmosphere. Largely because of that stray light, or sky glow, new giant telescopes are being built in the most remote corners of the planet. Yet even those sites are now threatened by artificial light from communities that may be located hundreds of kilometers away, as shown in Figure 22.1.

The projected growth of outdoor lighting, illustrated in Figure 22.2, paints a discouraging picture. Whereas the U.S. population is growing at an average rate of less than 1.5 percent per year, the amount of artificial light is increasing at an annual rate of 6 percent. Increases in population, standards of living, and isolation from the natural nighttime environment combine to lead communities and individuals to increase not only the number of situations in which outdoor lighting is deemed necessary but also the amount of light used for many applications. Witness the amount of

Figure 22.1 The sky over Mauna Kea, Hawaii, is affected by outdoor lighting in communities from Waimea 30 km away to Honolulu 300 km away. (Photo by Richard Wainscoat.)

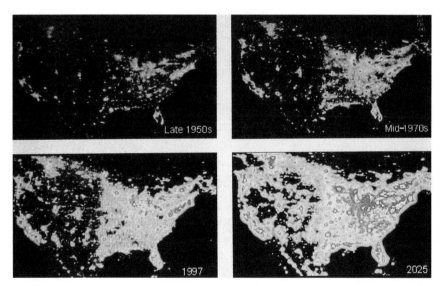

Figure 22.2 Sky glow over time is estimated by extrapolating 1997 satellite measurements forward and backward using a 6% annual growth rate. That average rate, determined from ground-based measurements of sky brightness, is applied equally to all points; no attempt has been made to model actual growth rates for different regions. (Adapted from [9], p. 39.)

light commonly seen at service stations today compared with that of only a few decades ago: Lighting levels comparable to or even brighter than those recommended for indoor office work are common. Parking lots are often illuminated five to ten times as brightly as they were twenty years ago. Bright light is becoming a form of advertising.

Although lighting is rarely installed with the purpose of brightening the sky, even in the best circumstances some fraction of outdoor lighting propagates upward by reflection from the illuminated area. And actual outdoor night lighting rarely represents the best circumstances. Inefficient, careless practices and poorly designed fixtures dramatically increase sky glow through direct upward emission, wasted light falling into areas that do not need illumination, and over-illumination.

Falling Back

Astronomers have long been retreating from encroaching lights. As planning began in the 1930s for a new 5-m telescope, planners realized that the Mount Wilson Observatory, home of the then largest telescope (at 2.5 m), already suffered from too much light pollution. A new site on Palomar Mountain was chosen, farther from the lights of the Los Angeles area.

The retreat continued. When the National Observatory was searching for a site in the early 1950s, another remote area was chosen, on Kitt Peak in Arizona, 80 kilometers from Tucson and 170 kilometers from Phoenix, whose populations at the time were about 125,000 and 330,000, respectively. From the 1960s to the present, new telescopes have been built on ever more remote sites in Chile, Arizona, Hawaii, and the Canary Islands. Although the most remote sites currently suffer insignificant light pollution, the prospects for the future are uncertain. There are few high-quality sites for further retreat. The choice of Kitt Peak for the National Observatory was based in part on a general confidence that the desert conditions would limit the growth of southern Arizona communities, but today the Tucson and Phoenix metropolitan areas have populations exceeding 1 million and 4 million, respectively. And populations are rapidly growing near most other observatory sites.

Many people suggest that the next stage of the retreat will follow the Hubble Space Telescope into orbit, or even to the moon. (See the article by Paul Lowman Jr. in *Physics Today* [November 2006]: 50.) But the enormous—yes, astronomical—costs associated with building and maintaining space-based facilities mean that ground-based telescopes must continue to provide the data for the vast majority of observational astronomical research. Hubble catches the eye and imagination of the public, but it catches a very small percentage of the photons that lead to discoveries in astronomy.

Running the Numbers

Although no distinct thresholds of observational capability are crossed as the sky is brightened by artificial lighting, the effectiveness of telescopes measuring faint sources gradually deteriorates. At the limit in which the source under study is negligibly brighter than the background against which it is observed, a 10 percent increase in that background means that astronomers need 10 percent more time to observe the same object with the same signal-to-noise ratio. If sky glow continues to increase, the faintest sources will eventually become unobservable within practical time constraints. During the dark phases of the lunar cycle, the sky over Palomar Observatory is now more than 50 percent brighter than it would be with no artificial light sources. The effectiveness of the 5-m telescope is thereby reduced to that of a 4-m telescope. (Even with no artificial lighting or moonlight, the sky is not perfectly dark. Natural sky glow in the visible spectrum results primarily from sunlight scattered by dust in the solar system and emission from upper-atmosphere oxygen atoms that were excited by daytime sunlight.)

Quantitative treatment of the relationship between lights and the sky glow they produce began in 1965 when Merle Walker, driven by increasing light pollution over the Lick Observatory on Mount Hamilton, California, undertook an effort to find a new site for observation of very faint objects. Seeking to identify a site with not only good current conditions but also the expectation that encroaching development would not unacceptably brighten the night sky in the foreseeable future, Walker developed a crude estimate of the distance a site must be from a city to keep the sky glow below a 10 percent increase in the natural condition at the zenith. He extended his work in 1977 to develop a general empirical relation, now called Walker's law, between sky brightness, population, and distance.[1] The law states that the sky-glow intensity from a light source is

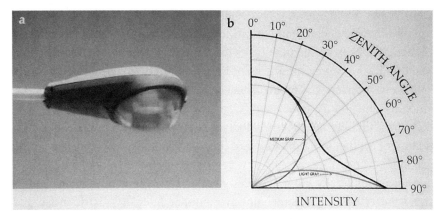

Figure 22.3 (a) A typical light fixture that allows some emission above the horizontal. (b) The angular intensity distribution used by Roy Garstang to represent light propagating upward into the sky. The light gray line represents light emitted directly from fixtures such as the one in panel a, the medium gray line represents light emitted downward and reflected off the ground, and the black line is the sum of the two. (Adapted from [5].)

approximately proportional to the distance raised to the −2.5 power. The intensity falls off more quickly than the inverse square primarily because of atmospheric absorption.

A more comprehensive approach was adopted in 1986, when Roy Garstang of the Joint Institute for Laboratory Astrophysics (now JILA) in Boulder, Colorado, published models that treated the scattering of light off molecules and aerosols in the atmosphere, including the variation in the density of molecules and aerosols with altitude.[2] The models also accounted for Earth's curvature. They have become the standard in the field and have successfully reproduced the variation of sky glow with position in the sky and distance from light sources. What they have not done is relate the sky glow produced by cities to the way lighting is actually used on the ground, such as the number, brightness, and optical characteristics of lighting fixtures. Instead, the models assume a particular angular distribution function based on the shielding of typical light fixtures (shown in Figure 22.3a), and the reflection off surfaces of light directed downward. The light output of cities is then empirically adjusted based on limited measures of sky glow.

Using Garstang's models, Pierantonio Cinzano (now at the Light Pollution Science and Technology Institute in Thiene, Italy), his coworkers, and

Christopher Elvidge of the National Oceanic and Atmospheric Administration have produced maps of artificial night-sky brightness,[3] such as in Figure 22.2; they created the maps by using U.S. Defense Meteorological Satellite Program measurements of light emitted by towns and cities.

Recent work by Luginbuhl connects ground-based surveys of lighting amounts and fixture types with Garstang's models to compare the predictions with detailed measurements of sky glow.[4] With the addition of a treatment for the partial blocking of light emissions due to objects near the ground,[5] excellent across-the-sky agreement between model predictions and measures of sky glow has been obtained. It is now possible to understand the effects of different upward angular intensity distributions and spectral characteristics of artificial lighting.

Two topics of current interest in outdoor lighting exemplify the importance of sky-glow models. Members of the lighting profession frequently point out that shielding lighting fixtures incompletely and thereby allowing a small percent of the light output to be directed just above the horizontal will also provide a wider distribution of light in a downward direction. Fixtures could be placed farther apart, and perhaps 10–15 percent less light could be used to accomplish a given lighting task. At first glance, the trade-off would appear to be favorable for astronomy. But it raises the question: Does the reduced amount of light from widely spaced fixtures decrease sky glow more than the small amount of light emitted upward increases it? There has also been much interest in recent years in broad-spectrum (white) lighting from metal halide and LED sources. What are the implications for astronomy? Answering such questions leads to insight into the nature of the processes that produce light pollution over observatories.

All Uplight Is Not Equally Polluting

It is qualitatively clear that light directed upward and toward an observatory site has a greater impact on the observatory sky than light directed toward the zenith or away from the observatory. Quantitative analysis, described in the sidebar, shows that light emission between zenith angles of 60° and 90° (0° to 30° above the horizontal plane) is far more harmful to observatory skies than light directed toward the zenith, even though on average much of the near-horizontal light is directed away from the

Direction Matters

To quantify the relationship between upward emission angle and sky glow, we have modeled a series of nine light sources, each emitting light upward into a 10°-wide zone spanning 0°–10°, 10°–20°, 80°–90° from zenith.[10] The light source's altitude is set at 1 km; the observatory altitude, 3 km; and the ratio of total aerosol scattering to molecular scattering, 3:1. That ratio corresponds to a low aerosol content and a very clear atmosphere, typical of world-class observatory sites.

Any light source brightens some parts of the sky more than others, but it is convenient to have a single number to represent the glow over the whole sky. One such measure, representative of the parts of the sky most commonly used in astronomical observation, is a weighted average of the sky glow at the zenith and at four points with zenith angle 60°—one toward the light source and the others at 90° intervals in azimuth with the zenith assigned twice the weight of the other points. That average sky glow is divided by the sky glow produced by an equal amount of light directed downward and reflected off a surface with 15 percent reflectivity, a typical value.

Of course, a real light source emits light over a range of angles, both above and below the horizontal. The uplight intensity distribution shown in Figure 22.3b, from Roy Garstang's models, can be used to represent the upward emission from real light fixtures. The table below shows the resulting sky-glow ratios for fixtures with 1 percent, 3 percent, and 10 percent direct uplight as measured at observatories from 50 to 200 km away. The 3 percent figure is representative of fixtures commonly discussed in the tradeoff between uplight and pole spacing, as described in the text. The minimum practical limit for partially shielded fixtures is about 1 percent. Though fractions lower than 1 percent can be optically designed, the accumulation of dirt and deterioration in the optical surfaces drives the uplight fraction toward 1 percent or higher as the fixtures age. And light pollution researchers, starting with Garstang, have found 10 percent to be representative of the average uplight proportion from all fixtures used for outdoor lighting.

	Sky-glow ratio		
Uplight	50 km	100 km	200 km
0%	1.0	1.0	1.0
1%	1.3	1.6	2.0
3%	1.8	2.7	3.9
10%	3.8	6.7	10.6

observatory. And the sky-glow increase from the near-horizontal rays is 6 to 160 times as great as that of an equal flux directed downward and reflected off the ground.

Because most of the upward light emission from incompletely shielded fixtures is directed just above the horizontal, such fixtures have a disproportionate effect on sky glow. From the table in the sidebar, a fixture with an unshielded fraction of only 3 percent produces between 80 and 290 percent more sky glow than a fully shielded fixture with the same light output, with the worst value occurring for the most distant light sources. Startlingly, for a typical community that emits 10 percent of its light directly upward, direct uplight causes almost three-fourths of the sky glow at an observatory 50 km away and more than nine-tenths at a site 200 km away. Even though the amount of direct uplight (10 percent) is similar to the amount of light reflected off the ground (90% × 0.15 albedo = 13.5%), direct upward emission produces the majority of artificial sky glow.

Those numbers don't account for the blocking of light by vegetation and structures near the ground. In a model of sky glow over the U.S. Naval Observatory near Flagstaff, Arizona, accounting for such blocking reduced the relative impact of upward emissions by 50–60 percent. Even so, direct uplight still produced much more sky glow than the same amount of light directed downward. Furthermore, the model did not account for the fact that direct upward emission usually arises from fixtures some distance above the ground, such as on buildings or poles, and may therefore be subject to less blocking than light reflected from the ground.

The answer to the lighting professionals' proposition is clear: The detrimental effect on observatory skies of even 3 percent direct uplight vastly outweighs the benefit of a 10–15 percent reduction in the total amount of light. Even if fixtures could be kept to just 1 percent direct uplight, the competing effects might approximately balance only for observatories located near cities; for the more distant observatories the detrimental effects still dominate.

A Spectrum of Light Sources

For decades astronomers have naturally favored lighting sources that confine emissions to as narrow a portion of the spectrum as possible. Low-pressure sodium lights are particularly preferred,[6] because most of their emission is in the sodium resonance doublet near 589 nm. Indeed, they are widely used in several regions near major astronomical observatories. High-pressure sodium lamps are second best; they emit mostly in the yellow portion of the spectrum, but through pressure broadening and the inclusion of other compounds in the discharge arc, they produce some light in the red and the blue.

About twenty years ago, heavy marketing pressures and improvements in lamp technology led to the more widespread use of broad-spectrum metal halide sources. More recently, white LEDs have begun to emerge as contenders in the outdoor lighting market, as described in the second sidebar. Their greater efficiency makes them especially attractive to municipalities seeking to use economic stimulus money tied to energy savings.

All such broad-spectrum sources interfere with astronomical observation at more wavelengths than do sodium sources, so they leave essentially no unpolluted windows in the visible spectrum. As a further complication, the shorter wavelengths they emit are much more strongly scattered by molecules in the atmosphere. The potential increase in sky glow from such sources is a concern, although the increased scattering leads also to increased attenuation with distance.

LEDs for Outdoor Lighting

The typical white LED used for outdoor lighting is made from a blue LED that emits light at about 450 nm and a phosphor that converts some of the blue light to green and red. White LEDs are typically characterized by their correlated color temperature (CCT), the temperature of the blackbody radiator that most closely resembles the appearance of the LED light.

The blue emission is particularly harmful to astronomers and to the environment. Rayleigh scattering, responsible for the daytime blue sky, has a λ^{-4} wavelength dependence: the 450-nm emission is nearly three times as strongly scattered as is the astronomers' preferred low-pressure sodium emission at 589 nm. Furthermore, wavelengths shorter than 500 nm interfere more strongly with circadian rhythms and melatonin production in humans and other animals. The higher the CCT of a white LED, the more strongly its light is scattered and the larger the perturbation to biological systems.

Fortunately for astronomers, few people like the appearance of the high-CCT LEDs (5000–6000 K, or daylight color), with many describing them as looking like welding torches. Even so, some municipalities prefer them because they are somewhat more efficient. Other communities, such as Anchorage, Alaska, have specified that white LEDs should not have a CCT higher than that of moonlight (4200 K), but even that approach does not properly account for the fact that the LEDs are more damaging than moonlight due to the blue peak, which is not present in moonlight, and because the moon is below the horizon during half of the nighttime. Low-CCT white LEDs (3000 K or lower, the color of typical incandescent lamps) are the least harmful to the environment and astronomical observation.

Damage Control

In 1973 Walker identified the critical issue of light pollution facing astronomy:

> At the time of their founding, the sites of the present major optical astronomical observatories in California and Arizona were among the best in the world. Now, however, work at all of these installations is either presently or potentially limited by the increase in the illumination of the night sky from nearby cities. . . . It is essential that immediate efforts be undertaken to: (1) Control outdoor illumination to lengthen the useful

life of existing observatory sites, and (2) Identify and protect the best remaining sites both within and outside the United States.[7]

Today, his words are as true for the remotest observatory sites as they were for California and Arizona thirty-six years ago.

Astronomers' efforts to address the issue have been ongoing since the late 1950s and are now having some effect. The effort has been aided in recent years by a broadening coalition of interests concerned about the many detrimental effects of artificial light at night: energy waste, poor visibility due to glare, disturbance of biological systems, and loss of starry skies for casual stargazers. A comprehensive study of lighting in Flagstaff shows that the growth rate of light pollution per person, added to the population, has been cut approximately in half since 1989, when a stringent outdoor lighting code was adopted that limits the total amount of light permitted. Sky glow continues to increase, but at a slower pace.

Lighting designers and manufacturers are increasingly aware of the many harmful effects of light pollution. Through extensive educational efforts led by the International Dark-Sky Association (http://www.darksky.org) and other similar organizations throughout the world, a greater selection of fully shielded lighting fixtures is becoming available. Trained lighting professionals are using more fully shielded fixtures, at least in areas where the sensitivity to light-pollution issues is high due to heightened environmental sensitivity or the presence of observatories.

Unfortunately, in most areas insufficient awareness of the problems that can arise from lighting at night still leads to poor control of upward emission and lighting amounts. In many places, particularly in small towns and rural areas, the majority of outdoor lighting is not designed by lighting professionals. And outdoor lighting is used for more situations and in greater amounts than it used to be. The best hope for progress is through continuing education, as described in the third sidebar, about the value of a starry sky—a value not just for astronomy and science but for everyone. Nobody ever seems to make the mistake of thinking that Yellowstone National Park and the Grand Canyon are protected just for geologists and rock hounds. Does the vista of a star-filled night sky matter only to astronomers?

The International Year of Astronomy and Dark Skies Awareness

Dark Skies Awareness is a Global Cornerstone Project of the United Nations–sanctioned 2009 International Year of Astronomy. Its goal is to raise the level of public knowledge about adverse effects of excess artificial lighting on local environments and to make more people aware of the ongoing loss of a dark night sky for much of the world's population. Toward that end, a range of programs and resource materials has been developed. One such program is GLOBE at Night, an international citizen-science event that takes place every March to encourage everyone—students, educators, dark-sky advocates, and the general public—to measure the darkness of their local skies and contribute their observations online to a world map. Everyone is invited to participate in GLOBE at Night and the other Dark Skies Awareness programs, offered as potential local solutions to a global problem. To learn more, visit http://www.darkskiesawareness.org.

Notes

*This chapter is adapted and reprinted with permission from C. B. Luginbuhl, C. E. Walker, and R. J. Wainscoat, Lighting and astronomy, *Physics Today* 62 (2009): 32–37. Copyright 2009, American Institute of Physics.

1. Walker M. *Publ. Astron. Soc. Pac.* 1977;89:405.
2. Garstang RH. *Publ. Astron. Soc. Pac.* 1991;103:1109.
3. Cinzano P, Falchi F, Elvidge C. *Mon. Not. R. Astron. Soc.* 2001; 328:689.
4. Luginbuhl CB, et al. *Publ. Astron. Soc. Pac.* 2009;121:185.
5. Luginbuhl CB, et al. *Publ. Astron. Soc. Pac.* 2009;121:204.
6. Luginbuhl CB. In: Cohen RJ, Sullivan WT III, eds. *Preserving the Astronomical Sky: Proceedings of the 196th Symposium of the International Astronomical Union, 12–16 July 1999.* San Francisco: Astronomical Society of the Pacific; 2001:81.
7. Walker M. *Publ. Astron. Soc. Pac.* 1973;85:508.

8. Rich C, Longcore T, eds. *Ecological Consequences of Artificial Night Lighting*. Washington, DC: Island Press; 2006.
9. Cinzano P. In: Schwarz H, ed. *Light Pollution: The Global View*. Dordrecht, the Netherlands: Kluwer; 2003:39.
10. Walker CE, Luginbuhl CB, Wainscoat RJ. In: *Proceedings of the CIE Light and Lighting Conference with Special Emphasis on LEDs and Solid State Lighting,* 27–29 May 2009, Budapest, Hungary (in press).

23

Green Living: Reducing the Individual's Carbon Footprint

Clinton J. Andrews and Robert H. Friis

Introduction

Widespread use of fossil fuels, rapid deforestation, and activities related to the production of goods and services are causing exponential increases in greenhouse gas emissions that are changing the earth's climate. The term *greenhouse gas* refers to "[a]ny gas that absorbs infra-red radiation in the atmosphere. Greenhouse gases include water vapor, carbon dioxide (CO_2), methane (CH_4), nitrous oxide (N_2O), halogenated fluorocarbons (HCFCs), ozone (O_3), perfluorinated carbons (PFCs), and hydrofluorocarbons (HFCs)."[1] These gases capture heat in the earth's atmosphere and cause the phenomenon known as global warming. When the levels of greenhouse gases increase excessively, they can cause the earth's temperature to heat up to potentially damaging levels.

As a result of global warming and its adverse consequences, people are now recognizing that they should change course and begin reducing anthropogenic greenhouse gas emissions. Human lifestyles that embrace low carbon outputs are thought to have numerous potential health benefits for high- and middle-income societies.[2] Moreover, reduction in carbon outputs also benefits the citizens of low-income nations by increasing available resources and limiting the amounts of toxic chemicals and hazardous wastes that are present in the environment. Reductions in greenhouse gases will require coordinated international efforts in view of the interconnectedness of today's world.

By adopting lifestyles and practices that result in reduction of greenhouse gases, individuals, commercial enterprises, and society contribute to a sustainable environment. The concept of environmental sustainability adheres to the philosophical viewpoint "that a strong, just, and wealthy society can be consistent with a clean environment, healthy ecosystems, and a beautiful planet" (p. 5383).[3] Three widely recognized dimensions of sustainable development are materials and energy use, land use, and human development. Environmental sustainability means that resources should not be depleted faster than they can be regenerated; the concept also specifies that there should be no permanent change to the natural environment. Critics of sustainable development argue that the definition of the term is not entirely clear and is open to interpretation.[4] Nevertheless, the concept of environmental sustainability is consistent with a low-carbon lifestyle and minimization of one's carbon footprint.

The present chapter is a companion piece to chapter 6, "Climate Change and Health," which covers the health-related implications of climate change. It examines general considerations regarding the components of people's carbon footprints (the levels of per capita emissions of CO_2 and other greenhouse gases) and individual actions such as ethical consumption in order to reduce the production of greenhouse gases; the chapter also provides examples of best practices for reducing carbon emissions by individuals and households well as by the various levels of government.

The Carbon Footprint

The production of greenhouse gases, carbon outputs, and carbon footprints of individual countries are highly asymmetrical when wealthy, developed countries are compared with poorer nations. Measurements of carbon footprints are one criterion for the assessment of the resource intensity of the modern consumer society. Estimation of the size of carbon footprints allows comparisons between groups of rich consumers (driven more by desires) and groups of poor ones (driven more by basic needs), within one country or across borders.

The *carbon footprint* refers to the level of greenhouse gases generated by a particular activity.[5] To simplify reporting, often the carbon footprint is stated as an amount of CO_2 generated by functions such as the manufacture and consumption of goods and services. "The carbon footprint is a measure of the exclusive total amount of carbon dioxide emissions that is directly and indirectly caused by an activity or is accumulated over the life stages of a product" (p. 4).[6] The carbon footprint encompasses the entire goods production chain, especially upstream linkages to coal, oil, natural gas, nuclear power, and renewables in addition to greenhouse gas emissions from use of the products themselves.

In the United States, fossil fuel consumption is the dominant contributor to CO_2 emissions. Between 1990 and 2007, total U.S. emissions of greenhouse gases increased by 17 percent, although the gross domestic product and population grew by 65 percent and 21 percent, respectively.[7] During this same period, emissions of CO_2 from fossil fuels increased by 21.8 percent, about the same as the population growth. However, emissions of methane declined by 5 percent and nitrous oxides by 1 percent; these decrements can be attributed to technological improvements, including capture of gases from landfills and better automobile emission controls. Accordingly, overall emissions of greenhouse gases lagged behind growth of the population and the gross domestic product. Nevertheless, in the United States emissions of greenhouse gases from anthropogenic sources are likely to continue to increase in the future; reductions can be achieved through the concerted actions of individuals, industry, government, and environmental organizations.

Both the benefits received from goods and services and responsibility for the corresponding volume of greenhouse gases produced vary greatly from one country to another. Friel et al. noted that "[t]he environmental and health outcomes of climate change impinge unequally across regions and populations. In 2000, climate change accrued to that point caused a conservative estimate of 150,000 deaths; although the poorest 1 billion people account for around 3% of the world's total carbon footprint, . . . the deaths were almost entirely confined to the world's poorest populations" (p. 1677).[8]

Examples of processes and activities that contribute to the production of greenhouse gases comprise eight general categories: "construction, shelter, food, clothing, mobility, manufactured products, services, and trade" (p. 6414).[9] Average national per person carbon footprints range from one ton of CO_2 emitted per year in some African countries to approximately thirty tons of CO_2 emitted per year in the United States. Household consumption of goods and services accounts for about three-fourths of total greenhouse gas emissions globally. Worldwide, the percentages of greenhouse gas emissions are 20 percent for food (production and consumption), 19 percent for shelter (operation and maintenance of residences), and 17 percent for mobility. The largest sources of greenhouse gases are mobility and manufacturing in affluent countries and production of food and services in developing countries.

Individual Actions for Climate Change

The components of the individual's carbon footprint include preferences for recreation, methods of commuting to work, styles of work, and choice of learning environments. Consumers are able to exercise responsibility in the selection of end-use goods and services—food, shelter, mobility, and modes of communication—and thereby help to reduce the output of greenhouse gases.

Ethical Consumption

One possibility for reducing one's carbon footprint is through ethical consumption practices. International comparisons of carbon footprints have sparked an interest in ethical consumption, so that thoughtful people are asking what are reasonable and fair levels and types of consumption. People can take a wide variety of actions in order to reduce their own carbon footprints and perhaps mitigate some of the causes of climate change. Individuals can influence climate change, either positively or negatively, in two broad ways—first, directly via their own behaviors and, second, indirectly through attempts to modify the behavior of others. An example of the first involves personal behaviors such as the choice of a fuel-efficient or gas-guzzling car or the simple decision of whether to turn off the lights when leaving a room. An example of the second (discussed later in this chapter) pertains to attempts to effect group change via social networks, educational programs, and lobbying the government to bring about policy changes that help to lower the carbon footprint.

What are some examples of behaviors that individuals can adopt to decrease global warming? Many consumers aspire to reduce their carbon footprints through greener living. Increasingly, many consumers' decisions about the purchase of goods and services are based on the desire to limit choices that are harmful to the environment. Consumers' purchase decisions have consequences for the entire society, for example, human rights including rights of workers, and animal welfare, particularly the treatment of farm animals used as food sources. "[E]thical trade [is defined] as any form of trade that consciously seeks to be socially and environmentally, as well as economically, responsible. Ethical consumers would, therefore, seek to purchase or use goods and services that can demonstrate social and/or environmental responsibility. In the natural resources sector, ethical trade includes fair-trade, trade in organic products, trade in products from sustainably managed resources such as forests, and ethical sourcing of fresh and processed produce following ethical codes of conduct" (p. 5).[10]

The economies of advanced industrialized countries already provide for the basic needs of their populations and now devote vast amounts of additional energy and materials to serving people's unbounded desires for material possessions and services. Contemporary examples of sought-after goods are fuel-inefficient luxury vehicles, expansive homes ("McMansions"), the latest state-of-the-art electronics, designer clothes, and foods and products derived from animals raised in crowded conditions or from endangered species; examples of services that may impact the environment adversely are luxurious travel, entertainment in energy-intensive venues (e.g., cruise ships and casinos), and the latest high-tech healthcare procedures.

Yet much of the leverage is in the hands of producers, not consumers. Consider how improvements in the healthcare industry might contribute to the reduction of greenhouse gases. English environmentalist Griffiths points out that the healthcare sector can adopt many efficient procedures that would aid in minimizing damage to the environment.[11] The National Health Service (NHS), the largest employer in England, can reduce its ecological footprint by designing ecologically friendly facilities, reducing water consumption, and managing waste effectively. Similar efforts could be adopted worldwide by the healthcare industry. Producers may respond to some consumer demands, but may also need "indirect direction" from individuals operating in their roles as citizens who encourage governments to enact environmental policies.

Ethical consumers believe that they share a global commons with the rest of humanity and feel an obligation not to take more than their fair share. This sentiment manifests itself in efforts such as Switzerland's

2,000 Watt Society vision, which urges rich nations to reduce their overall rate of energy use to 2,000 watts per person (17,250 kWh/person-year), the global average.[12] The United States currently averages about 12,000 watts per person, whereas Bangladeshis average less than 500 watts.

One of the objectives of ethical consumption is to consume less and conserve resources by changing behavior, that is, by driving less, flying less, eating less meat, and adjusting the thermostat so as to minimize energy consumption; such changes in behavior can aid in slowing the depletion of the earth's limited resources. Some ethical consumers hope that they can encourage systemic improvements by voting with their pocketbooks, and they favor ethical products, avoid unethical products, and boycott unethical companies. These individuals obtain help in distinguishing which products are ethical from those that are unethical by consulting organizations such as the Ethical Consumer (www.ethicalconsumer.org) and the GoodGuide (www.goodguide.com).

Exercising consumer choice can indeed help establish niche markets for less greenhouse gas–intensive products, provided innovative producers offer them.[13] Some ethical consumers also avoid sweatshop labor, factory farmed animals, nonorganic food production methods, and purchases associated with a variety of types of social irresponsibility.

The chapter by Griffiths presents an extensive list of actions that individuals can take to reduce their carbon footprints.[2] Examples of individual and household actions that influence a society's carbon footprint pertain to the following:

- **Regulation of home energy use.** Individual actions in this sphere can contribute significantly to the production of greenhouse gases. For example, overheating one's home during winter and using excessive air conditioning during summer increase demand for electric power. Individuals can reduce their carbon footprints by monitoring thermostats closely; purchasing updated, energy-efficient appliances; and switching to compact fluorescent lightbulbs (CFLs).[14]
- **Selection of foods.** Some food and beverage choices are associated with lower carbon output. Increasingly, consumers are interested in growing their own fruits and vegetables and purchasing those that are grown locally. Out-of-season and exotic produce may be transported

over long distances by polluting diesel-fueled trucks and jet aircraft. Also, consumption of tap water instead of bottled water and lowering the intake of meat may help to limit greenhouse gases.[15] Such reductions accrue from the reduced utilization of plastics, decreased release of methane from cattle and their wastes, and, most important, consumption of calories produced lower on the food chain using fewer natural resources.

- **Choice of transportation modes** (e.g., local transportation and long-distance air travel) that are energy efficient or energy conserving. Motorists can select more fuel-efficient automobiles or those powered by alternative energy sources. Other options include carpooling, use of public transportation, consolidation of trips, and bicycling and walking to work, if feasible.[14] Telecommuting spares the need to commute daily to work. Internet conferencing can provide many of the benefits of in-person meetings and reduces the need for long-distance business trips. Online courses offered by universities help minimize the number of trips to campus.

However, conserving behaviors and more energy-efficient products do not necessarily result in lower aggregate greenhouse gas emissions.[16] Jevons made a troubling observation about coal usage in industrial-age England: "It is wholly a confusion of ideas to suppose that the economical use of fuel is equivalent to a diminished consumption. The very contrary is the truth."[16] He and subsequent economists have demonstrated that improving fuel efficiency makes the end-use service cheaper, thereby stimulating demand (if demand is price-elastic); it also allows the economy to grow more rapidly, further stimulating demand; and it may also lead to step changes in production or consumption technology that further stimulate demand. Jevons's "efficiency paradox," in which efficiency improvements are overtaken by increases in aggregate consumption, shows that resource conservation depends not only on consumers, but also on producers and governments. Dick Cheney, former U.S. vice president, gained notoriety for saying that "conservation may be a sign of personal virtue, but it is not a sufficient basis for a sound, comprehensive energy policy," suggesting that ethical consumers cannot solve problems like climate change entirely on their own.[17]

Changing Social Networks and Lobbying Government

A second method for reducing the carbon footprint is through individuals' indirect effects on greenhouse emissions by influencing the consumption patterns of other people. Individuals are able to effect climate change by influencing members of their social networks and by demanding that governments change the incentives and constraints that are pushing individual decisions in a particular direction. Governments can operate and have influence at every level, from local to global. They can influence the pricing of goods and services though regulation and tax policy. They can pass laws that constrain individual options by taking certain goods off the market, or making certain activities illegal.

Individuals also can decide to exert influence within their social networks to influence the behavior of their peers ("don't forget to turn out the lights, honey"). Social networks have an impact because most people instinctively try to conform to the moral and cultural norms of their social groups. Individuals participate in social networks in many ways, from voting in federal and statewide elections to joining groups such as the garden club or softball league. Every individual has a different set of affinities and circumstances that will determine in which groups he or she participates.

Changing societal norms by encouraging exemplary behaviors via public education can help to reduce the production of greenhouse gases. Education and marketing campaigns that exhort consumers and citizens to make better choices for a sustainable environment and otherwise aid in creating awareness of issues related to the environment are a possible approach. "Public and private efforts to change individual behavior to more environmentally friendly practices usually rely on educating the public about the consequences of the climate change and what activities individuals can take to help prevent it."[18] An example of a behavioral change system is EcoIsland, which is used to persuade individuals and their family members to modify their behaviors in order to reduce CO_2 emissions.

Educational institutions such as universities can help further public education about the environment by imparting knowledge about ecological sustainability. "[U]niversities play a critical role in embedding sustainability principles and understanding in society, through the training of future leaders and professionals, cutting-edge research, and community outreach activities that empower local communities to implement sustainable principles and practices" (p. 2036).[19]

The impacts of individual decisions and the viability of strategies that rely on exhorting individuals to make better choices are widely debated. Proponents of this approach believe that government is incapable of leading on such issues, and change has to come from the bottom up. Of course, what individuals will do is heavily influenced by their government and social networks. Is gasoline subsidized by the government, thus negating the cost advantage of a fuel-efficient car? Will someone in your peer group frown at you if you forget to turn off the lights when leaving a room?

Role of Government and Global Organizations

Individual behavioral and consumption choices can help reduce greenhouse gas emissions, but many important levers are in other hands, specifically those of the business community and local, state, and national governments. Individual and household choices are nested within and constrained by these larger institutions. The remainder of this chapter covers the potential roles of three levels of government in the United States (local, state, and national) as well as the role of the global economy and international organizations in reducing the carbon footprint. It may be argued that activities to reduce greenhouse gases will require coordinated interactions among the various levels of governments within a single country and coordinated efforts among nations.

Global Actions to Reduce Greenhouse Gases

Given the growing interconnectedness and interdependence of the world's nations, efforts to tackle global environmental issues require creative partnerships among governments, international organizations, and policy actors.[20] The Kyoto Protocol is an example of an international and legally binding agreement that seeks to reduce the emissions of greenhouse gases by targeted percentages. Countries that affirmed the protocol agreed that between 2008 and 2012 they would reduce the percentages of their greenhouse gas emissions to levels below those of 1990. In contrast with the many developed nations that signed on to the Kyoto Protocol, the United

States did not ratify the agreement. Globally, the United States is among the largest producers of greenhouse gases.

In addition to the Kyoto Protocol, other international programs to reduce emissions of greenhouse gases aspire to improvements in energy efficiency. The Organization for Economic Cooperation and Development (OECD) nations have significantly reduced their carbon footprints since 1973 by encouraging energy efficiency.[21] Another example of an international program is the Prototype Carbon Fund (PCF), "[a] partnership between seventeen companies and six governments, . . . managed by the World Bank . . . [, which] became operational in April 2000. As the first carbon fund, its mission is to pioneer the market for project-based greenhouse gas emission reductions while promoting sustainable development and offering a learning-by-doing opportunity to its stakeholders. The Fund has a total capital of $180 million."[22]

Among the difficulties that impede global efforts to achieve a sustainable environment is lack of coordination, especially among nongovernmental organizations. For example, the United Nations maintains numerous international bodies that deal with water quality, but often the activities of the groups are not coordinated.[23]

National Actions in the United States

The United States did not ratify the Kyoto Protocol and does not have a national target for reduction of greenhouse gases, except for reductions in criteria air pollutants (e.g., carbon monoxide, nitrogen dioxide, and ozone). Nevertheless, many noteworthy initiatives are in place, including public and private partnerships that concentrate on improving energy efficiency, designing technological improvements, optimizing agricultural practices, and developing renewable energy resources.[24] The transportation system in the United States is the largest in the world and the largest contributor in the U.S. economy to CO_2 emissions. The use of technologies to raise the energy efficiency of transportation vehicles can reduce the emissions of greenhouse gases by one-quarter by 2030.[25]

Among U.S. activities to reduce greenhouse gases is the adoption of requirements for reduction of energy use in federal facilities. The Energy Independence and Security Act of 2007 mandates a 30 percent reduction in the intensity of energy use in federal facilities by the year 2015.[26]

Similarly, President Barack Obama has declared a target of a 28 percent reduction in greenhouse gas pollution from federal departments and agencies by 2020.[27] "Energy efficiency plays a critical role in the U.S. energy policy debate because future national energy needs can be met only by increasing energy supply or decreasing energy demand" (p. 162).[28]

The U.S. Environmental Protection Agency's (EPA) greenhouse gas initiatives encourage voluntary reductions in greenhouse gases by U.S. consumers, large corporations, and industrial and commercial facilities. A total of nine initiatives to reduce greenhouse gases include federal government partnerships with states and industry for the development of climate change strategies, use of green energy, and lowering the emissions of gases, for example, hydrofluorocarbons, that have high global-warming potential.[29] In 2008 the EPA implemented the Mandatory Reporting of Greenhouse Gases Rule (74 FR 5620). The rule requires that facilities that release 25,000 or more metric tons of greenhouse gases per year provide an annual report to the EPA.[30] Another EPA initiative is the ENERGY STAR program, a voluntary labeling program to promote energy-efficient products such as appliances. The ENERGY STAR designation is used by more than 1,400 manufacturers for more than forty product categories.

Taxation and support of research are techniques for limiting the nation's carbon footprint. The federal government has encouraged the reduction of greenhouse gases through income tax credits for energy efficiency and tax incentives for use of hybrid electric vehicles.[31] The U.S. federal government also regulates energy production and taxes production of carbon as a method for encouraging energy conservation and reduction of CO_2 emissions.[32] Another role of the federal government has been the sponsorship of basic research and development policies, including funding, that aid in creation of innovative clean technologies. Also, the government supports the development and commercialization of applied technologies that reduce the formation of greenhouse gases. On the downside, these research-related activities can take many years to yield tangible results and justify the investments made.

State and Local Governments

The federal government has taken on primarily a nonregulatory and voluntary stance regarding the control of greenhouse gases. In contrast, "states are important players in interpreting, applying, enforcing, and regulating beyond the scope of federal law" (p. 1015).[33] Possible actions at the state level are carbon and energy taxation and regulation; waivers of state sales taxes to encourage energy efficiency; income tax credits and deductions for reduction of greenhouse gases; and implementation of policies for increasing the energy efficiency of transportation, for example, allowing access by single drivers of fuel-conserving cars to high occupancy vehicle (HOV) lanes to encourage the purchase of automobiles that are fuel efficient.

Several states have formulated their own climate goals, which vary regarding the degree to which they are voluntary or required as well as their stringency.[34] In 2006 the California legislature passed Assembly Bill 32, the Global Warming Solutions Act, which was signed by Governor Schwarzennegger.[35] The law specifies a number of actions that will enable the state to reach goals that have been established to reduce greenhouse gases consistent with the limit that the state has set for 2020.

With respect to local governments, actions of municipalities are often top-down decisions driven by the beliefs of officials and staff members about actions that are good business or rational policy choices, for example, have potential for cost savings.[36] Nevertheless, despite the fact that decisions are mainly top-down, local jurisdictions can be strong players in efforts to lower greenhouse gas emissions. Regional settlement patterns, transportation systems, energy infrastructure adequacy, and architectural and building standards typically fall within the purview of local governments and their zoning ordinances. Sprawling settlement patterns that force residents into automobile dependence and encourage larger housing units are highly greenhouse gas intensive on a per capita basis, and they do not sequester carbon at a significant enough scale to offset those factors.[37] In these types of municipalities, architectural standards that require green commercial buildings can aid in reducing greenhouse gases. Also, cities can reduce pollution by increasing both walkability and the availability of public transportation to encourage commuters to abandon the use of their own vehicles.

Role of Corporations and Commercial Enterprises

During the current century, in commercial settings, ecological and social concerns have become salient forces that influence the behaviors of businesses. Some environmentalists have argued that corporations might spur the movement of heavy industries to developing countries and pollution havens, where environmental standards are less stringent.[38, 39] Despite these concerns, responsible environmental philosophies are an increasingly important concern for the business community, which is responding to market-based policy instruments that use signals from the marketplace, rather than directives, for controlling pollution levels.[40]

The term *green marketing* is defined as "[t]he holistic management process responsible for identifying, anticipating and satisfying the needs of customers and society, in a profitable and sustainable way" (p. 727).[41] *Green alliances* refer to cooperative arrangements between environmental organizations and businesses in order to link market goals with corporate environmental responsibilities. Such alliances "are an outgrowth of an emerging philosophy called 'Market-based Environmentalism[,]' which advocates making ecology attractive to businesses via market incentives" (p. 184).[42] In illustration, McDonald's and the Environmental Defense Fund have formed an alliance to enhance the positive environmental aspects of the chain's operations.[43] Real estate companies and stakeholders are increasingly including environmental sustainability as a business strategy for their operations.[43] Some builders are marketing homes with installed solar panels and promoting "green" homes and developments that optimize energy efficiency. With respect to commercial enterprises, sustainable economic development "includes issues of corporate social responsibility and citizenship along with improved management of corporate social and environmental impacts and improved stakeholder engagement" (p. 20).[44]

Leading corporations have begun to systematize their responses to social and environmental concerns. Illustratively, the Greenhouse Gas Protocol Initiative (GHG Protocol) was convened in 1998, organized by the World Business Council for Sustainable Development and the World Resources Institute.[45] The initiative is a partnership of stakeholders from business, organizations, and governments that seeks to achieve transparency in companies' reporting of emissions of greenhouse gases from various activities. The initiative specifies direct and indirect emission sources by defining three "scopes." Scope 1 concerns emissions that are produced directly by a reporting company; scope 2 involves emissions that are the result of imported electricity, heat, or steam; and scope 3 pertains to other indirect greenhouse gas emissions from a variety of business activities, for example, employee business travel and transportation of products.

The initiative "serves as the premier source of knowledge about corporate GHG accounting and reporting."[45] These metrics make corporate performance more transparent and help differentiate good and bad actors.

Conclusion

Reducing the individual's carbon footprint is one of the major environmental challenges of the twenty-first century. Among the associated challenges is the requirement for humanity to adapt expeditiously to new temperature and rainfall patterns and their associated effects on agriculture; other related challenges are the potential for spread of disease to previously nonendemic regions as well as the need to accommodate rising sea levels that may impinge upon the crowded coastal regions throughout the world. Human actions are a major driver of global warming and some of the associated conditions that have been observed in recent years. There is an urgent need to implement concerted mitigation efforts to reduce greenhouse gases and the carbon footprint to minimize the future pain of adaptation to a lessened output of goods and services.

In addition to the requirement of combining top-down and bottom-up solutions, meeting the challenges of global warming will demand coordinated actions at many levels: individual and family as well as international, national, state, and local. Individual actions to reduce global warming are limited primarily to the consumption side of the equation of global warming. Also, because behavioral change by individuals is ephemeral and often has limited impact, technical fixes are needed; however, technical fixes require appropriate taxation and innovative policies developed at federal and other levels of government. Among the impediments to reducing the emission of greenhouse gases in federal and international systems is the ju jitsu of policy making: as leaders try to shift the center, laggards resist, and inertia from higher levels of governance further constrains the extent to which leaders can diverge from the status quo. Efficient reductions in emissions of greenhouse gases require an appropriate allocation of responsibilities among the various governmental levels—local, state, national, and international. Yet initiative often comes from the "wrong" level, such as the locality, so that policy entrepreneurs must fight to advance their issues on higher-level institutional agendas. This push and pull—two steps forward and one backward—yields only incremental progress until a tipping point is reached at which enough active citizens and consumers coordinate their voices to precipitate systemic changes. Within limits, these voices can alter the directions of both political and economic decision making about the problem of worldwide climate change.

Notes

1. United States Environmental Protection Agency, Climate Change Kids Site, *Greenhouse effect* (2008), http://www.epa.gov/climate change/kids/greenhouse.html (accessed March 19, 2011).
2. J. Griffiths, Climate change and health, in *Praeger handbook of environmental health*, edited by R. H. Friis (Westport, CT: Praeger Publishing Company, 2012).
3. V. M. Thomas and T. E. Graedel, Research issues in sustainable consumption: Toward an analytical framework for materials and the environment, *Environmental Science and Technology* 37 (2003): 5383–5388.
4. J. Maddox, Positioning the goalposts: The best environmental policy depends on how you frame the question, *Nature* 403 (2000): 139.
5. National Geographic, *Sustainability—Our carbon footprint*, http://environment.nationalgeographic.com/environment/national-geographic-sustainability/carbon-footprint/ (accessed March 18, 2011).
6. T. Wiedmann and J. A. Minx, A definition of "carbon footprint," in *Ecological economics research trends*, edited by C. C. Pertsova (Hauppauge, NY: Nova Science Publishers, 2008), 1–11.
7. United States Environmental Protection Agency, Climate change—Greenhouse gas emissions, http://www.epa.gov/climatechange/emissions/index.html (accessed March 18, 2011).
8. S. Friel, M. Marmot, A. J. McMichael, et al., Global health equity and climate stabilisation: A common agenda, *Health Policy* 372 (2008): 1677–1683.
9. E. G. Hertwich and G. P. Peters, Carbon footprint of nations: A global, trade-linked analysis, *Environmental Science and Technology* 43 (2009): 6414–6420.
10. A. Tallontire, E. Rentsendorj, and M. Blowfield, *Ethical consumers and ethical trade: A review of current literature* (Chatham, UK: Natural Resources Institute, 2001).

11. J. Griffiths, Environmental sustainability in the national health service in England, *Public Health* 120 (2006): 609–612.
12. E. Jochem, D. Spreng, D. Favrat, et al., Steps towards a 2000 watt society: A white paper (input paper to an international workshop on September 9 and 10, 2002, Zürich, Switzerland), http://www.stadt-zuerich.ch/2000-watt-society (accessed April 3, 2011).
13. C. J. Andrews and D. DeVault, Green niche market development: A model with heterogeneous agents, *Journal of Industrial Ecology* 13 (2) (2009): 326–345.
14. Natural Resources Defense Council, *NRDC's guide to greener living: Save energy on the road*, http://www.nrdc.org/cities/living/gover.asp (accessed March 18, 2011).
15. Carbon Footprint Ltd., *Carbon footprint reduction*, http://www.carbonfootprint.com/minimisecfp.html (accessed April 3, 2011).
16. W. S. Jevons, *The coal question: An inquiry concerning the progress of the nation, and the probable exhaustion of our coal-mines*, 2nd ed. (London: Macmillan and Co., 1866), http://www.econlib.org/library/YPDBooks/Jevons/jvnCQ.html (accessed April 3, 2011).
17. D. Cheney, Speech at the annual meeting of the Associated Press, Toronto, Canada, 2001, quoted in R. Benedetto, Cheney's energy plan focuses on production, *USA TODAY*, May 1, 2001, http://www.usatoday.com/news/washington/2001-05-01-cheney-usat.htm (accessed April 3, 2011).
18. M. Shiraishi, Y. Washio, C. Takayama, et al., Using individual, social and economic persuasion techniques to reduce CO_2 emissions in a family setting, in *Persuasive '09* (Claremont, CA: April 26–29, 2009), http://www.hiit.fi/~vlehdonv/documents/Shiraishi-2009-persuasion-co2.pdf (accessed April 3, 2011).
19. M. L. Fleming, T. Tenkate, and T. Gould, Ecological sustainability: What role for public health education? *International Journal of Environmental Research and Public Health* 6 (2009): 2028–2040.
20. C. Streck, New partnerships in global environmental policy: The clean development mechanism, *Journal of Environmental Development* 13 (3) (2004): 295–322.
21. H. Geller, P. Harrington, A. H. Rosenfeld, et al., Polices for increasing energy efficiency: Thirty years of experience in OECD

countries, *Energ Policy* 34 (2006): 556–573. (doi:10.1016/j.enpol.2005.11.010).
22. The World Bank, Carbon Finance Unit, http://wbcarbonfinance.org/Router.cfm?Page=PCF (accessed March 26, 2011).
23. J. Jeffery, Governance for a sustainable future, *Public Health* 120 (2006): 604–608.
24. United States Environmental Protection Agency, *U.S. climate policy and actions*, http://epa.gov/climatechange/policy/index/html (accessed March 22, 2011).
25. D. L. Greene and A. Schafer, *Solutions: Reducing greenhouse gas emissions from U.S. transportation* (Arlington, VA: Pew Center on Global Climate Change, May 2003).
26. United States Department of Energy, Energy Efficiency and Renewable Energy, Federal Energy Management Program, *Energy management requirements by subject*, http://www1.eere.energy.gov/femp/regulations/requirements_by_subject.html (accessed April 3, 2011).
27. The White House, Office of the Press Secretary, President Obama sets greenhouse gas emissions reduction target for federal operations, Washington, DC, January 29, 2010, http://www.whitehouse.gov/the-press-office/president-obama-sets-greenhouse-gas-emissions-reduction-target-federal-operations (accessed March 22, 2011).
28. K. Gillingham, R. Newell, and K. Palmer, Energy efficiency policies: A retrospective examination, *Annual Review of Environmental Resource* 31 (2006): 161–192.
29. United States Environmental Protection Agency, *Current and near-term greenhouse gas reduction initiatives*, http://epa.gov/climatechange/policy/neartermghgreduction.html (accessed March 22, 2011).
30. United States Environmental Protection Agency, Greenhouse Gas Reporting Program, http://www.epa.gov/climatechange/emissions/ghgrulemaking.html (accessed March 21, 2011).
31. K. S. Gallagher and E. Muehlegger, *Giving green to get green: Incentives and consumer adoption of hybrid vehicle technology*, Faculty research working papers series (Harvard University, John F. Kennedy School of Government, February, 2008), http://www.hks.harvard.edu/fs/emuehle/Research%20WP/Gallagher%20and%20Muehlegger%20Feb_08.pdf (accessed April 3, 2011).

32. R. Boyd, K. Krutilla, and W. K. Viscusi, Energy taxation as a policy instrument to reduce co_2 emissions: A net benefit analysis, *Journal of Environmental Economics and Management* 29 (1995): 1–24.
33. K. Engel, State and local climate change initiatives: What is motivating state and local governments to address a global problem and what does this say about federalism and environmental law? *Urban Lawyer* 38 (2006): 1015.
34. M. A. Brown, F. Southworth, and A. Sarzynski, *Shrinking the carbon footprint of metropolitan America* (Washington, DC: The Brookings Institution, Metropolitan Policy Program, 2008).
35. California Environmental Protection Agency, Air Resources Board, Assembly Bill 32: Global Warming Solutions Act, http://www.arb.ca.gov/cc/ab32/ab32.htm (accessed March 22, 2011).
36. C. Kousky and S. H. Schneider, Global climate policy: Will cities lead the way? *Climate Policy* 3 (2003): 359–372.
37. C. J. Andrews, Greenhouse gas emissions along the rural-urban gradient, *Journal of Environmental Planning Management* 51 (6) (2008): 847–870.
38. D. Vogel, Trading up: Consumer and environmental regulation in a global economy, *Review of International Economics* 5 (2) (1997): 284–293.
39. J. M. Dean, M. E. Lovely, and H. Wang, Are foreign investors attracted to weak environmental regulations? Evaluating the evidence from China, *Journal of Development Economics* 90 (2009): 1–13.
40. R. N. Stavins, Experience with market-based environmental policy instruments, in *The handbook of environmental economics*, edited by K.-G. Mäler and J. Vincent (Amsterdam: North-Holland/Elsevier Science, 2001).
41. K. Peattie and M. Charter, Green marketing, in *The marketing book*, by M. J. Baker and S. J. Hart (Oxford: Butterworth-Heinemann, 2008), 726–728.
42. C. L. Hartman and E. R. Stafford, Green alliances: Building new business with environmental groups, *Long Range Planning* 30 (2) (1997): 184–196.
43. U. K. Vyas and S. E. Cannon, Shifting the sustainability paradigm: From advocacy to good business, *Real Estate Issues* 33 (3) (2008): 1–2.

44. M. J. Epstein, *Making sustainability work: Best practices in managing and measuring corporate social, environmental, and economic impacts* (San Francisco: Berrett-Koehler, 2009).
45. The Greenhouse Gas Protocol Initiative, About the GHG protocol, http://www.ghgprotocol.org/files/ghgp/public/ghg-protocol-2001.pdf (accessed March 21, 2011).

24

Educating the Environmental Health Workforce

Maureen Lichtveld and Christine Rosheim

Introduction

Environmental health encompasses aspects of human health, disease, and injury influenced by physical, chemical, biological, social, and psychosocial factors. Given this broad definition, environmental health professionals are employed in workplace settings and areas that are equally wide ranging. It is a complex and constantly evolving field that relies on a workforce from multiple disciplines to accomplish the ten essential services of environmental health (1):

1. **Monitor** environmental and health status to identify and solve community environmental health problems.
2. **Diagnose and investigate** environmental health problems and health hazards in the community.
3. **Inform, educate, and empower** people about environmental health issues.
4. **Mobilize** community partnerships and actions to identify and solve environmental health problems.
5. **Develop policies and plans** that support individual and community environmental health efforts.
6. **Enforce** laws and regulations that protect environmental health and ensure safety.

7. **Link** people to needed environmental health services and assure the provision of environmental health services when otherwise unavailable.
8. **Assure** a competent environmental health workforce.
9. **Evaluate** effectiveness, accessibility, and quality of personal and population-based environmental health services.
10. **Research** for new insights and innovative solutions to environmental health problems and issues.[1]

This chapter focuses on number 8, assuring a competent, well-educated and trained environmental health workforce. Environmental health professionals play a crucial role in decreasing and preventing illness in our communities; therefore, challenges to the competence and effectiveness of the environmental health workforce are critical concerns. Current challenges to assuring a robust environmental health workforce include an aging workforce (an estimated 40–50 percent of the environmental health workforce in state and local agencies may be eligible to retire in the next five years),[2] newly emerging environmental threats,[3] and the need for adequate environmental health professional education and training.[3, 4]

The recent economic downturn has also placed additional strain on an already vulnerable and fragmented environmental health system.[4] A survey by the National Association of County and City Health Officials (NACCHO) showed that in 2008 alone more than half of the nation's local

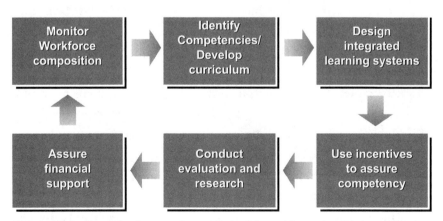

Figure 24.1 Strategic elements for public health workforce development. (Centers for Disease Control and Prevention.)

health departments (LHDs) had either laid off employees or lost them through attrition and have not replaced them due to budget limitations.[5]

This chapter addresses the strategic elements of the National Strategic Framework for Public Health Workforce Development: monitor workforce composition; identify competencies and develop curricula; design an integrated learning system; use incentives to assure competency; conduct evaluation and research; and assure financial support (see Figure 24.1).[6]

The Environmental Health Workforce

Public health has been defined as "what we as a society do collectively to assure conditions in which people can be healthy."[7] As a field within public health, environmental health plays an increasingly central role in this broad mission—concerned with all aspects of the natural and built environment that may affect human health; the three core governmental public health functions of assessment, policy development, and assurance; and practice as well as research. To carry out this expanding mission, environmental health professionals inspect restaurants; monitor air and water quality; measure radon levels in homes; identify patterns of disease; investigate sources of outbreaks; conduct research to prevent, diagnose, and treat environmentally induced illness; provide environment health education to those at risk; promote environmentally sound and sustainable business practices; and develop environmental policy and regulations.

Monitoring Workforce Composition: Environmental Health Workforce Enumeration

Enumerating the workforce is challenging because 1) the field of environmental health contains a multiplicity of different job titles and classifications, including but not limited to physicians, sanitarians, laboratorians, epidemiologists, scientists, engineers, educators, and managers; and 2) from a practice perspective, environmental health activities take place within and beyond the traditional public health settings; therefore environmental health practitioners in many agencies, including public health, environmental protection, agriculture, housing, and others, depend on state and local structures.[4]

The 1988 DHHS Report to the President and Congress estimated an environmental public health workforce of 235,000, with the majority having little or no formal training.[4] The report also foresaw the need for an additional 137,000 environmental health professionals on the basis of population growth and expanding responsibilities.[4] In 1996 the DHHS Risk Communication and Education Subcommittee (RCES) of the Environmental Health Policy Committee conducted a survey to update these data and review the status of the recommendations made in 1988.[8] The RCES found no current information on the actual number of environmental health specialists in the U.S. workforce, but found the need for more training of local and entry-level environmental health professionals to be critical.[8] The *Public Health Workforce: Enumeration 2000* reported that only 19,431 environmental health professionals could be identified—4,549 environmental engineers and 14,882 environmental scientists and specialists—but noted that undercounting was probable due to many environmental health activities being organizationally separated from other parts of public health.[9] Other recent sources estimate the number closer to 60,000.[10] Several documents report that over 90 percent of the current EH workforce has no formal education in public health or environmental health.[11-13] However, these documents did not provide an original source for this estimate.

Other Demographics of the Environmental Health Workforce

Similar to the difficulties with enumeration, other demographic information on the environmental health workforce is also challenging to find. Data from the Bureau of Labor Statistics indicate that as of 2006 only 23 percent of environmental engineers and 13 percent of environmental scientists are women.[14] The percentage of underrepresented minorities (Hispanics, African Americans, and American Indians/Alaska Natives [AI/ANs]) in environmental science and geoscience occupations varied between 4.7 and 8.1 percent between 2003 and 2008 and dropped to 2.2 percent in 2009.[15] (In all U.S. occupations, the underrepresented minority percentage varied between 23.2 and 25.0 in the years 2003–2008 and was 24.7 in 2009.[15]) Males currently represent a much great percentage than females enrolled in environmental sciences in Association of Schools of Public

Health (ASPH).[16] AI/ANs represent the largest percentage of minorities enrolled in ASPH programs in environmental sciences.[16]

Environmental Health Professionals and Professionals in Environmental Health

Larry Gordon, past president of the American Public Health Association (APHA), states that those in the environmental health workforce may be classified either as environmental health professionals or as professionals in environmental health, and regards both as essential components of a comprehensive environmental health effort.[17] Environmental health professionals are defined as

> those who have been adequately educated in the various environmental health technical disciplines as well as in epidemiology, biostatistics, toxicology, management, public policy, risk assessment and reduction, risk communication, environmental law, social dynamics and environmental economics; while professionals in environmental health include other essential personnel such as chemists, geologists, biologists, meteorologists, physicists, physicians, economists, engineers, attorneys, social scientists, public administrators, and planners.[17]

This chapter refers to environmental health professionals in the inclusive sense as all those who are employed to provide the essential services in environmental health.

Work Settings

Although this chapter focuses on those employed in the public sector, environmental health professionals work in a variety of settings, including federal, state, tribal, territorial, and local government; academia; private industry and consulting firms; the armed services; and community-based organizations, domestically and internationally. Environmental health includes a vast number of state, tribal, local, and territorial agencies working in collaboration with the federal government. This collaboration is guided by a financial resource framework, which allows the federal government to promote equality and minimum standards, a legal foundation that delegates specific regulatory authorities to states, and a practice base of local agencies as the point of contact for communities.[18]

- The federal role in environmental health is to provide funding through taxation, technical assistance, training, research, and regulation of international and interstate commerce.
- The state role in environmental health, in addition to regulatory and taxation roles, includes police powers, creation of political subunits, providing funding to LHDs, and implementing federal requirements.
- The local role in environmental health is to carry out state laws and local ordinances, deliver environmental health programs, and establish and collect local fees for services.[18]

Whether one is serving in a federal public health agency, a state or tribal environmental agency, or a local county health department, the core knowledge and skills used in the practice of environmental health are universal:[19]

Federal level: The Environmental Protection Agency (EPA) employs approximately 17,000 people;[20] about 1,000 staff are employed in the environment-focused centers at the Centers for Disease Control and Prevention (CDC) and the Agency for Toxic Substances and Disease Registry (ATSDR);[21] and an estimated 700 employees are engaged in research at the National Institute for Environmental Health Sciences (NIEHS).[22] The U.S. Public Health Service employs a commissioned force of about 400 environmental health officers among these federal employers.[19] An additional 7,000 active duty environmental health practitioners are employed in the uniformed services of the United States.[19]

State level: Environmental health professionals comprise approximately 9 percent (9,000) of the states' 100,000 public health employees.[23]

Tribal level: Gary Hartz, director of Environmental Health and Engineering of the Indian Health Service (IHS), estimates there are 1,200 federal environmental health employees at IHS in the regions or at headquarters, and 600–800 environmental health employees at compact/contract tribes.[24]

Local level: Environmental specialists or scientists represent approximately 10 percent or 15,500 of the estimated 155,000 LHD workforce.[5]

As of 2008, 80 percent of all LHDs employed environmental health specialists (sanitarians) and 20 percent of LHDs employed other environmental health scientists.[5]

Current Scope of Work

Local public health agencies are focused on providing services directly to the communities they serve. Local environmental health services and programs typically include indoor air quality, emergency response, food safety, lead screening and abatement, sewage disposal, vector control, surface water pollution, and private drinking water.[5] Local agencies may also inspect and license food and milk processing and distribution centers, public and private drinking water facilities, recreational water facilities, restaurants, and health-related facilities.[5]

More generally, an environmental health professional has the responsibility to

> recognize and assess the real or potential impact of environmental agents and conditions on human health; and develop preventive or corrective measures in order to reduce the risk of environmentally provoked disease, dysfunction, and premature death. Increasingly, the applied environmental health specialist is becoming more deeply involved in functions such as the interpretation of biological and environmental monitoring data; the creation of regulatory standards for environmental conditions; the conduct of environmental policy analysis; the design of risk management strategies within a social, economic, and political framework; and communicating about uncertain risk in a manner that both educates the public and improves the communication process.[25]

Environmental health positions at LHDs can be categorized generally into trainees, field inspectors, supervisor/managers, and directors, with most positions being in the field inspector area.[19] A field inspector can be either a generalist or a specialist. For instance, a generalist may work on sewage disposal, drinking water safety, food safety, and nuisance complaints, while a specialist may focus solely on food safety.[19] Possible job titles for field inspectors include environmental health scientist, environmental health technician, sanitarian, environmental control technician, or air quality specialist. Data from a NEHA survey show that 95 percent of environmental health professionals are required to have a minimum of a bachelor's degree[19] For supervisor/manager level positions, 24 percent of positions required a master's degree.[19]

Given the broad scope of environmental health, the range of environmental disciplines, and the reality that much of the environmental workforce does not have formal training in either environmental health or public health, there is a need to ensure that those in the environmental health workforce are competent to perform the ten essential services in the context of their work setting and level. This need has given rise to the development of educational and workforce competencies, discussed in the next section.

Identify Competencies: What Does the Environmental Workforce Need to Know?

Our nation is facing a serious shortage of qualified environmental health professionals,[2, 4, 19] and the field and profession of environmental public health face a number of challenges:

1. **Decreasing enrollment in environmental education programs.** Since 2004 enrollment in Association of Schools of Public Health (ASPH) public health programs other than environmental health has increased over the enrollment in environmental health sciences.[16] ASPH enrollment in environmental science in 1998 was 11.3 percent, and in 2008 was 7 percent, a decrease of 20 percent over a ten-year time frame.[16] Similarly, starting in 1994 there was a downturn in environmental science enrollment in undergraduate programs.[26] This decline worsened between 1996 and 2001, when the number of students enrolled in accredited environmental science programs declined from 2,257 to 1,130, a 50 percent decrease in a five-year time period.[26]
2. **Workforce issues.** These include recruitment shortfalls, inability to retain qualified staff, impending retirements, insufficient compensation, and lack of career advancement pathways.[3] Since graduates from accredited programs are in great demand in the public sector, many LHDs are hiring personnel from two- or four-year college programs without a degree or educational training in environmental health or

public health.[26, 27] These new employees may require significant additional training. Training costs, challenging daily workloads leading to employee burnout, and lack of capacity to fully protect public health are possible consequences of hiring inadequately prepared staff.[27, 28]

3. **Increasing scope and need for services.** Demand on the environmental infrastructure has increased, and the scope of services has recently broadened to include the role of the built environment, climate change, and disaster and terrorism preparedness.[2, 4, 12, 28]

4. **Increasing need for education and training.** Training is needed for new and existing environmental health professionals to maintain and enhance the technical knowledge required to adequately address essential environmental health services and issues such as terrorism, natural disasters, and other emerging environmental threats.[2, 4, 12, 28]

5. **Uneven educational foundation.** Enumeration studies to date indicate that a large majority of the environmental health workforce had little or no formal training in their field.[8, 4, 12]

6. **Aging workforce.** State health departments have been particularly concerned about their aging workforce and a shortage of environmental health workers, both of which may create critical challenges to the ability of states to effectively respond to threats to the public's health.[2, 4, 22]

Americans' health depends on knowledgeable and experienced environmental health professionals who can identify threats, mitigate or eliminate hazards, and offer assistance to those exposed or otherwise affected. For environmental health professionals to be effective in preventing and responding to environmental threats, they must be able to create innovative solutions. Hence, environmental health professionals, regardless of their area of specialty, must master basic public health competencies, including epidemiology, statistics, and communication skills, combined with critical thinking skills, and be thoroughly trained in a number of advanced technologies.[2] In addition, with fewer workers, there is a need to "work smarter" and strengthen capacity through the development of discipline-specific competencies.[2] The benefits of requiring a competency-based education are a highly skilled workforce, position uniformity, increased workforce pool mobility, and improved communication and coordination between sectors.[29]

From a workforce perspective, a competency is defined as "a cluster of related knowledge, skills, and attitudes that affect a major part of one's job (a role or responsibility), that correlates with performance on the job, that can be measured against some accepted standards, and that can be improved via training an development."[30] Competencies are used to develop, deliver, and evaluate education; identify job responsibilities and position descriptions; and assess individual and agency capacity.[29]

Recognizing a disconnect between public health education and public heath practice, in 1988 the Institute of Medicine (IOM) recommended the development of universal competencies for public health professionals and specific competencies for those practicing in specific areas such as environmental health.[7] As a response, in 2001 the Council on Linkages adopted a model set of core competencies for all public health professionals,[32] and the Association of Schools of Public Health (ASPH) defined sets of competencies across the five required core areas of master in public health curriculum, including environmental health.[29] Since then several other sets of competencies and performance standards have been developed. In fact, the proliferation of environmental competencies may create a challenge for those attempting to select appropriate competencies for targeting education and training.[29] However, environmental health college programs accredited by the National Environmental Health Science and Protection Accreditation Council (EHAC) are currently deemed the "gold standard."[19, 28]

EHAC is the accrediting body for undergraduate and graduate programs in environmental health.[33] (Refer to appendices A and B in this chapter for lists of EHAC-accredited schools.) The Association of Environmental Health Academic Programs (AEHAP) represents the thirty-eight schools currently accredited by EHAC.[34] AEHAP environmental health academic programs are developed with extensive consultation with employers, and periodic reviews of employer needs are conducted to ensure that graduates possess the right mix of science, math, public health, and communication skills.[34] The EHAC Web site lists curricula guidelines; all programs must have foundational courses in toxicology, biostatistics, and epidemiology, and prerequisite requirements include college courses in chemistry, biology, and physics.[33] EHAC expects that necessary knowledge will be gained through courses in environmental economics, environmental health management, environmental law and public policy development, risk assessment, and risk communication.[33] Every student must complete in-depth study in at least four of the following technical areas and be exposed to a majority of the following topics:[34]

Air quality control	Institutional health
Environmental chemistry	Noise control
Environmental epidemiology	Occupational heath and safety
Environmental health planning	Radiation health
Environmental microbiology	Recreational environmental health
Food protection	Soils
Global environmental health	Solid waste management
Hazardous material	Vector control
Housing	Wastewater
Hydrogeology	Water quality
Industrial hygiene	Water supply
Injury prevention	

All students at AEHAP schools must also complete a field practicum of at least 180 hours, which can be waived if the student has significant work experience.[34] Earning an undergraduate degree in environmental health typically takes four years.[34] The graduate program requirements are similar, but allow for additional specialization or management concentration.[34] EHAC-accredited program graduates currently have a 99 percent placement rate within six months of graduation.[19, 34]

Technical environmental health competencies required to pass NEHA's Registered Environmental Health Scientist/Registered Sanitarian exams include basic environmental health and protection; basic sciences—toxicology, physics, chemistry, geology, biology, epidemiology; environmental, occupational, communicable and chronic disease; environmental law—statutes, regulations; risk assessment; and risk management.[35]

Beyond these required technical competencies, the CDC and the American Public Health Association (APHA) developed fourteen core competencies (grouped into three functions) needed for environmental health practitioners to function effectively:[31]

Assessment Function—1) information gathering; 2) data analysis; 3) evaluation

Management Function—4) problem solving; 5) economic and political issues; 6) organizational knowledge and behavior; 7) project management; 8) computer and information technology; 9) reporting, documentation, and record keeping; 10) collaboration

Communication Function—11) education, 12) communication, 13) conflict resolution, and 14) marketing

Leadership Competencies

Training specifically for environmental health leaders should emphasize strategic visioning and direction setting skills, critical thinking and analysis, political effectiveness, and organizational and team development.[36] Such training should lead to 1) improved delivery of essential services to communities, 2) timely response to emerging environmental public health threats, and 3) effective and efficient coordination of service to areas affected by environmental disasters.[35]

Designing an Integrated Learning System: How Is This Learning Accomplished and Incentivized?

The educational pathways to become an environmental health professional are as diverse as the subspecialties. At the bachelor's level, the degrees held by the environmental workforce encompass an array of areas, including environmental health, biology, chemistry, physics, engineering, computer science, geography, geology, and communications. Graduate degrees include master of public health (MPH), master of science (MS), doctor of public health (DrPH), and doctor of philosophy (PhD).

A. Undergraduate Level

AEHAP was formed in 1999 to address decreasing enrollment in environmental health undergraduate programs and a severe workforce shortage of environmental professionals.[34] Its members include thirty-one undergraduate and seven graduate programs. Ten of its member institutions serve a large minority population of students: Alabama A&M; Benedict College; California State Universities at Fresno, Northridge, and San Berndardino; Mississippi Valley State University; New Mexico State University; North Carolina Central University; Spelman College; and Texas Southern University.[34] For the 2008–2009 academic year, a total of 327 undergraduate and 59 graduate students graduated from AEHAP member schools.[34]

B. Graduate Level

Schools of public health and programs in public health are accredited by the Council of Education in Public Health (CEPH).[37] To be CEPH-accredited and to be a member of the ASPH, schools of public health must offer graduate degrees in environmental health. A list of the forty-three ASPH members is included in appendix C in this chapter.[37] There are also sixty-five programs in public health accredited by CEPH.[39] CEPH-accredited programs in public health are not required to offer environmental health degree programs, but as of 2003, nineteen of them did.[40] Appendix D in this chapter is a list of CEPH-accredited programs in public health offering environmental health degrees. As of 2008, ASPH member schools with the highest number of students enrolled in environmental health science programs were Minnesota (141), UNC Chapel Hill (132), Tulane (128), and Michigan (115).[16] The total of environmental health students enrolled at all forty-three ASPH schools in 2008 was 1,810, which was 7.7 percent of the 23,357 total students enrolled.[16] Of the 1,810 ASPH students in environmental health programs, 67.3 percent were in master's degree programs and 32.7 percent in doctoral degree programs.[16]

C. On-the-Job Training, Continuing Education, Credentialing, and Certification

An environmental health program benefits from a workforce engaged in lifelong learning provided through continuing education, job-related training, and mentoring. Continuing education may encompass distance learning, workshops, seminars, national and regional conferences, and other activities intended to strengthen the professional knowledge and skills of employees.[41] Opportunities should be available for staff to work with academic and research institutions, particularly those connected with school and programs of public health and accredited undergraduate and graduate programs of environmental health.[41] Almost 70 percent of LHDs report that staff has taken public health classes or workshops offered by an accredited school or program of public health[5] Many AEHAP universities have distance learning programs, including mail and satellite courses, especially in large state university networks.[19] An updated list is maintained on the AEHAP Web site under "Student Resources."[34] Another training resource is IHS's Environmental Health Support Center (EHSC), offering training courses on a wide variety of subjects related to the programs of the IHS's Office of Environmental Health and Engineering.[42]

In addition to the REHS/RS credential through examination, NEHA offers eight additional credentials demonstrating mastery of specialized knowledge: Certified installer of onsite waste water treatment systems; Certified Healthy Homes Specialist; Certified Professional in Food Safety; Certified Environmental Health Technician; Registered Environmental Technician; Registered Hazardous Substances Professional; Registered Hazardous Substance Specialist; and Radon Proficiency Program.[35]

Aligned with the chapters in NEHA's REHS/RS Study Guide is a new package of comprehensive, online, on-demand courses. The Environmental Public Health Online Courses (EPHOC) were developed by NEHA, the Jefferson County Department of Health Environmental Health Services, the University of Alabama School of Public Health, and the CDC, and were designed for two main audiences: (1) new hires who want to learn about the broad field of environmental health practice and (2) more experienced staff who are ready to pursue professional credentialing.[43, 44]

Certification and credentialing are proven strategies to assure competence; however, both depend heavily on both educational as well as workplace incentives. Examples of effective incentives include expanded career trajectories, financial benefits, professional recognition, and access to membership benefits in professional associations. Without a sustainable incentive structure and defined job descriptions, strategic approaches to assuring competence may remain voluntary rather than mandatory.

D. Leadership Level

According to the CDC's Model Standard 8.4 (Environmental Health Leadership Development), an environmental health program should provide formal education programs or leadership institutes and informal coaching or mentoring opportunities for leadership at all organizational levels.[41] One example of such an opportunity is the CDC's Environmental Public Health Leadership Institute (EPHLI), launched in 2005.[36] EPHLI is a one-year program of seminars and workshops, special projects, and individual study.[36] Each year about thirty practicing environmental health professionals are admitted to the program, with applications for new cohorts accepted each fall.[36] In 2010 graduating presentation topics included air quality, biosolids, children's environmental health, climate change, community environmental health assessment, credentialing, drinking water, and emergency shelter standards.[36]

E. Health Professional Education in Environmental Health

The limited research available on environmental health in medical education indicates that medical students and residents are not adequately prepared in environmental health.[45] Beyond local continuing medical education (CME) offerings or course work at schools or programs of public health, the CDC and ATSDR offer educational programs and materials, including the *Case Studies in Environmental Medicine*, which can be completed for CME credits.[46] Other educational resources include the following:

The National Environmental Education Foundation (NEEF) is a private, nonprofit organization dedicated to advancing environmental education in its many forms, including strengthening the public health safety net through educating healthcare professionals.[47] NEEF supports several health professional and environmental educational activities, including a pesticide initiative and a pediatric asthma initiative.[47]

Since its founding in 1987, the Association of Occupational and Environmental Clinics (AOEC) has grown to a network of more than 60 clinics and more than 250 individuals committed to improving the practice of occupational and environmental medicine through information sharing and collaborative research.[48] The AOEC Educational Resource Library is a multifaceted resource available to both AOEC members and nonmembers.[48] Library holdings include videotapes, peer-reviewed slide presentations, CD-ROM presentations, non-peer-reviewed programs donated by members, and other instructive materials.[48]

The Pediatric Environmental Health Specialty Units (PEHSUs) were created to ensure access to special medical knowledge and resources for children faced with a health risk due to a natural or human-made environmental hazard.[49] Located throughout the United States, Canada, and Mexico, PEHSU professionals provide quality medical consultation for health professionals, parents, caregivers, and patients.[49] The PEHSUs are also dedicated to increasing environmental medicine knowledge among healthcare professionals about children's environmental health by providing consultation and training.[49] Finally, the PEHSUs, provide information and resources to school and community groups to help increase the public's understanding of children's environmental health.[49]

In addition to environmental health practitioners, there is a cadre of scientists engaged in environmental health research. Environmental health scientists are employed in academia and the public and private sectors and are specialized in both traditional as well as emerging disciplines—from water quality to climate change. Growing the environmental health workforce pipeline will require academicians to be closely attuned to what future graduates are interested in as well as what employers need.

F. Sustaining the Environmental Health Workforce

Recent advances in public health systems research has contributed significantly to a greater appreciation of the need for workforce development research.[6, 50, 51] Findings from quality improvement research and credentialing efforts by the National Association of County & City Health Officials (NACCHO) indicate that quality improvement practice among most local public health departments has been limited, and that local health department workers must complete on-the-job training programs to ensure that public health competencies are achieved and maintained.[52] The Multi-State Learning Collaborative, a Robert Wood Johnson Foundation initiative, has demonstrated that participating states are showing signs of progress with increased communication, information exchange, and health department accountability.[51, 53]

A growing body of research is also strengthening the evidence base for financing public health infrastructure in general and workforce development specifically to promote high performance and continuous quality improvement through a National Voluntary Accreditation Program. In 2005 NACCHO released the report, *Operational Definition of a Functional Local Health Department*, which incorporates standards for accountability and a framework for local health departments in preparing for voluntary national accreditation. The report's intentions include guiding local health departments in sustaining a quality public health workforce through the expansion of functions; enhancement of activities; and open communication channels with governing bodies, elected officials, and the community.[54] To date none of these efforts have resulted in a concrete extension of the environmental health workforce.

Future Directions and Recommendations

The authors believe that the following recommendations will lead to an enhanced ability of the environmental health workforce to address existing and emerging needs, through 1) better understanding of the scope of work, size, composition, and performance standards; and 2) increasing competencies of the environmental health workforce and its leadership, by increasing the number of professionals who engage in competency-based continuing education and training. The authors recommend the following:

a. Using new technology and new media for education, recruiting, retention, and on-the-job training to reach new audiences. In the 2008–2009 academic year, AEHAP reports that undergraduate enrollment actually increased 10.3 percent and graduate enrollment increased 6.5 percent, respectively.[34] Innovative methodologies for increasing enrollment need to be developed and shared.

b. Increasing the diversity of the environmental health workforce. Multicultural staff are increasingly essential to ensure an effective workforce and leadership capable of meeting environmental health challenges of the twenty-first century.[56, 57] As a field, environmental health must recruit, train, and retain a workforce from a broad range of disciplines and with a broad range of backgrounds and experience to allow maximum creativity in addressing health disparities and emerging issues.[55, 56] AEHAP reports that its support for innovative projects in diversity recruitment and retention have helped yield a 31.5 percent increase in the number of minority students enrolled in accredited environmental health programs over a four-year period.[56]

c. Marketing the image of environmental health to boost recruitment and retention. While needing to attract youth, the field faces an aging workforce, worker shortages, budget constraints, noncompetitive wages, and lack of visibility.[55] Innovative approaches; use of new media; and connecting environmental health to bigger picture "green" issues, such as climate change, the built environment, healthy homes, and sustainability, should promote its marketability.

d. Expanding and sustaining support for environmental health internships and fellowships. Internships and fellowships are effective mechanisms, allowing students to gain practical skills and apply academic theory to the on-the-job setting, and allowing agencies to have a preview of potential employees.[4] The authors support the establishment of federally funded public health workforce scholarships and loan repayment programs. Stable funding is also needed for Title VII of the Public Health Service Act, under which the Health Resources and Services Administration (HRSA) supports traineeships in fields where there is a severe shortage, including environmental health.[58]

e. Sustaining leadership programs. These should include the CDC's Leadership Institute and proposed National Environmental Health Service Corps.[4]

f. Encouraging transdisciplinary research, education, and training to broaden collaboration among epidemiology, laboratory, and environmental health students, faculty, and professionals.

g. Supporting workforce research to address issues of quality (competence) and quantity (shortage). For example, each year many of these highly trained and skilled military environmental health practitioners retire or are honorably discharged; however, only a small percentage of these practitioners select a postmilitary career working with state or local environmental health programs.[19] Research is needed to understand why this is so and how to encourage them to seek such positions.

h. Building environmental literacy in the general population. An environmentally literate populace, managing behaviors and ecosystems in order to live sustainably, could be predicted to support environmental quality and an effective environmental workforce.[59] With goals for educating schoolchildren, the adult public, health professionals and business managers, and other decision makers, NEEF's ultimate aim is to activate environmentally responsible behavior in the general public.[47] One method of doing this is finding innovative ways to integrate K–12 environmental education into existing school science and other curricula.[60]

Conclusion

"Hiring graduates from accredited environmental programs . . . will help ensure a highly skilled workforce that is articulate, adaptable, and better equipped to work in the field of public health. Graduates of accredited environmental programs receive a standardized education tailored specifically to meet the growing challenges in the field of environmental public health."[28] The authors agree and conclude the following:

1. Environmental health education should be competency-based. This allows benchmarking employees' knowledge, skills, and abilities. Educational and workforce competencies to be used should be based on continuing collaboration between employers and academia to prevent a schism between existing curricula and the needs of employers.

2. More qualified graduates and more accredited programs are clearly needed. "Large areas of the country, including whole states, are still without a single accredited environmental health program."[55]

3. Employers should hire from accredited programs when possible. Agencies should hire an environmental health workforce that is equipped to work effectively in the complex field of environmental health. Graduates of accredited programs should possess the essential knowledge, skills, and abilities to provide the ten essential environmental health services to their communities and to protect and promote environmental health.

4. There is a need for lifelong learning. New and existing staff will need additional training to meet current emerging health and environmental challenges. Every member of the environmental health workforce should demonstrate the core environmental health competencies regardless of the strategy employed to attain those competencies—via undergraduate, graduate, on-the job, or other training.

5. New strategies for recruitment and retention are urgently needed in environmental health and in public health overall. A survey being conducted by The Council on Linkages between Academia and Public Health Practice to determine how, when, and why individuals enter, stay in, and leave the public health workforce may inform future, more sustainable approaches.[32]

Notes

1. C. Osaki, D. Hinchey, and J. Harris, Using 10 essential services training to revive, refocus, and strengthen your environmental health programs, *Journal of Environmental Health* 70 (1) (2007): 12–15.
2. Association of State and Territorial Health Officials, *2007 state public health workforce survey results* (2008), http://www.astho.org/Programs/Workforce-and-Leadership-Development (accessed April 8, 2010).
3. F. Phillips, S. Kelly, T. Burke, et al., Enhancing the Maryland environmental public health workforce, *Journal of Environmental Health* 70 (1) (2007): 32–36.
4. Centers for Disease Control and Prevention, *A national strategy to revitalize environmental public health services* (2003), http://www.cdc.gov/nceh/ehs/Publications/Strategy.htm (accessed April 8, 2010).
5. National Association of County and City Health Officials, *2008 national profile of local health departments*, http://naccho.org/topics/infrastructure/profile/resources/2008report/index.cfm (accessed April 8, 2010).
6. M. Y. Lichtveld, J. P. Cioffi, E. L. Baker, et al., Partnership for frontline success: A call for a national action agenda on workforce development, *Journal of Public Health Management and Practice* 7 (4) (2001): 1–7.
7. Institute of Medicine, *The future of public health* (Washington, DC: The National Academies Press, 1988).
8. B. Johnson and T. Tinker, An assessment of federal environmental health training resources, *Journal of Environmental Health* 62 (5) (1999): 21–25.
9. Health Resources and Services Administration, Bureau of Health Professions, National Center for Health Workforce Information and

Analysis, *The public health workforce, enumeration 2000* (Washington, DC, 2000), http://bhpr.hrsa.gov/healthworkforce/reports/ (accessed June 4, 2010).

10. K. Gebbie, J. Merrill, and H. Tilson, The public health workforce, *Health Affairs* 21 (6) (November 2002): 57–67, http://content.healthaffairs.org/cgi/content/full/21/6/57 (accessed June 4, 2010).

11. M. Hering, Facts about the environmental health workforce shortage and the impact of accredited environmental health degree programs (presented at Building Environmental Health Workforce of the 21st Century: A Workshop for Minority Serving Institutions, Houston, Texas, April 16–18, 2008), http://www.aehap.org/MSIWorkshop proceedings08.pdf (accessed June 8, 2010).

12. Centers for Disease Control and Prevention, *Strategic options for CDC support of the local, state, and tribal environmental public health workforce* (2009), http://www.cdc.gov/nceh/ehs/Docs/Strategic_Options_for_CDC_Support.pdf (accessed June 8, 2010).

13. Environmental Health Office of Multnomah County, Oregon, Workforce development: Post-secondary internship program, 2007, http://www.mchealthinspect.org/documents/workforce_development.pdf (accessed June 8, 2010).

14. American Geological Institute, Participation of women in geoscience occupations, *Geoscience Currents* 33 (May 14, 2010), http://www.agiweb.org/workforce/currentshtml (accessed June 1, 2010).

15. American Geological Institute, Underrepresented minorities in the US workplace, *Geoscience Currents* 34 (May 21, 2010), http://www.agiweb.org/workforce/currentshtml (accessed June 1, 2010).

16. Association of Schools of Public Health, Annual data report: 2008, http://www.asph.org/UserFiles/2008%20Data%20Report.pdf (accessed April 15, 2010).

17. L. Gordon, Setting the context, environmental health practitioner competencies, *Journal of Environmental Health* 65 (1) (2002): 25–27.

18. B. J. Turnock, *Public health: What it is and how it works* (Sudbury, MA: Jones and Bartlett, 2004).

19. Uniformed Services Environmental Public Health Careers Work Group, *Career resource guide for uniformed services environmental health practitioners* (2007), http://www.cdc.gov/nceh/ehs/Docs/career_resources (accessed April 8, 2010).

20. U.S. Environmental Protection Agency Web site, http://www.epa.gov (accessed April 15, 2010).
21. Centers for Disease Control and Prevention Web site, http://www.cdc.gov (accessed April 15, 2010).
22. National Institute of Environmental Health Sciences Web site, http://www.niehs.gov (accessed June 3, 2010).
23. 2007 State Public Health Workforce Survey Results, 2008, http://www.astho.org/Programs/Workforce-and-Leadership-Development (accessed April 8, 2010).
24. Private conversation, May 7, 2010.
25. B. Walker, F. Blackwell, and V. Adrounie, Educating the workforce: Expectations and curricular needs of environmental health, *Journal of Environmental Health* 52 (4) (1990): 220–222.
26. T. Murphy and J. Neistadt, *Improving public health services and saving resources: Hiring graduates from accredited environmental health programs* (National Environmental Public Health Leadership Institute, 2006–2007 Fellows Project), 85–102 (also presented at NEHA Annual Education Conference, June 2007), http://www.heartlandcenters.slu.edu/ephli/FinalReport07/6jeff_tim.doc (accessed April 15, 2010).
27. J. Neistadt and T. Murphy, Are we really saving resources with current hiring practices at local health departments? *Journal of Environmental Health* 71 (6) (2009): 12–14.
28. National Association of Local Boards of Health, *Board of Health recommendations for hiring qualified environmental health practitioners* (2009), http://www.cdc.gov/nceh/ehs/Docs/NALBOH_EH_Workforce_Guide.pdf (accessed April 15, 2010).
29. D. Koo and K. Miner, Outcome-based workforce development and education in public health, *Annual Review of Public Health* 31 (March 2010): 253–269.
30. S. R. Parry, The quest for competencies, *Training* (July 1996): 50.
31. American Public Health Association and Centers for Disease Control and Prevention, *Environmental health competency project: Recommendations for core competencies for local environmental health practitioners* (May 2001).
32. Council of Linkages Web site, http://www.phf.org/link/index.htm (accessed June 2, 2010).

33. National Environmental Health Science and Protection Accreditation Council Web site, http://www.ehacoffice.org (accessed April 8, 2010).
34. Association of Environmental Health Academic Programs Web site, http://www.aehap.org (accessed April 8, 2010).
35. National Environmental Health Association, Credentials, http://www.neha.org/credential/index.shtml (accessed April 8, 2010).
36. Centers for Disease Control and Prevention Web site, CDC Environmental Public Health Leadership Institute, http://www.cdc.gov/nche/ehs/EPHLI/default.htm (accessed April 5, 2010).
37. Council on Education for Public Health Web site, http://ceph.org/i4a/pages/index.cfm?pageid=1 (accessed April 15, 2010).
38. Association of Schools of Public Health Web site, http://www.asph.org/ (accessed April 15, 2010).
39. Council of Graduate Programs of the Association for Prevention Teaching and Research (APTR) Web site, http://www.atpm.org/membership/Council.html (accessed April 12, 2010).
40. Association of Teachers of Preventive Medicine, *Directory of graduate programs in preventive medicine and public health* (Washington, DC, 2003).
41. Centers for Disease Control and Prevention, *Environmental health performance standards* (2010), http://www.cdc.gov/nceh/ehs/EnvPHPS/default.htm (accessed June 3, 2010).
42. Indian Health Service, Environmental Health and Engineering, http://www.oehe.ihs.gov/ (accessed May 7, 2010).
43. Centers for Disease Control and Prevention, Environmental Public Health Online Courses (EPHOC), http://www.cdc.gov/nceh/ehs/Workforce_Development/EPHOC.htm (accessed June 4, 2010).
44. W. Studyvin and T. Struzick, EPHOC: A new training resource for the environmental public health workforce, *Journal of Environmental Health* 72 (7) (2010): 28–29.
45. J. Roberts and R. Reigart, Environmental health education in the medical school curriculum, *Ambulatory Pediatrics* 1 (2) (2001): 108–111.
46. Agency for Toxic Substances and Disease Registry, *Case studies in environmental medicine*, http://www.atsdr.cdc.gov/csem/csem.html (accessed June 4, 2010).

47. National Environmental Education Foundation Web site, http://www.neefusa.org/health/index.htm (accessed May 4, 2010).
48. Association of Occupational and Environmental Clinics Web site, http://www.aoec.org (accessed June 2, 2010).
49. Association of Occupational and Environmental Clinics, Pediatric Environmental Health Specialty Units, http://www.aoec.org (accessed June 2, 2010).
50. F. D. Scutchfield, M. W. Bhandari, N. A. Lawhorn, et al., Public health performance, *American Journal of Preventive Medicine* 36 (3) (2009): 266–272.
51. C. J. Leep, *The local health department workforce: Findings from the 2005 national profile of local health departments study* (Washington, DC: National Association of County & City Health Officials, 2007), 1–11.
52. National Association of County & City Health Officials, *Statement of policy, workforce certification and credentialing* (Washington, DC: National Association of County & City Health Officials, March 2009).
53. B. M. Joly, G. Shaler, M. Booth, et al., Evaluating the multi-state learning collaborative, *Journal of Public Health Management and Practice* 16 (1) (2010): 61–66.
54. National Association of County & City Health Officials, *Operational definition of a functional local health department* (Washington, DC: National Association of County & City Health Officials, 2005).
55. R. Blake, Education and workforce development issues associated with invisibility, *Journal of Environmental Health* 70 (4) (2007): 4, 39.
56. D. Harper, A diverse environmental public health workforce to meet the diverse environmental health challenges of the 21st century, *Journal of Environmental Health* 69 (6) (2007): 52–53.
57. B. Walker and M. Spann, The need for diversity in the environmental health workforce, *Journal of Health Care for the Poor and Underserved* 19 (1) (2008): 19–25.
58. C. M. Perlino, The public health workforce shortage: Left unchecked, will we be protected? *American Public Health Association Issue Brief* (September 2006), http://www.apha.org/advocacy/reports/reports (accessed April 8, 2010).

59. J. Palmer, *Environmental education in the 21st century* (London and New York: Routledge, 1998).
60. M. Stone and Z. Barlow (eds.), *Ecological literacy: Educating our children for a sustainable world* (San Francisco: Sierra Club Books, 2005).

Acknowledgments

The authors thank Farah Arosemena for her contributions to this book chapter.

Appendix A: EHAC-Accredited Undergraduate Programs in Environmental Health

Alabama A&M
Benedict College
Boise State University
Bowling Green State University
California State University at Fresno
California State University at Northridge
California State University at San Bernardino
Colorado State University
East Carolina University
East Central University
Eastern Kentucky University
East Tennessee State University
Illinois State University
Indiana University of Pennsylvania
Indiana University–Purdue University Indianapolis
Lake Superior State University
Mississippi Valley State University
Missouri Southern University
New Mexico State University
North Carolina Central
Ohio University
Old Dominion University
Salisbury University
Spelman College
Texas Southern University
University of Georgia Athens
University of Washington
University of Wisconsin Eau Claire
West Chester University
Western Carolina University
Wright State University

Appendix B: EHAC-Accredited Graduate Programs in Environmental Health

California State University at Northridge
East Tennessee State University
Old Dominion University
University of Illinois Springfield
East Carolina University
Mississippi Valley State University
University of Findlay

Appendix C: CEPH-Accredited Schools of Public Health

Boston University School of Public Health
Columbia University Mailman School of Public Health
Drexel University School of Public Health
East Tennessee State University College of Public Health
Emory University Rollins School of Public Health
Florida International University College of Public Health and Social Work
George Washington University School of Public Health and Health Services
Harvard University School of Public Health
Instituto Nacional de Salud Publica Escuela de Salud Public de Mexico
Johns Hopkins University Bloomberg School of Public Health
Loma Linda University School of Public Health
Ohio State University College of Public Health
Saint Louis University School of Public Health
San Diego State University Graduate School of Public Health
Texas A&M University System Health Science Center School of Rural Public Health
Tulane University School of Public Health and Tropical Medicine

University of Medicine and Dentistry New Jersey, Rutgers, The State University of New Jersey, New Jersey Institute of Technology School of Public Health

University at Albany–SUNY School of Public Health

University at Buffalo–SUNY School of Public Health and Health Professions

University of Alabama at Birmingham School of Public Health

University of Arizona Mel and Enid Zuckerman College of Public Health

University of Arkansas for Medical Sciences Fay W. Boozman College of Public Health

University of California, Berkeley School of Public Health

University of California, Los Angeles School of Public Health

University of Florida College of Public Health and Health Professions

University of Georgia College of Public Health

University of Illinois at Chicago School of Public Health

University of Iowa College of Public Health

University of Kentucky College of Public Health

University of Louisville School of Public Health and Information Sciences

University of Massachusetts Amherst School of Public Health and Health Sciences

University of Michigan School of Public Health

University of Minnesota School of Public Health

University of North Carolina Gillings School of Global Public Health

University of North Texas Health Science Center School of Public Health

University of Oklahoma College of Public Health

University of Pittsburgh Graduate School of Public Health

University of Puerto Rico Graduate School of Public Health

University of South Carolina Arnold School of Public Health

University of South Florida College of Public Health

University of Texas Health Science Center at Houston School of Public Health

University of Washington School of Public Health

Yale University School of Public Health

Appendix D: CEPH-Accredited MPH Programs with Degrees (or Concentrations) in Environmental Health

University of Arkansas, College of Public Health

California State University, Long Beach, Dept. of Health Science

UC-Davis, Dept. of Epidemiology and Preventive Medicine

Florida A&M, College of Pharmacy and Pharmaceutical Sciences

University of Miami, Dept. of Epidemiology and Public Health

Indiana University/Purdue University, School of Medicine

Eastern Kentucky University, Dept. of Health Promotion, Administration, and Environmental Health Sciences

Western Kentucky University, Dept. of Public Health

Tufts University School of Medicine, Dept. of Family Medicine and Community Health

University of New Mexico, Dept. of Family and Community Medicine

Hunter College of the City University of New York, Dept. of Urban Public Health

New York Medical College, School of Public Health

East Carolina University, Dept. of Family Medicine

Bowling Green University, Medical College of Ohio, and University of Toledo

Drexel University, School of Public Health

Temple University, Dept. of Public Health

Thomas Jefferson University, College of Graduate Studies

West Chester University, Dept. of Health

Medical College of Wisconsin, Division of Public Health

About the Editor and Contributors

Editor Robert H. Friis, PhD, FRSPH, is Professor Emeritus and Chair Emeritus of the Department of Health Science at California State University, Long Beach (CSULB), and Director of the Department of Veterans Affairs/CSULB Joint Studies Institute. Dr. Friis has been a Visiting Professor at the Max Planck Institute and Dresden University of Technology, both in Germany, and a Visiting Researcher at the Karolinska Institute in Sweden. He retired as a faculty member from the Department of Medicine, University of California, Irvine, and continued for several years as a voluntary Clinical Professor in the Department of Community and Environmental Medicine. Earlier, he was an Associate Professor at the Albert Einstein College of Medicine and an Assistant Professor at the Columbia University School of Public Health. A fellow of the Royal Society of Public Health, he has also served as President of the Southern California Public Health Association and has provided editorial service to scientific journals in the field of public health. An epidemiologist by profession, he authored the widely adopted textbook *Essentials of Environmental Health*, now in its second edition, as well as three other textbooks on epidemiology and biostatistics. He is a recipient of research funding from the University of California's Tobacco-Related Research Program. His two major research interests are tobacco control and the health effects of air pollution in the Los Angeles Basin. As a health department epidemiologist, he conducted investigations into the adverse effects of exposures to various environmental hazards.

Margaret Alkon, JD, is an environmental attorney and Adjunct Assistant Professor of Political Science at Carthage College, Kenosha, Wisconsin.

About the Editor and Contributors

Clinton J. Andrews, PhD, is a Professor of Urban Planning and Policy Development at Rutgers, the State University of New Jersey, New Brunswick, New Jersey.

Wolfgang Babisch, PhD, is an associate of the Federal Environmental Agency, Department of Environmental Hygiene, Berlin, Germany.

Grant Baldwin, PhD, MPH, is an associate of the National Center for Injury Prevention and Control, Centers for Disease Control and Prevention, Atlanta, Georgia.

Brian C. Black, PhD, is Head, Division of Arts and Humanities, and Professor of History and Environmental Studies at Penn State University, Altoona, Pennsylvania.

Eva Cecilie Bonefeld-Jørgensen, PhD, is a professor at the Center for Arctic Environmental Medicine, School of Public Health, University of Aarhus, Denmark.

A. L. Brown, PhD, is Professor of Environmental Planning at the Griffith School of Environment, Griffith University, Nathan, Australia.

Joanna Burger, PhD, is Professor of Cell Biology and Neuroscience in the Division of Life Sciences, Consortium for Risk Evaluation with Stakeholder Participation, and the Environmental and Occupational Health Sciences Institute, Rutgers University, Piscataway, New Jersey.

Colin D. Butler, PhD, is an associate professor at the National Centre for Epidemiology and Population Health (NCEPH), Australian National University, Canberra, Australia.

Vincent T. Covello, PhD, is Director, Center for Risk Communication, New York, New York.

Alison F. Dabney is Senior Environmental Scientist, Division of Drinking Water and Environmental Management, California Department of Public Health, Sacramento, California.

Keith B. G. Dear, PhD, is Senior Fellow, National Centre for Epidemiology and Population Health, Australian National University, Canberra, Australia.

About the Editor and Contributors

T. Bella Dinh-Zarr, PhD, is Road Safety Director, FIA Foundation, and North American Director, MAKE ROADS SAFE, Washington, DC.

Reid Ewing, PhD, is Professor of City and Metropolitan Planning, University of Utah, Salt Lake City, Utah.

John E. Fa, DrPhil, is Chief Conservation Officer, Durrell Wildlife Conservation Trust, Jersey, United Kingdom, and a visiting professor, Imperial College London, Division of Biology, Berkshire, UK.

Richard E. Flarend, PhD, is Associate Professor of Physics, Penn State University, Altoona, Pennsylvania.

Cindy A. Forbes, PE, is Chief, Southern California Drinking Water Field Operations Branch, California Department of Public Health, Fresno, California.

Mandana Ghisari, PhD, is an associate of the Center for Arctic Environmental Medicine, School of Public Health, University of Aarhus, Denmark.

Robert M. Gould, MD, is President, San Francisco Bay Area Chapter, Physicians for Social Responsibility, Berkeley, California.

Jenny Griffiths, OBE, MSc, FFPH, FRSPH, is a member of the Climate and Health Council and Non-Executive Director of the National Institute of Health and Clinical Excellence, United Kingdom.

Kristen Gunther, MPH, is an associate of the School of Public Health, Loma Linda University, Loma Linda, California.

Mark B. Horton, MD, MSPH, is the former director of the California Department of Public Health, Sacramento, California.

Barry L. Johnson, PhD, FCR, is Adjunct Professor, Department of Environmental and Occupational Health, Rollins School of Public Health, Emory University, Atlanta, Georgia, and Editor in Chief/Managing Editor of the *Journal of Human and Ecological Risk Assessment*.

Patrick Kennelly is Chief, Food Safety Section, California Department of Public Health, Sacramento, California.

About the Editor and Contributors

Richard A. Kreutzer, MD, is Chief, Division of Environmental and Occupational Disease Control, California Department of Public Health, Richmond, California.

Tanja Krüger, PhD, is an associate of the Center for Arctic Environmental Medicine, School of Public Health, University of Aarhus, Denmark.

Gregg W. Langlois, MS, is Senior Environmental Scientist, Division of Drinking Water and Environmental Management, California Department of Public Health, Richmond, California.

Maureen Lichtveld, MD, MPH, is a professor in the Department of Environmental Health Sciences, Tulane University School of Public Health and Tropical Medicine, New Orleans, Louisiana.

Carl Lischeske, PE, is Chief, Northern California Drinking Water Field Operations Branch, California Department of Public Health, Sacramento, California.

Manhai Long, MD, PhD, is an associate professor at the Center for Arctic Environmental Medicine, School of Public Health, University of Aarhus, Denmark.

Christian B. Luginbuhl is an astronomer at the U.S. Naval Observatory Flagstaff Station in Arizona.

Anthony J. McMichael, PhD, is a professor and NHMRC Australia Fellow at the National Centre for Epidemiology and Population Health, Australian National University, Canberra, Australia.

Rebecca Naumann, MSPH, is an epidemiologist at the National Center for Injury Prevention and Control, Centers for Disease Control and Prevention, Atlanta, Georgia.

Mihalis I. Panayiotidis, PhD, is an associate of the School of Community Health Sciences, University of Nevada at Reno, and the Department of Pathology, University of Ioannina Medical School, Ioannina, Greece.

Aglaia Pappa, PhD, is an assistant professor in the Department of Molecular Biology and Genetics, Democritus University of Thrace, Alexandroupolis, Greece.

About the Editor and Contributors

Christine Rosheim, DDS, MPH, is an associate of the Centers for Disease Control and Prevention in Atlanta, Georgia.

Robert Schlag, MSc, is Chief, Division of Food, Drug, and Radiation Safety, California Department of Public Health, Sacramento, California.

Karen E. Setty, MS, is an associate of the Southern California Coastal Water Research Project Authority, Costa Mesa, California.

Derek G. Shendell, DEnv, MPH, is Assistant Professor, Department of Environmental and Occupational Health Sciences, University of Medicine and Dentistry of New Jersey, School of Public Health, and an associate of the Environmental and Occupational Health Sciences Institute, Exposure Measurement and Assessment Division.

Ryan G. Sinclair, PhD, MPH, is Assistant Professor, Department of Environmental Health, School of Public Health, Loma Linda University, Loma Linda, California.

David A. Sleet, PhD, is Associate Director for Science at the National Center for Injury Prevention and Control, Centers for Disease Control and Prevention, Atlanta, Georgia.

Rhonda Spencer-Hwang, DrPH, is Assistant Professor, Department of Environmental Health, School of Public Health, Loma Linda University, Loma Linda, California.

Marc A. Strassburg, DrPH, is an Adjunct Professor, Department of Epidemiology, UCLA, Los Angeles, California, and Professor of Health Science at Trident University International online.

Irene van Kamp, PhD, is an associate of the National Institute for Public Health and the Environment, Bilthoven, The Netherlands.

Richard J. Wainscoat, PhD, is an astronomer at the University of Hawaii and Chairman of the International Astronomical Union Commission 50 Working Group on Controlling Light Pollution.

Constance E. Walker, PhD, is an associate scientist and education specialist at the National Optical Astronomy Observatory in Arizona, and Director of both the International Year of Astronomy 2009 Dark Skies Awareness Cornerstone Project, and GLOBE at Night.

Stephen B. Weisberg, PhD, is Director of the Southern California Coastal Water Research Project Authority, Costa Mesa, California.

Gary H. Yamamoto, PE, is Chief, Division of Drinking Water and Environmental Management, California Department of Public Health, Sacramento, California.

Dominique Ziech, MSc, is an associate of the Department of Student Success Services, University of Nevada at Reno.

Index

Note: Page numbers followed by *f* and *t* indicate figures and tables, respectively.

AAP. *See* American Academy of Pediatrics (AAP)
Abernethy, Virginia, 221
Absolute risk, 167
Absorption, 37
Abundance, 55
 definition of, 55
Accountability, for environmental health policy, 315
Acid aerosol particles, 33
Acoustic environment
 context of, and responses to, 83
 response to, factors affecting, 82–83
 restorative effects of, 83
 social inequalities and, 86–87
Acta Biotheoretica, 102
Active Living, 430–431
Active travel
 health benefits of, 118
 in low-carbon society, 120–121
 promotion of, 124
Adaptation
 of assemblage structure and function, 14
 to disturbances, 14
 genetic variability and, 45–46
 of individuals, 14
 of species, 14
 to threat of climate change, 191

Adipate(s), as endocrine disruptor, 260
Adriatic Sea, 137
 eutrophication in, 137
Adsorption, 37
Advocacy
 for action on climate change and health, 122–124
 definition of, 122
 for environmental health policy, 309
 health professionals and, 392
Africa, 217, 219–220
 health effects of climate change in, 113, 116
African Academy of Science, 219
Age, as vulnerability factor, 40–42
Agency for Toxic Substances and Disease Registry (ATSDR), workforce of, 480
Agenda 21, 237
Agent(s), definition of, 162
Agent Orange, 20
Agent transport (agents from sources), 36–37
Agriculture
 and biofuels, 420
 climate change and, 114, 116, 414–415
 energy use in, and costs, 412
 technological improvements, benefits of, 118

Index

AhR. *See* Aryl hydrocarbon receptors (AhR)
Air pollution, 156
   adverse health effects of, 160
   and children's cognitive functioning, 82
   and noise
      cardiovascular effects of, 80–82
      combined effects of, 80–82
   reduction of, benefits of, 118, 119
   significant federal statutes regarding, 327–329
   the "London Fog" (1952), 183–184
Air quality
   climate change and, 113
   in human exposure assessment, 34
   indoor, 34
   outdoor, 34
   processes affecting, 37
Air quality standards
   primary, 328–329
   secondary, 329
Albedo, 42
Albrecht, G., 190
Algae, 136
   and biofuel production, 422
   overabundance of, 136
   toxins, 136
   toxins from, 136
Algae beds and reefs
   primary productivity in, 8*t*
   transient time in, 8*t*
Algal blooms, 360
   climate change and, 113
   as ocean pollutants, 136
Alliance of Nurses for Healthy Environments, 399
Alpha diversity, 49
Alzheimer's disease, epigenetics and, 96
AMA. *See* American Medical Association (AMA)
American Academy of Pediatrics (AAP)
   and advocacy for environmental health, 398
   Green Book, 398
American Medical Association (AMA), and advocacy for environmental health, 400–401
American Public Health Association (APHA)
   core environmental health competencies, 485–486
   and nuclear freeze movement, 398
Amnesic shellfish poisoning (ASP), 360
Anchovies, 9
Andersen, H. R., 265
Androgen receptor(s) (AR)
   effects of persistent organic pollutants on, in Arctic people, biomarkers for, 268–272
   endocrine disruptors and, 259, 261–262
   transactivation assays, 262
Angel's trumpet, hallucinogenic seeds of, 104
Animal(s)
   number of species of, 47
   in toxicology studies, 286–287
Animal study(ies), in dose-response assessment, 289–290, 290*t*
   acute, 289, 290*t*
   chronic, 289, 290*t*
   date extrapolation from, 291–292, 292*f*
   specialized, 290, 290*t*
   subacute, 289, 290*t*
   subchronic, 289, 290*t*
Annapolis Protocol (1999), 144–145
Annoyance, noise and, 69–70, 72
Antarctica, 146
AOEC. *See* Association of Occupational and Environmental Clinics (AOEC)
Aoetoroa (New Zealand), 217
APHA. *See* American Public Health Association (APHA)
Apple seeds, cyanogenic glycosides in, 104
Aquaculture, 146
   invasive species from, 146

Aquatic Nuisance Species Task Force, 146
Arbitration, binding *versus* nonbinding, 241
Arctic fox, 19
Arctic Monitoring and Assessment Programme (AMAP), 268–273
Area of distribution (AD), 56
Argentina, 241–242
  and bilateral management of River Uruguay, 242
Aristotle, theory of epigenesis, 95
Arita, H., 56
Aromatase activity
  assay, 263
  endocrine disruptors and, 259, 261
Arsenic, 32
Aryl hydrocarbon receptors (AhR)
  effects of persistent organic pollutants on, in Arctic people, biomarkers for, 268–272
  endocrine disruptors and, 259, 261–262
  transactivation assays, 262
Asbestos, 20, 33
ASP. *See* Amnesic shellfish poisoning (ASP)
ASPH. *See* Association of Schools of Public Health (ASPH)
Associated specialists, in estimation of biodiversity, 47
Association of Environmental Health Academic Programs (AEHAP), 484–488
Association of Occupational and Environmental Clinics (AOEC), 489
Association of Schools of Public Health (ASPH)
  accredited schools, 487, 500–501
  competencies for environmental health, 484
  enrollment statistics, 482
Asthma
  climate change and, 113
  epigenetics and, 96

Astronomical observations, outdoor lighting and, 441–443, 442f, 449–450
Astronomical observatories, sites for, light pollution and, 443–446, 450–451
ATDSR. *See* Agency for Toxic Substances and Disease Registry (ATSDR)
Atomic bomb survivors, studies of, 160, 175
Attention, adverse effects of noise exposure on, 77
Attributable risk, 167
Ausimmune case-control study, 185
Australia
  indigenous population, sub-incision among, 222
  Ross River virus disease rates in, 191
  use of subincision among indigenous populations, 222
Autism, epigenetics and, 96
Automobile battery(ies), recycling of, 209
Automobile travel, safe environments for, 431–434, 434f
Autotrophs, 7, 9
Azacitidine, epigenetic therapy with, 105

Baby Boomers, aging, road safety enhancements for, 433–434
Bahia Province, 49
Balanced Menus, 403
Ballast water
  invasive species in, 146
  management of, 146
Baltic Sea, 137
  eutrophication in, 137
Bangladesh, 201
  municipal solid waste production in, 201, 202t
  sea-level rise in, health impacts of, 191–192
Barcelona Convention for the Protection of the Marine Environment and the Coastal Region of the Mediterranean, 234

Barro Colorado Island (Panama), 47, 55
  bird extinctions in, 55
Basel Convention (1989), 236
BBP. *See* Benzyl butyl phthalate (BBP)
Beaches and Environmental Assessment and Coastal Health (BEACH) Act, 330–331
Beetles
  number of species of, 47
  specializing in single-tree species, 47
Benzyl butyl phthalate (BBP), as endocrine disruptor, 266–267
Bhopal, India, pesticide disaster, 175
Bias(es), 164–165, 285
  definition of, 164
  types of, 164
Bicycle rental(s), 431, 432f
Bicycle safety
  incorporation into transportation planning, 430–431
  promotion of, environmental strategies for, 429–430
  threats to, 428, 430
Bierce, Ambrose, definition of "dawn," 164–165
Bioaccumulation, 37
Bioconcentration, 37
Biodegradation, 37
Biodiversity, 45, 48–50
  conservation of, reasons for, 61–63
  consumptive value, 62
  distribution of, 48–50
  estimation of, 46–48
  existence value, 62
  global patterns of, 48
  habitats supporting, 48
  loss of, 50–53, 63–64
    from habitat destruction and fragmentation, 51–52
    introduced species and, 52–53
    from overkill, 51
    See also Extinction(s)
  nonconsumptive value, 62

option value, 62
  productive value, 61
  See also Biodiversity, loss of
2010 Biodiversity Indicators Partnership (BIP), 63
Biodiversity loss, 218
  climate change and, 115–116
2010 Biodiversity Target, 62–63
Biofuels, 420–422
Biogeophysical systems, 183
Biological resources, value of, determination of, 61–62
Biological rules, in estimation of biodiversity, 47
Biomagnification, 37
Biomarkers, 30, 187, 260
  definition of, 260
  of effects, 260, 261f
  of exposure, 260, 261f
  in exposure assessment, 288
  ex vivo serum xenohormone activity, 267–268
  in human exposure assessment, 35
  for POP and metal exposures, 268
  for receptor effects, 268, 270–273
  of susceptibility, 260, 261f
  types of, 260
Biomass, 7–8, 10, 11
  definition of, 3t, 8
  noncrop, and biofuel production, 421–422
Biome, definition of, 3t
Biomonitoring, 260, 261f
Bipolar disorder, epigenetics and, 96
Bird(s)
  extinctions of, 57f
  marine, oil pollution and, 148
  new species of, 48
  number of species of, 47
Bis-ethyl-hexyladipate (DEHA), as endocrine disruptor, 266–267
Bisphenol A (BPA), 260
  as endocrine disruptor, 260

Index 513

in vitro bioassays of, 266–267
epigenetic effects, in mice, 103–104
Bis(2-ethylhexyl)phthalate. *See*
Diethylhexyl phthalate (DEHP)
Black-legged tick. *See Ixodes scapularis*
Black rats *(Rattus rattus),* effect on insular ecosystems, 53
Black Sea, eutrophication in, 137
Blood pressure, noise exposure and, 74–75
Blowback world, 228
Blumberg, Bruce, 102
Body fluid(s), in human exposure assessment, 35
Bolger, D. T., 53
Boreal forest, 6
biomass in, 8*t*
primary productivity in, 8*t*
transient time in, 8*t*
Borlaug, Norman, 224–225
Boron, 32
Botero, Giovanni, 222
Boundary Waters Treaty (1909 [BWT]), 233–234, 241
Bowen, E., 223
BPA. *See* Bisphenol A (BPA)
Breastfeeding, 221–222
effects on fertility, 221–222
Brown pelican, DDT accumulation in, 139
Brundtland Commission, 115
Built environment
design, approaches to, 430–431
and injury prevention, 427
and traffic injury prevention, 429
Burden of disease (BoD), of environmental noise, 78–80, 79*f*, 80*f*
Bush, George W., 342, 413, 420
BWT. *See* Boundary Waters Treaty (1909 [BWT])
Bygren, Lars Olov, 102–103

CA. *See* Concentration addition (CA)
CAA. *See* Clean Air Act (CAA)
Cadmium, 32

as ocean pollutant, 138
rice contaminated with, 160
Caldwell, J. C., 222
CalFERT. *See* California Food Emergency Response Team (CalFERT)
California
climate goals formulated by, 466
environmental enforcement in, roles of federal, state and local government in, 348, 350*t*–351*t*
Global Warming Solutions Act, 466
solar power production in, 422
California Food Emergency Response Team (CalFERT), 357
response to *Salmonella* Rissen outbreak, *case study,* 357–358
California Medical Association (CMA), and advocacy for environmental health, 399–400
Canada
transboundary pollution from, 241
and United States, treaties, 233–234
Cancer
diet and, 118
epigenetic pathways and, 100–101
linked with environmental factors, 159–160, 160*t*
Cancer clusters, analysis of, 162
Cancer risk analysis, 297–298
CAPs. *See* Criteria Air Pollutants (CAPs)
Carbon
personal allowances, 120
See also Low-carbon
Carbon counting, 413
Carbon dioxide (CO_2), 32, 141–142, 143*f*
atmospheric concentrations, historical perspective on, 142, 143*f*
emissions
production-side accounting for, 416
See also Carbon footprint; Greenhouse gas(es)
environmental, increased, 113
from fossil fuels, 414

Index

Carbon footprint, 456–458
  average national per-person, 457
  reduction of, individual actions for, 458–459
  See also Greenhouse gas(es)
Carbon monoxide, 32
  See also Carbon monoxide exposure
Carbon monoxide exposure, dose–effect relationship, 158
Carbon monoxide poisoning. *See* Carbon monoxide exposure
Carbon outputs, reduction of, 455
Carbon tax, 416
Carcinogen(s), dose–response curves for, 290–291, 291*f*
Carcinogenicity
  linear multistage model for, 293–294, 294*f*
  multihit model for, 293
  one-hit model for, 292–293
  weight of evidence categories for, 287, 287*t*, 292, 297
Cardiovascular disease
  climate change and, 113
  combined effects of noise and air pollution and, 80–82
  diet and, 118
  epigenetics and, 96
  noise exposure and, 69, 73, 75
Carnivore(s), 9, 10
  definition of, 3*t*
  energy efficiency of, 11
Carrying capacity, 12, 225
  human, 225–227
Carson, Rachel, *Silent Spring*, 391
Carter, Jimmy, 225, 424
Cascading effects, 13
  definition of, 3*t*
Case, T. J., 53
Case-comparison study(ies). *See* Case-control study(ies)
Case-control study, 284, 285
  in dose-response assessment, 289, 290*t*
  and risk, 168
Case-control study(ies), 173–175
  advantages and disadvantages of, 174
Case-referent study(ies). *See* Case-control study(ies)
Case reports, 171
  in dose-response assessment, 289, 290*t*
Case series, 171
Case Studies in Environmental Medicine, 489
Case study(ies), 171
Causation
  and epidemiology, 162
  establishing, 163–164
CBD. *See* United Nations Convention on Biological Diversity (CBD)
CCT. *See* Correlated color temperature (CCT)
Cd. *See* Cadmium
CECs. *See* Contaminants of emerging concern (CECs)
Cellulose, and biofuel production, 421–422
Centers for Disease Control and Prevention (CDC)
  core environmental health competencies, 485–486
  Environmental Health Leadership Institute, 489
  National Center for Health Statistics (NCHS), 36
  workforce of, 480
CEPH. *See* Council of Education in Public Health (CEPH)
CERCLA. *See* Comprehensive Environmental Response, Compensation, and Liability Act (1980) (CERCLA)
CFCs. *See* Chlorofluorocarbons (CFCs)
CFR. *See* Code of Federal Regulations (CFR)
Cheetah *(Acinonyx jubatus)*, lack of genetic variability in, 55

Chemical(s)
  dirty dozen, 140
  emissions of, sources of, 140
  existing, 341
  industrial, as ocean pollutants, 138
  mixtures, analyses of, 263
  monitoring, in environment, 159
  new, 341
  nonorganic, as exposure agents, 32–33
  as ocean pollutants, 138
  organic, as exposure agents, 32
  safety criteria for, 159
  Toxic Substances Control Act and, 340–341
  See also Endocrine-disrupting chemicals (EDCs)
Chemical contamination, effects on human health, 17*t*, 20
Chemical risk assessment, 282
Cheney, Dick, 461
Chernobyl nuclear reactor, 394
Chesapeake Bay, eutrophication in, 137
Children
  cognitive functioning of, air pollution and, 82
  noise exposure
    annoyance response to, 72
    cardiovascular effects of, 75
    cognitive and academic effects of, 70, 76–77
  susceptibility/vulnerability of, 40–41
China
  food imports from, inspection of, 234
  municipal solid waste production in, 201, 202*t*
Chivian, E., et al., *Critical Condition*, 396
Chlordane, 140
Chlorination, of water, 329
Chlorofluorocarbons (CFCs), 141
4-Chloro-3-methyl phenol (CMP), as endocrine disruptor, 266
Chlorpyrifos, as endocrine disruptor, 265–266

Cholera, 159, 351
Chromium, 32
Chronic daily intake, 294, 297–298
Cinzano, Pierantonio, 445–446
City Bike Systems, 431
Clean Air Act (CAA), 236, 282, 327–329, 391
  compliance with, monitoring for, 30
Clean Water Act (CWA), 140, 145, 150, 330–331, 336, 391
  compliance with, monitoring for, 30
Climate, health effects of, 190
Climate change, 141–142, 143*f*, 414
  American Medical Association advocacy about, 400
  coordinated approach to, 468–469
  deaths caused by, 112–113, 457
  diseases caused by, 112–113
  epidemiology of, 189–192
  global environmental threats and, 114–116
  Health Care Without Harm and, 402–403
  health effects of, 111–114, 190–191
  in lower-income countries, 113–114
  indicators and effects of, monitoring, 31
  individual actions affecting, 458–459
  injuries caused by, 112–113
  management of, treaty regime for, 245
  Physicians for Social Responsibility and, 395–397
  taking action against, 119–127
Clinical trial(s), 176, 284, 286
  in dose-response assessment, 289, 290*t*
Cluster investigations, in environmental epidemiology, 162
CMA. *See* California Medical Association (CMA)
CME. *See* Continuing medical education (CME)
CMP. *See* 4-Chloro-3-methyl phenol (CMP)
CO. *See* Carbon monoxide

Coal, sulfur in, 413
Coal Ash: Toxic Threat to Our Health and Environment (Gottleib et al.), 397
Coastal Zone Management Act, 150
Cod, 217
Code of Federal Regulations (CFR), 324
Cohort study, 284–285
 in dose-response assessment, 289, 290t
 and risk, 168
Cohort study(ies), 175
Coitus interruptus, 221
Collaborative on Health and the Environment (CHE), 403
Colorectal cancer
 diet and, 118
 epigenetic pathways and, 100
Combined sewer overflows (CSOs), 144
Command and control, 326, 416
Communication. *See* Risk communication
Community
 definition of, 3t
 See also Sustainable community(ies)
Community trial, 176
Compensation requirements, 325
Competency(ies)
 definition of, 484
 environmental health, 484–486
Competition
 definition of, 3t, 12
 effects on competing organisms, 12–13
 and extinctions, 57–58
Complexity science, epidemiology and, 192
Composting
 definition of, 209
 mechanism of action of, 210
 organisms important for, 210
Comprehensive Environmental Response, Compensation, and Liability Act (1980) (CERCLA), 316, 337–338
Concentration addition (CA), 263
Confidence interval, 168
Confidence limit(s), 162

Confirmatory bias, 378
Confounding, 165–166, 188
 in study of climate and health, 191
Confounding factors, potential, 40
Consumer Product Safety Act (1972), 318, 341–342
Consumer Product Safety Commission (CPSC), 317, 341–342
Contaminants of emerging concern (CECs), 138–139
Continuing medical education (CME), in environmental health, 489
Contraception, historical perspective on, 221–222
Copper, 32
 as ocean pollutant, 138
Coral reef bleaching, 142
Corn, prices and production of, biofuels and, 421
Correlated color temperature (CCT), 449–450
Cost-benefit analysis, for environmental health policy, 315
Council of Education in Public Health (CEPH), 487
 accredited MPH programs with degrees (or concentrations) in environmental health, 502
 accredited schools, 487, 500–501
Covariates, 40
CpG islands, 99
 hypermethylation, in cancer, 100–101
Crete, composting program in, 210
Crisis communication, 376–383, 379f–380f
 questions commonly asked by journalists and, 384f–385f
 recommendations for, 300, 300t
Criteria Air Pollutants (CAPs), 328
Critical Condition (Chivian et al.), 396
Critical thinking, 163–164
Crop failures, climate change and, 114
Cross-sectional study(ies), 171–172, 284, 285–286

in dose-response assessment, 289, 290t
CSOs. *See* Combined sewer overflows (CSOs)
Customary law, 240–241
CWA. *See* Clean Water Act (CWA)
Cycling, health benefits of, 118
Cyclones, 112

DALY. *See* Disability adjusted life years (DALY)
Darfur, Sudan, 114
Dark Skies Awareness, 452
Darwin, Charles, 221
Datura plants, hallucinogenic seeds of, 104
dBA. *See* Decibels, A-weighted (dBA)
DDT, 32, 140
  as endocrine disruptor, 265
  as ocean pollutant, 138
  in sediments, and marine pollution, 139
Dead Reckoning (Physicians for Social Responsibility), 394
Death(s)
  caused by climate change, 112–113, 457
  from environmental risks (2004), 156, 157f
  linked to air pollution, 160
  risk of, for common activities, 283t
Death by Degrees (Physicians for Social Responsibility), 397
Debris
  marine
    reduction/prevention of, 148
    sources of, 147–148
    types of, 148
  as ocean pollutant, 135t, 146–148, 147f
Decibels, A-weighted (dBA), 33
Decomposers, 10
  definition of, 3t
Deepwater Horizon oil spill, 149
Deer, overabundance of, 19
Deforestation, and extinctions, 59–61
DEHA. *See* Bis-ethyl-hexyladipate (DEHA)

Deltamethrin, as endocrine disruptor, 265
Dengue fever, climate change and, 114
Denver, CO, bicycle rental program, 431, 432f
Department of Agriculture
  Division of Chemistry, 333
  meat safety authority, 335–336
Department of Health and Human Services, as environmental health policy-maker, 307
Department of Transportation, U. S., planning process, incorporation of bicycle and pedestrian safety into, 431
Desert scrub
  biomass in, 8t
  primary productivity in, 8t
  transient time in, 8t
Detritivore(s), definition of, 3t
Diabetes mellitus, type 2
  diet and, 118
  epigenetics and, 96
Diamond, J. M., 53
Diarrheal disease
  climate change and, 113–114
  linked to environmental causes, 160t
2,4-Dichlorophenol (2,4-DCP), as endocrine disruptor, 266
Dichlorophenyltrichloroethane. *See* DDT
DIDP. *See* Di-isodecyl phthalate (DIDP)
Dieldrin, as endocrine disruptor, 265
Diet
  epigenetic effects of, 102–103
  and health, 118
Diethylhexyl phthalate (DEHP)
  carcinogenicity in rodents, 287
  as endocrine disruptor, 266–267
Differential misclassification, 164
Di-isodecyl phthalate (DIDP), as endocrine disruptor, 266–267
Di-isononyl phthalate (DINP), as endocrine disruptor, 266–267
Di-n-octyl phthalate (DnOP), as endocrine disruptor, 266–267

518 Index

DINP. *See* Di-isononyl phthalate (DINP)
Dioxin(s), 20, 32
  in medical waste incineration, 401
  as ocean pollutants, 138
Disability adjusted life years (DALY)
  climate change and, 113
  from environmental risks (2004), 156, 157*f*
  noise exposure and, 78, 80*f*
Disease(s)
  adult, epigenetics and, 96
  causal web for, 163
  causation of, epidemiological concepts of, 162–163
  ecosystem disruption and, 20
  epigenetic pathways and, 100–104
  fecal waste and, 143–144
  frequency, measures of, 166–167
  incidence of, 166–167
  linked with environmental factors, 160, 160*t*
  prevalence of, 166–167
  transfer among ecosystems, 18
DNA (deoxyribonucleic acid)
  hypermethylation, in cancer, 100
  hypomethylation, in cancer, 100
  methylation
    in cancer, 100–101
    and development, 101
    and disease, 101
    as epigenetic pathway, 99–100
Doll, Richard, 158
Dominican Republic, 226
Domoic acid, *case study,* 360–361
Dose, 159
  biologically active, 39
  internal, 39
  units for, 289
  See also Reference dose (RfD)
Dose–effect relationship, 158
Dose response, 27, 282, 283*t*
Dose–response assessment, 289–297
  animal studies in, 289–290, 290*t*
  concepts used for, 293–297
Dose–response curves, 290–291, 291*f*
  based on extrapolation, 291–292, 292*f*
  for carcinogens *versus* noncarcinogens, 290–291, 291*f*
Dose–response relationship, 158, 185, 289
Double-blind study, 176
Draper, William, 224
Drinking water
  bottled, regulations governing, 334–335
  environmental enforcement regarding, in California, roles of federal, state and local government in, 350*t,* 351–355
  significant federal statutes regarding, 351–352
  uncovered reservoirs, *case study,* 353
Drinking water standards, 331
Drought, climate change and, 113–114, 414–415
Dust, in human exposure assessment, 35

Earth Day, 391
Earth Summit (1992), 235, 237
East Africa, population growth in, 223
Eastern Mediterranean, health effects of climate change in, 113
Eco-cultural attributes, 18, 19
  definition of, 3*t*
Eco-cultural dependency webs, definition of, 3*t*
EcoIsland, 462
Ecological evaluations, definition of, 3*t*
Ecologically extinct species, 50
Ecological risk, definition of, 3*t*
Ecological study(ies), 172–173
Ecologic fallacy, 186
Ecology
  definition of, 2, 3*t,* 181
  terminology used in, 2, 3*t*–4*t*
Ecosystem(s)
  definition of, 2, 3*t,* 46
  energy flow in, 10–11, 10*f*
  functions of, 6–7

healthy, characteristics of, 2
interconnectedness of, 21
marine. *See* Marine ecosystem(s)
protective value of, 19
recovery from disturbance, 14
structure of, 5–6
sustainability of, 21
Ecosystem degradation
　effects on goods and services, 18
　and human health, 15–21, 15*t*–17*t*
　in response to stressors, 14
Ecosystem diversity, 46
Ecosystem health, defining, 2–5
Ecosystem integrity, definition of, 5
Ecosystem services, 215
Ecosystem stressor(s), 1, 13–15
　anthropogenic, 14, 18
　biological, 13–14
　chemical/radiological, 14
　ecosystem vulnerability to, 13–15
　natural, 13–14, 18
　physical, 14
EDCs. *See* Endocrine-disrupting chemicals (EDCs)
Education
　and encouragement of exemplary behaviors, 462
　See also Environmental health workforce, education and training
Egg Products Inspection Act, 336
EHAC. *See* National Environmental Health Science and Protection Accreditation Council (EHAC)
Ehrlich, Paul, *The Population Bomb,* 224
Elderly
　road safety enhancements for, 433–434
　susceptibility/vulnerability of, 41–42
Electromagnetic field(s), 33
"Elephant's Child, The" (Kipling), 155
Elvidge, Christopher, 445–446
Emergency Planning and Community Right-to-Know Act of 1986, 338
Emission standards, 326

Endangered Species Act, 150
Endocrine-disrupting chemicals (EDCs), 398
　adverse health effects of, 259
　biomarkers for, 260, 261*f*
　biomonitoring for, 260, 261*f*
　definition of, 259
　in vitro bioassays of, 264
　man-made, 260
　mechanism of action of, 259
　natural, 260
Endocrine disruptors, 20
Endocrine Society, The, and advocacy for environmental health, 398
Endosulfan, as endocrine disruptor, 265
Energy
　alternative sources of, 409–410
　future paradigm for, 411
　production of, environmental accounting for, 415–416
　usage
　　in home, regulation of, 460
　　as watts per person-year, 460
　See also Renewable energy
Energy cycling, in ecosystems, 6–7
Energy efficiency
　actions for, in U.S. government, 464–465
　as business strategy, 467
Energy flow, in food webs, 10–11, 10*f*
Energy Independence and Security Act (2007), 464
ENERGY STAR program, 465
Energy transfer
　among trophic levels, 9
　definition of, 3*t*
　in ecosystems, 7
Energy transition(s), 409–410
Energy use, household, reduction, health benefits of, 117–118
Engels, F., 185
Environment, definition of, 181
Environmental accounting, 413, 415–416

Environmental cleanup site(s), questions commonly asked at, 386f–389f
Environmental enforcement
 in California, roles of federal, state and local government in, 348, 350t–351t
 definition of, 347
 goal of, 347
 and public health, 348–351, 349t
 scope of, 347
Environmental epidemiology, 182–183
 advances in (future directions for), 189–192
 challenges of, 186–187, 189–192
 and complexity science, 192
 data sources for, 187–188, 187f
 definition of, 156
 functions of, 156
 in high- *versus* low-income countries, 183–184
 historical perspective on, 159–160
 levels of observation and effect in, 184–186
 outcome measures used in, 161
 scope of, 156
 statistics and statistical modeling used in, 161–162
 study endpoints used in, 161
 timescales and, 188–189, 193
 toxicological concepts relevant to, 158–159
Environmental health, 181
 advocacy for, health professional associations and, 397–401
 definition of, 2, 475
 and ecosystem integrity, differentiation of, 5
 EHAC-accredited graduate programs in, 500
 EHAC-accredited undergraduate programs in, 499
 functions of, 477
 health professional education in, 489–490
 and infectious disease, 184
 mission of, 477
 need for, 483
 Physicians for Social Responsibility and, 395–397
 and politics, 306
 professionals in, 479
 risk communication and, 368
 science of, 181–182
 scope of services, 483
 significant federal statutes regarding, 326–344, 391
 ten essential services of, 475–476
Environmental health policy
 advocacy for, 309
 definition of, 306
 domestic economic policies and, 310f, 313
 experts' input and, 310f, 313
 factors affecting, 310–314, 310f
 form of governance and, 310f, 314
 government policy-makers for, 307–308
 Internet and, 310f, 311–312
 news media and, 310f, 311–312
 policy-makers for, 307–308
 policy-making process for, 309
 PACM model, 319–321, 319f
 private-sector policy-makers for, 308
 public concerns and, 310–311, 310f
 public expectations about, 314–318
 social network communications and, 310f, 311–312
 special interest groups and, 310f, 311
 trade policies and, 310f, 313–314
Environmental health professionals, 479
Environmental health workforce, 475
 advances in (future directions for), 491–494
 aging of, 476, 483
 certification, 487–488
 competencies required, 482–486
 composition of, 477–479
 credentialing, 487–488

Index 521

demographics of, 478–479
development of, strategic elements of, 476f, 477
education and training
  continuing education for, 487–488
  at graduate level, 487, 500
  integrated system for, 486
  lack of, 483
  at leadership level, 488–489
  need for, 483
  on-the-job, 487–488
  programs for, enrollment statistics, 482
  at undergraduate level, 486–487, 499
enumeration of, 477–479
at federal level, 480
issues involving, 482–483
leadership competencies required, 486
at local level, 480–481
recommendations for, 491–494
scope of work, 481–482
at state level, 480
sustaining, 490–491
at tribal level, 480
work settings of, 479–481
Environmental justice, and environmental health policy, 315–316
Environmental law, international, 233
  See also specific law
Environmental media
  in indoor microenvironments, factors affecting, 42–43
  in outdoor microenvironments, factors affecting, 42–43
  relevant to human exposure assessment, 34–36
Environmental monitoring, 29–30
Environmental Noise Directive (END), 71, 72
Environmental Protection Agency, U.S. (EPA), 318, 324–325, 391
  Carcinogenic Risk Assessment Guidelines, 282
  chemical regulatory authority of, 341
  classification of CO_2 as pollutant, 142
  creation of, 199
  definition of biomarkers, 260
  and drinking water safety, 352
  as environmental health policy-maker, 307
  greenhouse gas initiatives, 465
  hierarchy for management of municipal solid waste, 204
  and pesticide use/sales, 339–340
  risk assessment by, 282
  water quality criteria for nutrient pollution, 138
  weight of evidence categories for carcinogenicity, 287, 287t, 292, 297
  workforce of, 480
EPA. *See* Environmental Protection Agency, U.S. (EPA)
Epidemiological study(ies)
  causal processes and, 188
  data sources for, 187, 187f
  design of, 168–169, 170–176, 284–286
  determination of exposure status for, 169–170
  in dose-response assessment, 289, 290t
  methods of, 168–170
  sampling frames for, 169
Epidemiology, 182–183
  and complexity science, 192
  definition of, 155–156, 184
  historical perspective on, 184–185
  models of causation used in, 162
  observational *versus* experimental, 161
  See also Environmental epidemiology
Epigenetic pathway(s), 98
  and disease, 100–104
  DNA methylation as, 99–100
  histone modification as, 98–99
Epigenetics
  of adult diseases, 96
  definition of, 95–97
  historical perspective on, 95–96

Epigenetic therapies, 104–106
Epigenome, 97
Equatorial Guinea, bushmeat harvests, and species declines, 51
Ergonomics, 34
Erwin, T. L., calculation of beetle species, 47
Espoo Convention on Environmental Impact Assessment in a Transboundary Context, 242
Estrogen receptor(s) (ER)
  effects of persistent organic pollutants on, in Arctic people, biomarkers for, 268–272
  endocrine disruptors and, 259, 261–262
  transactivation assays, 262
Estuaries
  primary productivity in, 8t
  transient time in, 8t
Ethanol, as biofuel, 420–422
Ethical Consumer, 460
Ethical consumption, 458–461
EU. *See* European Union (EU)
Europe
  noise exposure from road traffic in, 71–72, 71f
  noise mapping in, 71–72
European Union (EU)
  and environmental health policy, 313, 316
  as environmental health policy-maker, 308
Eutrophication, 135t, 136–137, 143
Evil quartet, of species decline, 51
Experimental study(ies), 176
Exposure
  to climate change, 190
  definition of, 28
  direct measurement of, 170
  pathways of, 169–170
  routes of, 288
  sources of, 169–170
  total personal, contributions to, 28, 29f
  types of, 158–159
  See also Human exposure
Exposure agent(s), 27, 28, 31–34
  biological, 31–32
  carcinogenic, 31
  chemical, 32
  noncarcinogenic, 31
  physical, 33
  radiological, 33–34
  sources of, 31
Exposure assessment, 27–28, 282, 283t, 287–289
  biomarkers in, 288
  challenges of, 156
  data sources for, 187, 187f
  direct measures for, 288
  predictive measures for, 288
  principles of, 170
  reconstructive measures for, 288
Exposure–effect relationship, 185
Exposure measurement(s), 28–30
  direct, under controlled conditions, 29
  direct field, 29–30
  laboratory methods for, 29
  models for, 28–29
  qualitative, 30
  quantitative, 30
  uses of, 30–31
Exposure pathway(s)
  assessment of, 36–37
  biological processes and, 37
  chemical processes and, 37
  definition of, 36
  determinants of, 36
  physical processes and, 37
Exposure–response relationship, 158
Exposure science
  direct field measurements in, 29–30
  direct measurements under controlled conditions in, 29
  laboratory experiments in, 29
  measurements in, 28–30
  models used in, 28–29

Exposure status, determination of, for epidemiological studies, 169–170
Extinction(s)
  ancient, 50
  chains of, 51, 53
  competition and, 57–58
  definition of, 50
  deterministic relationships and, 50
  evil quartet and, 51
  future, 58–59, 59f
  from habitat destruction and fragmentation, 51–52
  intentional, 217
  introduced species and, 51–53
  local, 50
  mass, 50
  of megafauna, 217–218
  overkill and, 51, 217
  predation and, 57–58
  present-day rates of, 59–61, 115–116
  rates of, historical trends in, 58–59, 59f
  recent, 56–59, 57f
  species committed to, 60–61
  species threatened with, as of 2010, 63, 116
  stochastic processes and, 50–51
  vulnerability of species to, 54–55
  in wild, 50
Exxon Valdez oil spill, 149

Family Smoking Prevention and Tobacco Control Act (FSPTCA), 333
Farr, W., 185
FCTC. *See* Framework Convention on Tobacco Control (FCTC)
Fecal waste
  coastal ecosystems affected by, 144, 145f
  as ocean pollutant, 135t, 143–145
  sources of, 144
Federal government (U.S.)
  regulations, 325–326
  regulatory programs, 324–325
  rulemaking by, 325
  significant environmental health statutes, 326–344
  standards, 325–326
Federal Insecticide, Fungicide, and Rodenticide Act (FIFRA), 339–340
Federalism, and environmental health policy, 316
Federal Meat Inspection Act (FMIA), 335–336
Federal Register, 324, 325
Fenarimol, as endocrine disruptor, 265
Fertility
  human, factors affecting, 221–222
  male, environmental toxins and, 101
  persistent organic pollutants and, comparison of Inuit and Europeans, 272–273
Fertilizer(s), inorganic, environmental effects of, 136–137
FIFRA. *See* Federal Insecticide, Fungicide, and Rodenticide Act (FIFRA)
Fine particles, 33
Fish
  consumption of, 116
  guidelines/advisories for, 139
  mercury-contaminated, consumption of, 160
  new species of, 48
  supplies of, 116
Fishery(ies), critical species and, 9
Flame retardants, brominated, as endocrine disruptors, 260
Flannery, Tim, 217
Floods, 112
  climate change and, 414–415
FMIA. *See* Federal Meat Inspection Act (FMIA)
Follow-up study(ies). *See* Cohort study(ies)
Food
  and health, 118
  in human exposure assessment, 35

prices, energy costs and, 412
selection of, for low carbon footprint, 460–461
shortages, climate change and, 113–114
Food, Drug, and Cosmetic Act, 332–335, 339, 356
Food and Agriculture Organization (FAO), International Code of Conduct on the Distribution and Use of Pesticides, 238
Food and Drug Administration (FDA), 324–325, 355
  new drug procedures, 335
  public health authority of, 333–334
Food chain, 9
Food labeling, regulations governing, 334
Food poisoning, climate change and, 113
Food production, greenhouse gas emissions in, reduction of, 118
Food Quality Protection Act of 1996 (FQPA), 340
Food safety
  American Medical Association advocacy about, 400–401
  environmental enforcement regarding, in California, roles of federal, state and local government in, 350t, 355–359
  historical perspective on, 355–356
  significant federal statutes regarding, 332–336, 356
Food Safety Modernization Act, 334
Food web(s), 9
  bottom-up effects on, 13
  critical species and, 9
  definition of, 4t
  dynamics, 6–7
  energy flow in, 10–11, 10f
  top-down effects on, 13
Forest(s)
  boreal, 6
  temperate, 6
  tropical, 6

Fossil fuel, 410–411
  combustion, 218
  carbon dioxide emissions from, 142, 457
  costs of, 412–415
  environmental impact of, 412–415
Fourth National Report on Human Exposure to Environmental Chemicals, 159
FQPA. *See* Food Quality Protection Act of 1996 (FQPA)
Fragmentation
  definition of, 4t
  effects on human health, 16t
  and species loss, 51–52, 54
Framework Convention on Tobacco Control (FCTC), 250–251
Franklin, Benjamin, 222
Freiburg, Germany, waste management program in, 209
FSPTCA. *See* Family Smoking Prevention and Tobacco Control Act (FSPTCA)
Fuel poverty, 120
Fukushima Daiichi reactor, 419
Funding bias, 165

Gamma diversity, 49
Garstang, Roy, 445–446
GATT. *See* General Agreement on Tariff and Trade (GATT)
Gene(s), 45–46
Gene–environment interactions, 39
Gene expression, 95–97
  histone modification and, 98–99
General Agreement on Tariff and Trade (GATT), 245–246
Generations at Risk (Schettler et al.), 396
Gene silencing, 97, 99–100
Genetic diversity, 45–46
Genome, 97
Geographic information systems (GIS), 173
Germany, recycling programs in, 209

GIS. *See* Geographic information systems (GIS)
Giuliani, Rudolf, risk communication skills of, 381, 383
Global destabilization (warming), effects on human health, 17*t*
Global environmental threats, 183
Global Humanitarian Forum (2009), report on health effects of climate change, 112
Global Invasive Species Program, 146
Global warming, 455
  decreasing, individual actions for, 458–459
  See also Climate change
GLOBE at Night, 452
GLWQA. *See* Great Lakes Water Quality Agreement (1978) (GLWQA)
Gold, mining, and mercury contamination, 141
GoodGuide, 460
Gordon, Larry, 479
Government
  and environmental accounting for energy production and use, 415–416
  role in reducing carbon footprint, 463
Government Performance and Results Act (GPRA), 315
GPRA. *See* Government Performance and Results Act (GPRA)
Grassland
  biomass in, 8, 8*t*
  primary productivity in, 8, 8*t*
  transient time in, 8, 8*t*
Great Britain. *See* United Kingdom
Great Lakes Water Quality Agreement (1978) (GLWQA), 234
Greece, municipal solid waste production in, 201, 202*t*
Green alliances, 467
Green Book (American Academy of Pediatrics), 398

Green Chemistry, 141
Green consumerism, 424
Greenhouse gas(es), 455–456
  and climate change, 414, 456–457
  from fossil fuels, 414, 457
  as ocean pollutants, 135*t,* 141–142
  reduction of
    benefits of, 117, 118–119
    commercial enterprise initiatives for, 466–468
    corporate initiatives for, 466–468
    global actions for, 463–464
    governmental role in, 463
    national actions for, in United States, 464–465
    state and local government initiatives for, 465–466
  sources of, 117, 142, 457–458
  See also Carbon footprint
Greenhouse Gas Protocol Initiative (GHG Protocol), 467–468
Greenland, human POP markers for exposure in, 268, 269*f*
Green marketing, 467
Green Revolution, 224–225
Green spaces, beneficial effects of, 116–117
Griffiths, J., 459, 460
Guam, introduced (alien) species in, and species loss, 52
Guinea pig(s), in toxicology studies, 286
Gulf of Mexico
  Deepwater Horizon oil spill in, 149
  eutrophication in, 137

Habitat
  definition of, 4*t*
  loss of, and extinctions, 51–52, 59–61
Haiti, 226–227
HAPs. *See* Hazardous Air Pollutants (HAPs)
Hartz, Gary, 480
Hawaiian honey creepers, 53

526 Index

Hazard(s), 27–28
  direct, 183
  forms of, 158–159
  indirect, 183
Hazard identification, 27, 282, 283*t*, 284–287
  animal data for, 286
Hazard index, 298
Hazardous Air Pollutants (HAPs), 328
Hazardous and Solid Waste Amendments (HSWA), 200
Hazardous Substances Emergency Events Surveillance (HSEES), 238, 241
Hazardous waste site(s), questions commonly asked at, 386*f*–389*f*
Hazard quotient, 298
HCWH. *See* Health Care Without Harm (HCWH)
Health
  factors affecting, 114, 115*f*
  green space and, 116–117
  individual responsibility for, 126
  in low-carbon society, 120–121
  right to, 237
Healthcare systems
  ecological footprint of, 459
  low-carbon, 125–126, 459
Health Care Without Harm (HCWH), 401–403
Health co-benefits, in action on climate change, 123
Health impact assessment (HIA), 78
Health professional(s)
  as advocates, 392
  and environmental health movement, 392
Health professional associations, and advocacy for environmental health, 397–401
Health promoters, 299
Health risk assessment
  components of, 282, 283*t*, 284–298
  definition of, 281–282

historical perspective on, 282
Healthy Food in Health Care, 403
Hearing loss, noise exposure and, 69–70
Heat, 10*f*, 11
Heat islands, 116
  urban, 43
Heat waves, 112, 191
Heavy metals
  adverse health effects of, 101–102
  as exposure agents, 32
Heptachlor, 140
Herbivore(s), 9, 10
  definition of, 4*t*
  energy efficiency of, 11
Heterostyly, 53
Hippocrates, 329
  on water, 155
Histone(s), definition of, 98
Histone code, 98
Histone modification
  in cancer, 101
  as epigenetic pathway, 98–99
Hitler, Adolf, 223
Home energy use, regulation of, 460
Hong Kong
  living standard in, factors affecting, 227
  municipal solid waste production in, 201, 202*t*
Hopkins, Rob, 125
Horseshoe crabs, 9
Host, definition of, 162
HSEES. *See* Hazardous Substances Emergency Events Surveillance (HSEES)
HSWA. *See* Hazardous and Solid Waste Amendments (HSWA)
Human ecology, 181
Human exposure
  activity patterns and, 38–39
  acute, 38
  biological processes and, 37
  chemical processes and, 37

chronic, 38
continuous, 38
and dose, 29f
episodic, 38
estimated, for defined geographic area/region, 29–30
inter-individual differences and, 39
at intervals (intermittent), 38
intra individual differences and, 39
measurement of, 28–30
monitoring, 29–30
physical processes and, 37
qualitative measures of, 30
quantitative measures of, 30
routes for, 37–38
space and, 39
susceptibility factors and, 40–42
time and, 38
vulnerability factors and, 40–42
Human exposure assessment, environmental media relevant to, 34–36
Hung Liang-Chi, 217
Hurricane Katrina, 112–113
Hydromodification, 134
Hydrophobic agents, 37
Hydrothermal vent communities, new species in, 48
Hypertension
combined effects of noise and air pollution and, 81
noise and, 69–70
noise exposure and, 75

IHRs. See International Health Regulations (IHRs)
IHS. See Indian Health Service (IHS)
IJC. See International Joint Commission (IJC)
Immigrant(s). See Migration
Incidence, definition of, 166–167
Incidence density, 167
Incidence study(ies). See Cohort study(ies)

India, municipal solid waste production in, 201, 202t
Indian Health Service (IHS), 480
Environmental Health Support Center, 488
Office of Environmental Health and Engineering, 488
Induction period, 159
Industrial Revolution, ecological/ecosystem effects of, 1
Infant(s), psychosocial stress on, epigenetic effects of, 104
Infectious disease, and environmental health, 184
Infrasound, definition of, 77
In Harm's Way (Schettler et al.), 396
Injury prevention
built environment and, 427
public health and, 427–428
Insect(s), number of species of, 47
Insecticide(s), organochlorine. See DDT
Integrated personal exposure (IPE), 170
Integrated Risk Information System (IRIS), 282, 294, 297
Intergovernmental Panel on Climate Change, 414
report on health effects of climate change, 111–112
International Coastal Cleanup, 148
International Convention for the Prevention of Pollution from Ships (MARPOL), 150
International Convention on Oil Pollution Preparedness, Response and Cooperation, 149
International Court of Justice, and environmental dispute resolution, 241–242
International Dark-Sky Association, 451
International Health Regulations (IHRs), 248–250
International Human Epigenome Consortium (IHEC), 104

528 Index

International Joint Commission (IJC), 241
International law
  general (customary), 240–241
  treaties as. *See* Treaty(ies)
International Physicians for the Prevention of Nuclear War (IPPNW), 392, 394–395
International Road Assessment Program (iRAP), 433, 434*f*
International trade, 234, 245–247
International Year of Astronomy (2009), 452
Interstate Quarantine Act of 1893, 351–352
Interstate Shellfish Sanitation Conference, 359
Inuendo project, 272–273
Invasive species
  aquatic, transport of, 146
  control of, 146
  effects on human health, 17*t*
  as ocean pollutants, 135*t*, 145–146
Ionization, 37
Ionizing radiation, 33
IPCC. *See* Intergovernmental Panel on Climate Change
IPE. *See* Integrated personal exposure (IPE)
IPPNW. *See* International Physicians for the Prevention of Nuclear War (IPPNW)
Iprodion, as endocrine disruptor, 265–266
iRAP. *See* International Road Assessment Program (iRAP)
Ireland, potato famine in, 226
IRIS. *See* Integrated Risk Information System (IRIS)
Irruption, of human populations, 220
Ischemic heart disease
  diet and, 118
  noise exposure and, 70, 74–75
Islands, introduced (alien) species on, and species loss, 52–53

Itai-Itai disease, 160
Ixodes scapularis, host-seeking behavior, seasonal shifts in, 30

Japan, recycling programs in, 209
Jevons, W. S., 222, 461
Jevons paradox, 216, 461
Johnson, Lyndon B., 224, 225

Karr, J. R., 55
Kennedy, John F., 225
Keynes, John Maynard, 223
Kipling, Rudyard, "The Elephant's Child," 155
Kyoto Protocol, 142, 236, 244–245, 463–464

Lactation, effects on fertility, 221
LaHood, Ray, 431
Lakes
  primary productivity in, 8*t*
  transient time in, 8*t*
Landfills, closings of, 201
Landscape scale, definition of, 4*t*
La Selva Forest Reserve (Costa Rica), biodiversity in, 48
Latitudinal gradients
  in geographic range size, 49
  in species inventory, 48–49
Laurence, W. F., 54
Law(s), 323–324
  customary, 240–241
  See also Environmental law; Soft law; *specific law*
Lead, 32
  effects on human health, 20
  emissions of, reduction of, 140
  and neurocognitive development, 185
  as ocean pollutant, 138
  serum levels, in Arctic people, biomarkers for, 268, 269*f*
Lead exposure
  dose–effect relationship, 158

pathways for, 169
Learned helplessness, health effects of, 86
Lebensraum, 223
LEDs, for outdoor lighting, 449–450
Legacy contaminants, as ocean pollutants, 138
Legionnaires' disease, 160
Legislation, 323–324
  authorization, 325
  enabling, 325
LHDs. *See* Local health departments (LHDs)
Life cycle, definition of, 4*t*
Life stages, definition of, 4*t*
Lifetime risk, 294
Light, as exposure agent, 33
Light fixture(s), light distribution from, 445, 445*f*
Lighting, outdoor
  effects on astronomical observations, 441–443, 442*f*
  increased use of, 441–443, 442*f*
  LEDs for, 449–450
  sky-glow models, 445*f*, 446–448
  sources of, 441–443
Light pollution
  and siting of astronomical observatories, 443–446, 450–451
  uplight and, 446–449
Light propagation, into sky, 445–448, 445*f*
Limits to Growth, 227
Limulus polyphemus. See Horseshoe crabs
Lincoln, Abraham, 355
Linear multistage model, for carcinogenicity, 293–294, 294*f*
Lipophilic agents, 37
Lister, Adrian, 116
Livestock, fecal pollution caused by, 144
Local density (LD), 56
Local health departments (LHDs)
  quality improvement efforts, 490–491
  scope of work, 481–482
  workforce in, 476–477, 480–481

Logical fallacy(ies), 163
Long Island Sound, eutrophication in, 137
Longitudinal study(ies). *See* Cohort study(ies)
Loss aversion, 375
Love Canal, 337–338
Low-carbon healthcare systems, 125–126
Low-carbon healthy lifestyle, steps toward, 126–127, 127*t*
Low-carbon society, 119–121
  community-based approach for, 124–125
Lowest-observed-adverse-effect-level (LOAEL), 294–295, 295*f*
Low frequency noise (LFN)
  health effects of, 77–78
  sources of, 77
Luginbuhl, C.B., 446
Lung cancer, smoking and, risk statistics for, 167–168

Magnesium, 32
Malaria, climate change and, 114
Malthus, Thomas, 217
Malthus, Thorium, 222
Malthusianism, 222–224
Mammal(s)
  extinctions of, 57*f*
  marine, oil pollution and, 148
  Neotropical, distribution and abundance, and vulnerability to extinction, 56
  new species of, discovery of, 48
  number of species of, 47
Mandatory Reporting of Greenhouse Gases Rule (EPA), 465
Manganese, 32
Mangroves, protective value of, 19
Mao Tse Tung, 223
Marine ecosystem(s), stressors on, 134
Marine Mammal Protection Act, 150
Market-based environmentalism, 467
Marsh. *See* Swamp and marsh
Marx, Karl, 223

Massachusetts, composting program in, 210
Maximum contaminant levels (MCLs), 352–355
May, R. M., 47
McDonald's, and Environmental Defense Fund, green alliance of, 467
MCLs. *See* Maximum contaminant levels (MCLs)
Meat
  consumption
    and food availability, 11
    health effects of, 118
  federal statutes regarding, 335–336
  in hospital meals, 403
  prices, energy costs and, 412
Meat Inspection Act of 1906, 356
Media
  questions commonly asked by, in environmental emergency or crisis, 384*f*–385*f*
  reporting on risks, 376–377, 378, 379*f*–380*f*
Medical waste
  disposal, Health Care Without Harm and, 401
  environmental enforcement regarding, in California, roles of federal, state and local government in, 350*t*–351*t*, 363
  improper disposal of, *case study*, 363
Meningitis, meningococcal, 113
Mental health, environmental noise and, 76
Mercury, 32
  effects on human health, 20
  in fish, 113, 139
  in healthcare supplies, 402
  as ocean pollutant, 138
  serum levels, in Arctic people, biomarkers for, 268, 269*f*
  as water contaminant, 141
Metal(s)
  adverse health effects of, 101–102
  as exposure agents, 32
  See also Heavy metals; *specific metal*
Methane, 141–142
  from fossil fuels, 414, 457
Methiocarb, as endocrine disruptor, 265
Methyl tertiary butyl ether (MTBE), in water, *case study,* 354–355
Mexico, municipal solid waste production in, 201, 202*t*
Mice
  agouti strain, diet, and epigenetic modifications, 103
  in toxicology studies, 286
  See also Rodent(s)
Microbial risk assessment, 282
Microbial volatile organic compounds (MVOCs), 31
Microenvironment(s)
  indoor, exposure measurements in, factors affecting, 42–43
  outdoor, exposure measurements in, factors affecting, 42–43
  as vulnerability factor, 41–42
Migration(s), human, 217
Mill, John Stuart, 227
Millenium Development Goals, 223, 238
Millennium Development Goal(s), 63
  MDG 7, 63
2010 Millennium Goals Report, 63
Minamata disease, 160, 175
Mine Safety and Health Act (1977), 343
Mining, and mercury contamination, 141
Mississippi River, nutrient delivery by, and dead zone in Gulf of Mexico, 137
Missouri, recycling programs in, 209
Moa, 217
Monaco declaration (1992), 142
Montreal Protocol, 236, 242–245
Mosquito-borne disease, climate change and, 114
Mother-infant bonding, epigenetic effects of, 104

Index   531

Motorcycle travel, safe environments for, 431–434, 434f
MSW. *See* Municipal solid waste (MSW)
MTBE. *See* Methyl tertiary butyl ether (MTBE)
Multihit model, for carcinogenicity, 293
Municipal solid waste (MSW)
  by-products of, 203–204
  components of, 202–203, 203f
  current status in US, 200–201
  definition of, 197
  disposal of, historical perspective on, 197–199, 198f
  generation and recovery of materials from, by recycling/reuse, 206–209, 207t, 208t
  global production of, 197, 201, 202t
  management of, 203–204
  "pay-as-you-throw" (PAYT) programs for, 204–206
  recycling of, 206–209
  source reduction, 204–206
  volume produced in US, 200–201
  worldwide volume of, trends in, 197, 201, 202t
MVOCs. *See* Microbial volatile organic compounds (MVOCs)
mVOCs. *See* Microbial volatile organic compounds (mVOCs)
Myocardial infarction, combined effects of noise and air pollution and, 81–82

NAAQS. *See* National Ambient Air Quality Standards (NAAQS)
NAFTA. *See* North American Free Trade Agreement (NAFTA)
National Ambient Air Quality Standards (NAAQS), 328
National emission standards for hazardous air pollutants (NESHAP), 329
National Environmental Education Foundation (NEEF), 489

National Environmental Health Association (NEHA)
  credentials offered by, 488
  technical environmental health competencies, 485
National Environmental Health Science and Protection Accreditation Council (EHAC), 484–485
  accredited graduate programs in environmental health, 500
  accredited undergraduate programs in environmental health, 499
National Health and Nutrition Examination Survey (NHANES), 36
National Health Service (NHS), ecological footprint of, 459
National Institute for Environmental Health Sciences (NIEHS), workforce of, 480
National Institute for Occupational Safety and Health (NIOSH), 343
National Invasive Species Act, 146
National Oil and Hazardous Substances Pollution Contingency Plan, 338
National Pollutant Discharge Elimination System (NPDES), 331
National Primary Drinking Water Regulation, 331–332
National Report Card on Human Exposure to Environmental Chemicals, 36
National Research Council, *Risk Assessment in the Federal Government: Managing the Process*, 282
National Secondary Drinking Water Regulation, 331–332
National Shellfish Sanitation Program, 358, 359–360
Native Americans, traditional lifestyle, ecosystem degradation and, 20–21
Natural disasters, in low-income countries, increases in adverse effects of, 112
Natural environment, and health, 114, 115f

Natural resources
  consumption of, in low-carbon society, 120
  price of, in low-carbon society, 120
  sustainable use of, 121
Nature, access to, beneficial effects of, 116–117
NEEF. *See* National Environmental Education Foundation (NEEF)
Negative dominance theory, 374–375
NEHA. *See* National Environmental Health Association (NEHA)
NESHAP. *See* National emission standards for hazardous air pollutants (NESHAP)
Netherlands, The
  burden of disease attributable to noise in, 78–80, 80*f*
  renewable energy in, 423
Neurodevelopmental disorders, epigenetics and, 96
Neurological disease, linked to environmental causes, 160*t*
New England Journal of Medicine (NEJM), "The Medical Consequences of Thermonuclear War," 393
New Urbanism, 430–431
New Zealand, introduced (alien) species in, and species loss, 52
NHANES. *See* National Health and Nutrition Examination Survey (NHANES)
Niche(s)
  definition of, 4*t*
  ecosystem structure and, 6
Night Noise Guidelines, 74
NIOSH. *See* National Institute for Occupational Safety and Health (NIOSH)
Nitric oxide, 33
Nitrogen, as ocean pollutant, 136–138
Nitrogen dioxide, 33
Nitrogen oxides, 33
Nitrous oxide, 33, 141–142
  from fossil fuels, 413, 414, 457
Nixon, Richard, 225, 391
Noise
  air traffic
    and cognitive functioning in children, 76–77
    exposure-response relation for, 72
  characteristics of, and responses to, 69–70
  environmental
    burden of disease attributable to, 78–80, 79*f*
    health effects of, 69, 70*f*
    management, soundscape approach for, 82–85, 83*t,* 84*f*
    sources of, 70–72
  as exposure agent, 33
  metrics for, 71
  road traffic
    and cognitive functioning in children, 76–77
    exposure-response relation for, 72
    exposure to, in Europe, 71–72, 71*f*
  societal aspects of, 85–86, 86*f*
  See also Low frequency noise (LFN)
Noise control, 83*t,* 85
Noise exposure
  and air pollution, combined effects of, 80–82
  cardiovascular effects of, 74–75
  of children, cognitive and academic effects of, 70, 76–77
  chronic, health effects of, 69, 70*f*
  clinical effects of, 69
  context of, and responses to/health effects of, 83–85
  endocrine response to, 74
  and mental health, 76
  at night, guidelines for, 74
  physiological effects of, 74–75
  in school, and cognitive functioning in children, 77

stress response to, 69
well-being effects of, 69
Noise mapping, in Europe, 71–72
Nonindigenous Aquatic Nuisance Prevention and Control Act, 146
Nonionizing radiation, 33
n-Nonyl phenol, as endocrine disruptor, 266
No-observed-adverse-effect-level (NOAEL), 294–296, 295f, 296t
North American Agreement on Environmental Cooperation, 234
North American Free Trade Agreement (NAFTA), 234, 245–247, 313–314
North Sea, eutrophication in, 137
Notification requirements, 325
NPDES. *See* National Pollutant Discharge Elimination System (NPDES)
Nuclear age, health impacts of, 392–395
Nuclear Freeze movement, 393
health professional associations and, 398
Nuclear power, reemergence of, 418–419
Nuclear reactor accident(s), 419
Nuclear waste, 33
reprocessing of, 419
Nursing
and advocacy for environmental health, 398–399
and Health Care Without Harm, 402–403
Nursing, Health and Environment, 398
Nutrient(s), as ocean pollutants, 135t, 136–138
Nutrient cycling
in ecosystems, 7, 10, 10f
humans' effects on, 136
Nutritional status, epigenetic effects of, 102–103

Obama, Barack, 333, 334, 400, 413, 424, 465
Obesity, epigenetics and, 96, 102
Obesogens, 102

Occupational Safety and Health Act (1970), 317–318, 343–344
Occupational Safety and Health Administration (OSHA), 318, 324–325, 343–344
Occupational Safety and Health Review Commission, 343
Ocean(s)
acidification of, 141
dead zones in, 116, 136, 137
primary productivity in, 8t
temperature of, 142
transient time in, 8t
See also Oceanic pollution
Ocean Dumping Act, 150
Oceanic pollution
classes of, 134–149, 135t
effects of, 133–134
legislation addressing, 149–151
nonpoint sources of, 133, 150, 330–331
point sources of, 133, 150, 330–331
sources of, 133–134, 149–150
synergistic effects of, 134
Ocean pollutant(s)
acute toxicity of, 139
bioaccumulation of, 139
chronic toxicity of, 139
concentrated deposits of, 139
lipophilic, 139
Ocean pollution, management of, advances in (future directions for), 151–152
Ocean Pollution Act (1990), 150
n-Octyl phenol, as endocrine disruptor, 266
Odds ratio, 168
Odocoileus virginianus. See Deer
Oil
land-based runoff of, 149
as ocean pollutant, 135t, 148–149
Oil Pollution Act (1990), 149
Oil spills, 149
Omnivore(s), 9, 10
definition of, 4t

One-hit model, for carcinogenesis, 292–293
Onset of effect, 159
OptiSolar, 422
Organization for Economic Cooperation and Development (OECD), and reduction of carbon footprints, 464
Organochlorines
 as endocrine disruptors, 260
 in vitro bioassays of, 265–266
 See also DDT
OSHA. *See* Occupational Safety and Health Administration (OSHA)
Overhunting, 217
Oxidation/reduction reactions, 37
Oysters, raw, illness and death caused by, reduction of, *case study*, 358–359
Ozone-depleting substances (ODSs), 243–244
Ozone layer, 116
 stratospheric, treaty regime to protect, 242–245

Pacific islands, *Pomarea* monarch flycatchers, decline, with introduction of black rat, 53
PACM model, for environmental health policy-making, 319–321, 319*f*
PAHs. *See* Polycyclic aromatic hydrocarbons (PAHs)
Paralytic shellfish poisoning (PSP), 360
 case study, 361–363
Particulate matter
 as exposure agent, 32–33
 size fractions, 33
"Pay-as-you-throw" (PAYT) programs, for municipal solid waste, 204–206
PAYT. *See* "Pay-as-you-throw" (PAYT) programs
PCBs. *See* Polychlorinated biphenyls (PCBs)
Pedestrian safety
 incorporation into transportation planning, 430–431
 promotion of, environmental strategies for, 429–430
 threats to, 428, 430
Pediatric Environmental Health Specialty Units (PEHSUs), 490
Pediatric Environmental Health Toolkit (Physicians for Social Responsibility), 396
PEHSUs. *See* Pediatric Environmental Health Specialty Units (PEHSUs)
Pennsylvania
 coal production in, 409
 wind energy production in, 409
Perfluorinated compounds, as endocrine disruptors, 260
Period prevalence, 166
Pernambuco, Brazil, rain forest fragmentation in, and species loss, 52
Persistent organic chemicals, 32
Persistent organic pollutants (POPs), 184
 in Arctic people, biomarkers for, 268–273, 269*f*
 effects on fertility, comparison of Inuit and Europeans, 272–273
 as endocrine disruptors, 260, 261
Personal monitoring, 30
Pesticide(s)
 adverse health effects of, 101
 as endocrine disruptors, 260
 in vitro bioassays of, 265–266
 as ocean pollutants, 138
Pesticides and Human Health: A Resource for Health Professionals (Solomon), 396
Pet(s), fecal pollution caused by, 144
PF. *See* Potency factor (PF)
Phenol(s)
 as endocrine disruptor, 260
 as endocrine disruptors, *in vitro* bioassays of, 266–267
2-Phenylphenol, as endocrine disruptor, 266

Phosphorus, as ocean pollutant, 136–138
Photolysis, 37
Phthalate(s)
  as endocrine disruptor, 260
  as endocrine disruptors, *in vitro* bioassays of, 266–267
Physician(s)
  as advocates, 392
  and environmental health movement, 392
Physicians for Social Responsibility (PSR), 392–397
  and antinuclear movement, 392–395
  and climate change, 395–397
  Coal Ash: Toxic Threat to Our Health and Environment, 397
  Dead Reckoning, 394
  Death by Degrees, 397
  environmental health advocacy by, 395–397
  Generations at Risk, 396
  In Harm's Way, 396
  Pediatric Environmental Health Toolkit, 396
  Pesticides and Human Health: A Resource for Health Professionals, 396
Phytoplankton, toxic, 360–361
Pickens, T. Boone, 417
Pig(s), in toxicology studies, 286
Pimentel, David, 420
Plant(s)
  biomass of, 8
  distribution and abundance in British Isles, and vulnerability to extinction, 55–56
  energy efficiency of, 11
  extinctions of, 57
  number of species of, 47
Plastic(s)
  as marine debris, 146–148, 147*f*
  See also Phthalate(s)

Plasticizers, as endocrine disruptor, *in vitro* bioassays of, 266–267
Point prevalence, 166
Polar bear, vulnerability to extinction, 55
Policy
  definition of, 305
  See also Environmental health policy
Policy making
  definition of, 305–306
  for environmental health policy, 309
Politics, definition of, 306
Pollen, climate change and, 113
Pollinators, loss of, 53
Pollution, consequences of, polluters' payment for, 316
Polybrominated diphenyl ethers (PBDEs), 140
Polychlorinated biphenyls (PCBs), 32, 140
  in Arctic people, biomarkers for, 268, 269*f*
  effects on human health, 20
  as endocrine disruptors, 260
  in vitro bioassays of, 265
  in fish, 139
  as ocean pollutants, 138
  in sediments, and marine pollution, 139
Polychlorinated dibenzofurans (PCDFs), as endocrine disruptors, 260
Polychlorinated dibenzo-*p*-dioxins (PCDDs)
  as endocrine disruptors, 260
  See also Dioxin(s)
Polycyclic aromatic hydrocarbons (PAHs), as ocean pollutants, 138
Polyvinyl chloride (PVC), in healthcare supplies, 402
Pomarea monarch flycatchers, decline, with introduction of black rat on Pacific islands, 53
Pontiac fever, 160
POPs. *See* Persistent organic pollutants (POPs)

Population
  definition of, 4*t*
  and environment, human understanding of, 217–218
  global human, 215, 216, 216*f*
  See also Population dynamics
Population(s)
  ecological processes and, 6
  increases in, 11–12
  leveling off, 12
  limiting factors, 12
Population Bomb, The (Ehrlich), 224
Population dynamics, 11–12
  bottom-up effects on, 13
  deterministic relationships and, 50
  stochastic processes and, 50–51
  top-down effects on, 13
Population growth
  food availability and, 222
  historical perspective on, 216
  irruptive trajectory, 220
  natural checks on, 216, 217, 222
  popular attitudes toward, 218–219
  warnings about, 218–219
Population rates, and exposure–outcome relationships, 158
Port Pirie cohort study, 185
Post hoc fallacy, 163
Post-traumatic stress disorder (PTSD), epigenetic markers in, 104
Potency factor (PF), 293–294, 294*f*
Pott, Percival, 159
Poultry Products Inspection Act, 336
Power plant(s), economics of, 413–414
Precautionary principle, 124, 391–392
Predation
  cascading effects of, 13
  definition of, 4*t*, 12
  effects on food web, 12–13
  and extinctions, 57–58
Pre-Harvest Shellfish Protection and Marine Biotoxin Monitoring, 359–360

Prevalence
  definition of, 166
  period, 166
  point, 166
Prevalence surveys, 171–172
Prevention, and environmental health policy, 317
Primary productivity
  in different ecosystems, 7–9, 8*t*
  estimation of, 7–8
  factors affecting, 8–9
  in typical biomes, 8*t*
Primate(s), in infectious disease studies, 286
Process requirements, 325
Prochloraz, as endocrine disruptor, 265–266
Producers, 9, 10
  energy efficiency of, 11
Product controls, 325
Productivity
  definition of, 4*t*
  of ecosystems, 6–7
  lower, with ecosystem degradation, effects on human health, 15*t*
  See also Primary productivity
Product safety, 317
  significant federal statutes regarding, 338–342
Prospective study(ies). *See* Cohort study(ies)
Prosperity
  redefinition of, in healthy society, 121
  sustainable development and, 121–122
Prostate cancer, epigenetic pathways and, 101
Prototype Carbon Fund, 464
PSP. *See* Paralytic shellfish poisoning (PSP)
p statistic, 162, 168
Psychosocial stress, 34, 374
Publication bias, 165
Public education, and encouragement of exemplary behaviors, 462

Public health
 definition of, 477
 and injury prevention, 427–428
 ten essential services of, 348–351, 349*t*
Public health goal (PHG), 354–355
Public health nurses, and advocacy for environmental health, 398
Public Health Service, U.S., workforce of, 480
Public's right to know, and environmental health policy, 317–318
Pure Food and Drug Act, 355–356
Purvis, Nigel, 237
PVC. *See* Polyvinyl chloride (PVC)

q_1*. *See* Potency factor (PF)
QRA. *See* Quantitative risk assessment (QRA)
Quality standards, 326
Quantitative risk assessment (QRA), 27

Rabbit(s), in toxicology studies, 286
Rabinowitz, D., 55–56
Radiation
 ionizing. *See* Ionizing radiation
 nonionizing. *See* Nonionizing radiation
Radiological contamination, effects on human health, 17*t*, 20
Radon, 33–34
RANCH study, 72, 76–77, 82
Randomized controlled trials (RCTs), 176
Rao, Mala, 112
Rapoport's rule, 49
Rat(s)
 in toxicology studies, 286
 See also Rodent(s)
Raup, D. M., 54
RCRA. *See* Resource Conservation and Recovery Act (RCRA)
REACH, 101
Reagan, Ronald, 225, 393
Recall bias, 164

Recreation, ecosystem disruption and, 18–19
Recycling, 206–209
 generation and recovery of materials through, 206–209, 207*t*, 208*t*
Red Book, 282
Red List Index, 63
Redox reactions. *See* Oxidation/reduction reactions
Red tides, 136
Reference dose (RfD), 295–297, 295*f*, 296*t*
Refrigerants, 141
Regressive fallacy, 163
Regulation(s), 323–324, 325–326
 federal, 324–326
Regulatory agencies, 325
Relative risk, 158, 167–168
Remediation, and environmental health policy, 317
Renewable energy, public attitudes toward, 411
Report card(s). *See* National Report Card on Human Exposure to Environmental Chemicals
Reproductive outcomes, linked to environmental causes, 160*t*
Reproductive potential, 12
Research, on reduction of carbon footprint, in U.S., 465
Resorcinol, as endocrine disruptor, 266–267
Resource(s)
 and human carrying capacity, 225–227
 and human conflict (war), 220
 nonrenewable, consumption of, 217–218
Resource Conservation and Recovery Act (RCRA), 199
 historical perspective on, 199–200
Resource Conservation and Recovery Act of 1976 (RCRA), 336–337
Respirable particles, 33
 coarse fraction of, 33
Respiration, 10*f*, 11

Respiratory disease, 20
  air pollution and, 119
  climate change and, 113
  linked to environmental causes, 160*t*
Response requirements, 325
Retrospective study(ies). *See* Case-control study(ies)
RfD. *See* Reference dose (RfD)
Rhinitis, climate change and, 113
Riggs, Arthur, 96
Rio Declaration on Environment and Development, 237
Risk
  absolute, 167
  attributable, 167
  and case-control study, 168
  and cohort study, 168
  definition of, 158
  measures of, 167–168
  media reporting on, 376–377, 378, 379*f*–380*f*
  misattribution of, 188
  public misperceptions of
    factors creating, 377–378
    strategies for overcoming, 380–383
  underestimation of, 188
Risk analysis, 281
  cancer, 297–298
  noncancer, 298
  See also Health risk assessment
Risk assessment, 282–284, 285*f*
  for cancer, 297–298
  components of, 282, 283*t*, 284–298
  and environmental health policy, 318
  for noncarcinogenic substances, 298
Risk characterization, 27, 282, 283*t*, 297–298
Risk communication, 27, 284, 285*f*, 299–303, 300*t*
  anxiety-provoking factors and, 368–372
  challenges to, 376–378
    strategies for overcoming, 378–383
  definition of, 367
  effective, 367

and environmental health, 368
goals of, 368
importance of, 367
mental noise model for, 374
models for, 373–376
negative dominance model for, 374–375
principles of, 368
research/publications/references on, 368, 369*t*–372*t*
risk perception model for, 373–374
seven cardinal rules for, 372–373
trust determination model for, 375–376
types of, 368
Risk management, 27, 182–183, 283–284, 285*f*, 299–303
  and environmental health policy, 318
Risk mapping, 301–302, 301*f*, 302*f*
Risk perception, 300–303
River Uruguay, bilateral management of, 242
Road(s), and ecosystem disruption, 14–15
Roadway design, advances in (future directions for), 434–436
Rodent(s), in toxicology studies, 286–287
Roosevelt, Franklin D., 333
Rose, Geoffrey, 184–185
Ross River virus, 191
Rotterdam Convention, 238
Route of entry, 158–159
Rulemaking, 325
Rwanda, 224, 226
  genocide in (1990s), 219–220

Safe Chemicals Act (2010), 159
Safe Drinking Water Act (SDWA), 282, 331–332, 352, 391
  compliance with, monitoring for, 30
Safety factor, 295
Sagan, Carl, 398
Salmonella Rissen outbreak, CalFERT response to, *case study*, 357–358
Salmonellosis, climate change and, 113
Salt marshes, protective value of, 19
Sampling frames, for epidemiological studies, 169

Sandman, Peter, 300
San Francisco Bay, invasive species in, 146
School, noise in, and cognitive functioning in children, 77
Science, 181–182
SDWA. *See* Safe Drinking Water Act (SDWA)
Seafood, contamination by fecal pollution, 144
Sediments
　capping of, 139
　contaminated, and marine pollution, 139
　dredging of, 139
　in human exposure assessment, 35
　natural remediation of, 140
Selection bias, 164
Selenium, serum levels, in Arctic people, biomarkers for, 268, 269*f*
Semivolatile organic compounds (SVOCs), 32
Serum
　endocrine disruptors in
　　biomarkers for, 267–273
　　extraction and fractionation methods for, 264
　xenohormone activity biomarkers in, 267–273
Seto Inland Sea, eutrophication in, 137
Shellfish
　environmental enforcement regarding, in California, roles of federal, state and local government in, 350*t*, 359–363
　See also Oysters
Silent Spring (Carson), 391
Silver, mining, and mercury contamination, 141
Silvermann, Lewis, 105
Sleep disturbance
　adverse health effects of, 72
　noise and, 69–70, 73–74
Smart Growth, 430–431
Smith, Adam, 222
Smoking. *See* Tobacco/tobacco smoke
Snow, John, 159, 351

Social change, community-based, 124–125
Social networks, and encouragement of exemplary behaviors, 462
Social support, and environmental health policy, 318
Socioeconomic status, as vulnerability factor, 41–42
Soft law, 237–238
Soil(s), in human exposure assessment, 35
Solar power, 422–423
Solastalgia, 190
Solid waste, components of, 202–203, 203*f*
Solid Waste Disposal Act (1965) (SWDA), 199, 336–337
Sorption, 37
Sound, societal aspects of, 85–86, 86*f*
Soundscapes, 82–85, 83*t*, 84*f*
South-East Asia, health effects of climate change in, 113
Soybeans, prices and production of, biofuels and, 421
Spatial mapping, 173
Species
　abundance, and vulnerability to extinction, 55–56
　introduced (non-native), and extinctions, 51–53
　inventory of, latitudinal gradients in, 48–49
　keystone, 53
　new (undescribed), 48
　number of, worldwide, 46–48
　rare, 55–56
　vulnerable to extinction, characteristics of, 55
Species diversity, 45, 46
　definition of, 4*t*
　and ecosystem stability, 5
　estimation of, 46–48
　geographic differences in, 6
　global patterns of, 48
　loss of, effects on human health, 16*t*
　measurement of, 46
　protective value of, 19

Sperm, DNA damage, serum xenobiotic markers and, comparison of Inuit and Europeans, 273
Standard(s), 325–326
Starvation, epigenetic effects of, 102–103
Statement of Forest Principles, 237
Statistics, and environmental epidemiology, 161–162
Statute(s), 323–324
Statute of the River Uruguay, 242
Stockholm Convention, priority contaminants of, 140
Stockholm Declaration (1972), 237–238
Stork, N. E., 47
Stress, and human illness, 20–21
Stroke, epigenetic therapy for, 105
Sub-incision, 222
Subsistence, ecosystem disruption and, 18–19
Sulfur, in fossil fuels, 412–413
Sulfur dioxide, 413
Sun exposure, and neural demyelination, 185
SunPower Corporation, 422
Superfund Amendments and Reauthorization Act of 1986, 337–338
Susceptibility factors, 40–42
Sustainability
  definition of, 4t, 21
  education about, 462–463
  environmental, 455–456
  as business strategy, 467
Sustainable community(ies), 124–125
Sustainable development, 121–122, 455–456
SVOCs. *See* Semivolatile organic compounds (SVOCs)
Swamp and marsh
  primary productivity in, 8, 8t
  transient time in, 8t
SWDA. *See* Solid Waste Disposal Act (1965) (SWDA)
Switchgrass, and biofuel production, 421–422

Switzerland, 2,000 Watt Society, 459–460
Synergistic effect, 166

Taxation, and reduction of carbon footprint, in U.S., 465–466
Telecommuting, and low carbon footprint, 461
Temperate areas, biodiversity in, 48–50
Temperate forest(s)
  biomass in, 8t
  layers of, 6
  primary productivity in, 8t
  transient time in, 8t
Temperate grassland
  biomass in, 8t
  primary productivity in, 8t
  transient time in, 8t
Temperature, environmental, and food poisoning, 113
Tert-octyl phenol, as endocrine disruptor, 266–267
Tetrachlorobenzene, as ocean pollutant, 138
Texas, wind-energy production in, 417
Texas sharpshooter fallacy, 162
Three Mile Island, 393
Threshold, for toxic effect, 290, 291f, 294
Thyroid hormone(s), endocrine disruptors and, 259, 261
Tick-borne disease, climate change and, 114
Tin, adverse health effects of, 101–102
Tipping fees, 201
TMDL. *See* Total maximum daily load (TMDL)
Tobacco/tobacco smoke
  control, as model for action on climate change, 123–124
  epigenetic effects of, 102
  exposure to, dose–effect relationship, 158
  federal statutes regarding, 333–334
  and lung cancer, risk statistics for, 167–168

See also Framework Convention on Tobacco Control (FCTC)
Tolclofos-methyl, as endocrine disruptor, 265
tOP. *See* Tert-octyl phenol
Top-level predators, 9, 10
 definition of, 4*t*
Torrey Canyon oil spill, 150
Total maximum daily load (TMDL), 330
Total suspended particles, 33
Toxaphene, 140
 as endocrine disruptor, 265
Toxic Chemicals Safety Act (2010), 159
Toxicity
 acute, 139
 chronic, 139
Toxicokinetics, 159
Toxic substances
 as ocean pollutants, 135*t*, 138–141
 regulation/control of, 140
Toxic Substances Control Act (1976) (TSCA), 159, 340–341, 399, 402
Toxin(s)
 adverse health effects of, 101–102
 as ocean pollutants, 135*t*, 138–141
Trace elements, as ocean pollutants, 138
Traffic injury(ies)
 environmental planning and, 429
 epidemiology of, 428
 prevention of, 428–429
 and public health, 428–429
Traffic safety, advances in (future directions for), 434–436
Transactivation assays, 262
Transient time, 8, 8*t*
Transition Towns, 125
Transportation
 eco-friendly, advances in (future directions for), 434–436
 modes, selection of, for low carbon footprint, 461
Travel
 future of, in low-carbon society, 120–121

See also Active travel
Treaty(ies)
 criteria for, 235
 definition of, in international law, 235
 and dispute resolution, 241–242
 types of, 233–235
Tributylin, as ocean pollutant, 138
Trophic level, definition of, 4*t*, 9
Tropical areas, biodiversity in, 48–50
Tropical forest
 biodiversity in, 48–49
 biomass in, 8–9, 8*t*
 layers of, 6
 primary productivity in, 8*t*
 species diversity in, 6
 transient time in, 8*t*
TSCA. *See* Toxic Substances Control Act (1976) (TSCA)
T-screen assay, 262–263
Tundra, 6, 19

Udall, Tom, 400
Uganda, population growth in, 223
Ultrafine particles, 33
Ultraviolet light. *See* Decibels, A-weighted (dBA)
Ultraviolet radiation, exposure to, climate change and, 113
Uncertainty factors, 295–296, 296*t*
Underground storage tank (UST) program, 200
UNFCCC. *See* United Nations Framework Convention on Climate Change
United Kingdom
 health inequality in, green spaces and, 116–117
 household energy use in, reduction, health benefits of, 117–118
United Nations, as environmental health policy-maker, 308
United Nations Convention on Biological Diversity (CBD), 62, 235

United Nations Convention on the Law of the Sea (UNCLOS), 150–151
United Nations Environment Program (UNEP), 114–115, 242, 243
  London Guidelines for the Exchange of Information on Chemicals in International Trade, 238
United Nations Environment Programme-World Conservation Monitoring Centre (UNEP-WCMC), 63
United Nations Framework Convention on Climate Change, 124, 142, 235, 245
United Nations Millenium Declaration (2000), 238
United Nations system, 239–240
United States
  activities to reduce greenhouse gases, 464–465
  and Canada, treaties, 233–234
  costs of municipal solid waste disposal in, 201
  executive agreements, 236–237
  invasive species in
    economic and ecological damage caused by, 145–146
    numbers of, 146
  municipal solid waste disposal in, 201
  municipal solid waste production in, 200–201, 202*t*
  nutrient pollution in, 136–138
  ocean pollution legislation, 150
  population exposure to environmental chemicals, monitoring of, 159
  renewable energy strategy, advances in (future directions for), 423–425
  transboundary pollution from, 241
  transportation environments in, incorporation of bicycle and pedestrian safety into, 430–431
  treaty approval process, 236–237
United States Code, 324
Universal Declaration of Human Rights, 237

Uranium, 419
Urbanization, and environmental noise, 70–72
Uruguay
  and bilateral management of River Uruguay, 242
  tobacco policies, Philip Morris' action against, 250
UST. *See* Underground storage tank (UST) program

Vector-borne disease(s), climate change and, 114
Venn diagram, of exposure and dose, 29*f*
Vertebrate(s), terrestrial, extinctions of, 57–58
Vibrio vulnificus, 359
Vibro-acoustic disease, 78
Vidaza, epigenetic therapy with, 105
Vinclozolin, 101
Virchow, R., 185
VOCs. *See* Volatile organic compounds (VOCs)
Volatile organic compounds (VOCs), 32
Volatilization, 37
Vulnerability factors, 40–42

Waddington, Conrad, 95
Walker, Merle, 444, 450–451
Walker's law, 444
Walking, health benefits of, 118
Wallace, Alfred Russell, 48
War, climate change and, 114
Waste management, significant federal statutes regarding, 336–338
Waste recycling, definition of, 204
Waste reduction, definition of, 204
Wastewater
  endocrine disruptors in, 261, 267
  extraction and fractionation methods for, 264
  fecal pollution caused by, 144, 145*f*
  sanitation, 144

treatment of, 144, 150
Water
  coastal, monitoring for pollution, 143–145
  fecal contamination of, adverse health effects of, 143
  recreational, monitoring for pollution, 144–145
  runoff, pollutants in, 134
Waterborne disease outbreak, *case study*, 352–353
Water pollution
  nonpoint sources of, 133, 150, 330–331
  point sources of, 133, 150, 330–331
Water quality
  climate change and, 113–114
  processes affecting, 37
  significant federal statutes regarding, 329–332
Water resources
  climate change and, 116
  in human exposure assessment, 35
Weather
  extreme, 112
  health effects of, 190
White-tailed deer. *See* Deer
WHO. *See* World Health Organization (WHO)
Wiley, Harvey W., 333
Williams, C. B., 47
Wind energy, 409–410, 416–418
  in The Netherlands, 423
Wingspread Conference (1998, Racine, WI), 392

Wood, and biofuel production, 421–422
Workplace safety and health, significant federal statutes regarding, 343–344
World Commission on Environment and Development, 115
World Health Organization (WHO), 239
  calculation of environmental burden of disease, 156, 157*f*
  classification of environmental impacts on health, 183
  as environmental health policy-maker, 308
  Framework Convention on Tobacco Control, 250–251
  governance structure of, 247–248
  and International Health Regulations, 248–250
  new directions for, 247–248
World Scientists' Warning to Humanity, 219, 224
World Trade Organization (WTO), 245–247
WTO. *See* World Trade Organization (WTO)

Xenohormone transactivity(ies), as biomarker for persistent organic pollutants exposure
  in Arctic people, 268–272
  comparison of Inuit and Europeans, 272–273
X-ray(s). *See* Ionizing radiation

Yusho disease, 175